Nature's Benefits in Kenya

An Atlas of Ecosystems and Human Well-Being

World Resources Institute
Department of Resource Surveys and Remote Sensing, Ministry of Environment and Natural Resources, Kenya
Central Bureau of Statistics, Ministry of Planning and National Development, Kenya
International Livestock Research Institute

World Resources Institute: Washington, DC and Nairobi

PHOTO CREDITS

Cover
Rural landscape in Olepolos, Rift Valley with acacia tree and maize plot.
Dreamstime.com / Josiah J. Garber

Young Maasai with goat.
ILRI / Stevie Mann

A farmer in Nyeri holding a bunch of carrots in her hands.
IFAD / G. Pirozzi

Schoolchildren at elephant sanctuary.
Fred Hoogervorst

A boy in western Kenya holding his baby brother.
IFAD / S. Nimeh

Chapter 1 Ecosystems and Ecosystem Services
New farmland in former wetland close to Amboseli National Park.
Jeffrey S. Worden

Chapter 2 Patterns of Poverty and Human Well-Being
A boy in western Kenya holding his baby brother.
IFAD / S. Nimeh

Chapter 3 Water
Tegla Loroupe, world-renowned athlete, is very happy with a sponge full of water after the Lewa Safaricom race.
Fred Hoogervorst

Chapter 4 Food
A farmer in Nyeri holding a bunch of carrots in her hands.
IFAD / G. Pirozzi

Chapter 5 Biodiversity
A portrait of two Grevy's zebras.
Dreamstime.com / Marinacano

Chapter 6 Tourism
Schoolchildren at elephant sanctuary with Nairobi National Park in the background.
Fred Hoogervorst

Chapter 7 Wood
A woman collecting fallen tree branches for fuelwood.
IFAD / S. Nimeh

Chapter 8 The Upper Tana: Patterns of Ecosystem Services and Poverty
Woman walking in the foothills of the upper Tana River, Kirinyaga District.
Norbert Henninger

Tributary of the Tana River in the foothills of Embu District.
Norbert Henninger

Cropland in Ndunduri in the foothills of Embu District.
Jenny Olson

Tea plantation at the edge of the Mount Kenya Forest Reserve.
Jenny Olson

Back Cover
A woman collecting fallen tree branches for fuelwood.
IFAD / S. Nimeh

New farmland in former wetland close to Amboseli National Park.
Jeffrey S. Worden

A portrait of two Grevy's zebras.
Dreamstime.com / Marinacano

Tegla Loroupe, world-renowned athlete, is very happy with a sponge full of water after the Lewa Safaricom race.
Fred Hoogervorst

Woman walking in the foothills of the upper Tana River, Kirinyaga District.
Norbert Henninger

Cite as: World Resources Institute; Department of Resource Surveys and Remote Sensing, Ministry of Environment and Natural Resources, Kenya; Central Bureau of Statistics, Ministry of Planning and National Development, Kenya; and International Livestock Research Institute. 2007. *Nature's Benefits in Kenya, An Atlas of Ecosystems and Human Well-Being.* Washington, DC and Nairobi: World Resources Institute.

Published by: World Resources Institute, 10 G Street NE, Washington, DC 20002, USA
The full report is available online at www.wri.org

ISBN: 978-1-56973-642-5
Library of Congress Control Number: 2007924864

This publication is produced collaboratively by four institutions: the World Resources Institute; the Department of Resource Surveys and Remote Sensing, Ministry of Environment and Natural Resources, Kenya; the Central Bureau of Statistics, Ministry of Planning and National Development, Kenya; and the International Livestock Research Institute. The views expressed in this atlas are those of the authors and contributors.

Printed on recycled paper.

Planting a Seedling for Better Decision-Making

When we started the Green Belt Movement some 30 years ago, Kenyan women identified the lack of water, fuelwood, and nutritious food as their major challenges. They were also concerned about the growing poverty they were experiencing in the communities. This was in preparation for the first meeting focusing on the status of women in the world that was to be held in Mexico in 1975. As I sat and listened to the women, it occurred to me that what we needed to do was heal the land. Why can't we plant trees? I thought. It was easy and doable. Trees would restore the land and produce fuel, fodder, or fruit. Planting trees could help break the cycle of deprivation and malnutrition. These were the beginnings of the Green Belt Movement.

Ever since, the Green Belt Movement has advocated for the protection of Kenya's environment. It has demonstrated time and again that there is a link between a healthy environment, good governance, and peace, and shown that they are the necessary precursors for sustainable development. Our environment, our economic prosperity, and our democratic systems are all inextricably linked. Where environmental destruction takes place, poverty is sure to follow. By contrast, a well-functioning environment—healthy soils and a vibrant wildlife sector—is the foundation of agriculture and tourism, the envisioned engines of Kenya's economic future. Through their function as water catchment areas, our forests deliver more than half of the electrical power generation capacity for our country and supply most of the water for our cities, our farms, and our wildlife areas.

Improving human well-being and safeguarding the environment from which Kenya derives so many benefits are now top priorities in national plans and strategies such as the *Economic Recovery Strategy* and the implementation plan for the Millennium Development Goals. The challenge is in implementing these plans—moving from words to concrete actions on the ground. But how do we build the knowledge base, reform institutions, and formulate policies to achieve these multiple development targets? How do we convince policymakers that investing in nature and environmental management will yield strong returns for poverty reduction and improve the livelihoods of Kenyans?

To manage our resources more wisely and more fairly, we have to base our decisions on knowledge and analysis rooted in sound science. I believe that to formulate better environmental management policies requires investing in a more comprehensive knowledge base on the state of the environment and of human well-being. It requires better information on the value and contributions of environmental services to livelihoods and the national economy. And it requires better understanding of the relationships between resource use and poverty.

Kenyan policymakers and Kenyan voters need to know:

- How does the location of poverty compare to the distribution of key environmental resources and services?
- Which areas provide critically important environmental services, and how do the supply areas for various services overlap?
- Who has access to environmental resources, and who benefits?
- Who bears the cost of environmental depletion and degradation?
- What is the impact of resource depletion on the economy and livelihoods?
- Where could we restore ecosystems and create economic opportunities?

Planting trees has been a way to break the cycle of diminishing resources for the women of the Green Belt Movement. I see the ideas and maps in this atlas to be much like a small seedling. If nurtured, if further developed and grown, and if used by both government and civil society, this seedling carries the promise of breaking the cycle of unenlightened decision-making—decision-making that takes environmental resources for granted; that ignores the deep poverty and hardships of people; that does not fairly disclose the cost and benefits of different choices; that is not accountable to the people most affected by economic or environmental changes; that does not consider the impact on our children and grandchildren.

It is for these reasons that I commend the production of *Nature's Benefits in Kenya: An Atlas of Ecosystems and Human Well-Being* and the contribution it can make to sound decision-making and good governance. As a result of this type of work, we will never be able to claim that we did not know. Rather, using this knowledge, we can move forward to protect our environment, provide economic opportunity for everyone, and build a strong democracy.

Wangari Maathai
Nobel Peace Laureate, 2004
Member of Parliament, Tetu Constituency

March 2007

Table of Contents

Executive Summary

Nature's Benefits in Kenya: An Atlas of Ecosystems and Human Well-Being integrates spatial data on poverty and the environment in Kenya, providing a new approach to examining the links between ecosystem services (the benefits derived from nature) and the poor. This publication focuses on the environmental resources most Kenyans rely on to earn their livelihoods, such as soil, water, forest, rangeland, livestock, and wildlife. The atlas overlays georeferenced statistical information on population and household expenditures with spatial data on ecosystems and their services (water availability, wood supply, wildlife populations, and the like) to yield a picture of how land, people, and prosperity are related in Kenya.

In Kenya's national development plans, improving the health and prosperity of Kenyan families while also safeguarding the natural environment and the many important economic and spiritual benefits it provides are identified as top priorities. Attaining these multiple development goals means that policymakers and civil society groups need to access information and analysis on the numerous interconnections among environmental resources, human well-being, and economic expansion. The maps and analyses presented in this atlas are a first attempt to provide such information.

This information can be used in developing poverty reduction programs and in designing policies for water resources management, agriculture production, biodiversity preservation, and charcoal production, among others. The maps and analyses presented here will not provide easy answers to questions concerning the causes of poverty in Kenya and how ecosystems can best be managed to increase economic growth and improve livelihoods. But they are a first step toward stimulating more informed dialogue and provoking questions for which answers may be found. With up-to-date data and additional analyses, the implementation of Kenya's *Economic Recovery Strategy* (and its successor strategy) can be targeted to specific geographic areas of the country, focusing on the poor, and making better use of Kenya's natural resources.

CHAPTER 1: ECOSYSTEMS AND ECOSYSTEM SERVICES

Kenyans—like all people on Earth—depend on nature to sustain their lives and livelihoods. Not only does it provide the basic goods needed for survival such as water, food, and fiber, people also rely on nature to purify air and water; produce healthy soils; cycle nutrients; and regulate climate. Collectively, these benefits derived from nature's systems are known as *ecosystem services*.

About 80 percent of Kenyans derive their livelihoods from agricultural activities; agriculture contributes, directly and indirectly, about 53 of the nation's Gross Domestic Product. Other contributions of ecosystem services to the economy come from tourism based on Kenya's natural endowment of wildlife, mountains, rangelands, beaches, and coral reefs; as well as timber production from forests; and fish catches from lakes, rivers, and the Indian Ocean.

For a given ecosystem service, the supply is often concentrated in specific areas. Understanding where such key resource areas are located, the ecosystem processes operating to create and maintain these areas, and the services produced and valued by the community is essential for managing resources for improved livelihoods and sustained use.

The maps in Chapter 1 give an overview of Kenya's physical geography; rainfall patterns; major ecosystem types; and densities of wildlife, livestock, and people. They provide a synoptic view of Kenya as a context for the subsequent chapters on poverty and selected ecosystem services.

Savanna and grassland ecosystems, and bushland and woodland ecosystems cover 39 and 36 percent of Kenya, respectively. Agroecosystems extend over another 19 percent and closed forests make up about 1.7 percent of Kenya's land area. Urban ecosystems cover only about 0.2 percent of the country.

CHAPTER 2: SPATIAL PATTERNS OF POVERTY AND HUMAN WELL-BEING

Chapter 2 presents a geospatial profile of poverty and human well-being in Kenya. Human well-being has many elements, including: sufficient income to obtain adequate food and shelter; security; good health; social acceptance; access to opportunities; and freedom of choice. Poverty is defined as a lack of these elements. Human well-being relies fundamentally on the ability to access a wide variety of ecosystem services.

Official Kenyan poverty statistics are based on surveys of household expenditures. They use a rural poverty line of Ksh 1,239 per month (US$ 0.59 per day) and an urban poverty line of Ksh 2,648 per month (US$ 1.26 per day). Given these poverty lines, about 53 percent of rural and 50 percent of urban Kenyans were poor in 1997.

Poverty rate (i.e., the percentage of the population below the poverty line) and *poverty density* (the number of poor in a given area) provide two distinct ways to depict the spatial distribution of poverty. Understanding the relationship between *poverty rate* and *poverty density* is important for designing and implementing poverty reduction interventions.

The *poverty gap* measures how far below the poverty line the poor in a given area are. On average, each rural Kenyan would require an additional Ksh 239 (US$ 3.41) per month to move out of poverty.

Measures of *inequality* look at the distribution of economic welfare across the population. Areas of highest inequality in Kenya are found near urban areas. Inequality is low in rural areas with the highest poverty rates. *Housing quality*, a measure reflecting overall wealth of a household, is higher in the central regions of the country.

The maps indicate some convergence in spatial patterns of poverty, showing that a great number of administrative areas in central Kenya are better off than the rest of the country. The maps also highlight the exceptions to this trend: some areas with low poverty rates nonetheless retain a significant density of poor people. At the same time, not all areas with high poverty rates and high poverty densities have high levels of poor housing or high inequality. A careful analysis of the spatial patterns of *multiple* indicators of well-being is therefore needed to describe and understand the poverty situation and to design effective poverty reduction interventions.

CHAPTER 3: WATER

Water is the lifeblood of Kenya's ecosystems; the hydrological cycle sustains all life. Kenyans use water for drinking, energy generation, livestock production, agriculture, tourism, industry, and many other essential activities. Unfortunately, water is not always plentiful, and Kenya is characterized as *water scarce* based on average per capita water availability.

This chapter's maps show that Kenya's annual rainfall is distributed unevenly over the land: about 15 percent of the country receives sufficient rain to grow maize and other non-drought-resistant crops; another 13 percent has more marginal rainfall sufficient only to grow selected drought-resistant crops; and the remaining 72 percent has no agronomically useful growing season. Rainfall amounts also show distinct seasonal patterns. Areas east of the Rift Valley have two rainy seasons per year, but neither is quite long enough to allow high crop yields. Rainfall amounts vary greatly from year to year as well. Major droughts and floods have occurred regularly in each decade over the past 30 years.

Open surface water is the major source of drinking water for 29 percent of Kenyan households, almost all of them in rural areas. About 32 percent of households rely on groundwater for their drinking water. The same proportion uses piped water (71 percent of urban households and 19 percent of rural households). Families using untreated surface water are relying completely on the regulating services of ecosystems to provide uncontaminated water in sufficient quantities.

Hydropower is the largest source of electricity providing 55 percent of the total installed grid capacity. A number of new hydropower facilities are either under construction or in the planning stages. Ninety-eight percent of Kenya's cropping is rainfed; just 2 percent is irrigated and only 19 percent of potentially irrigable land is currently being irrigated. In almost all of the subdrainage areas in Kenya's rangeland Districts, water demand for livestock is significantly greater than for wildlife. Water demand from livestock is projected to increase as demand for livestock products rises, and may comprise 15 percent of national water demand by 2010.

Decision-makers will face an increasingly difficult challenge in allocating the nation's water resources to accommodate the multitude of demands for agriculture, hydropower, tourism, industry, and drinking water, while still supporting plant and animal life. It will also be increasingly important to address the links between poverty and lack of access to improved water supply and sanitation services.

CHAPTER 4: FOOD

Obtaining food, the most basic human need, is an activity that is always closely linked to natural resources. This chapter covers four dominant sources of food and livelihoods in rural Kenya: crop production, livestock, fishing, and hunting-gathering, and explores how different livelihood strategies are influenced by ecosystems and the resources they provide.

In terms of total area and numbers, smallholders dominate Kenya's rainfed agriculture. Most rural households grow maize to help feed their families *and* rely on the market for food security (between 25 and 70 percent of smallholder income is from non-farm sources). Maps of cropping intensities show that Kenya's rainfed agriculture reflects the country's rainfall patterns, with a significant proportion of farmers being exposed to the risks of unreliable rainfall or prolonged drought.

A mix of dairy cattle, food, and cash crops dominates high-potential agricultural lands in central and western Kenya, where 90 percent of croplands occur. Similar mixed farming along Lake Victoria and large parts of Laikipia, Machakos, Mwingi, Kitui, Makueni, Taita Taveta, Kwale, Kilifi, and Malindi Districts is more marginal. Here rainfall is more erratic or soils are less fertile, resulting in lower yields and incomes.

Livestock production in Kenya also displays distinct spatial patterns: high dairy output and surpluses primarily in central Kenya; milk deficits in large parts of Nyanza and Western Provinces; and pastoral and agropastoral livestock rearing in the arid and semi-arid lands.

Nearly 40,000 people fish for a living—sometimes combined with livestock raising or food cropping—in selected areas along Lake Victoria, Lake Turkana, and the Indian Ocean. About 92 percent of the fish landed in Kenya is from Lake Victoria.

Gathering nuts, fruits, and tubers; collecting honey; and hunting wildlife—including rodents, guinea fowl, and other birds, as well as larger animals such as antelope—are also important sources of food.

CHAPTER 5: BIODIVERSITY

Biodiversity—the full variety of plants, animals, and microorganisms found on Earth—provides the underlying conditions necessary for the delivery of ecosystem services. The maps in this chapter depict both the breadth of Kenya's biodiversity and current pressures and trends affecting it.

Of the 60 Important Bird Areas set up to ensure the survival of local and migratory bird species, half were in decline, about a quarter were improving, and eight were stable, as indicated in a 2003-04 assessment.

Agriculture is a significant factor in Kenya's biodiversity decline, but not all cropping is detrimental to biodiversity. In large parts of Kenya's agroecosystems, farmers' fields are interspersed with patches of forests, woodlands, and other vegetation types. This suggests that farmers could manage their lands in ways that support biodiversity. Average field size, extent of tree cover in croplands, and average number of crops grown represent important components of agrobiodiversity in a landscape. Maps of these three indicators show the following: Throughout central and western Kenya, field sizes are small (less than 2 hectares). Croplands with high levels of tree cover are east of the Aberdare Range; south of Mount Kenya; as well as in Gucha, Central Kisii, and Nyamira Districts. Kirinyaga, Meru Central, and Gucha are the Districts where farmers grow the greatest number of crops at one time.

Kenya's rangelands support primarily livestock and grazing mammals such as gazelle, wildebeest, zebras, and other wildlife species—an important source of tourism revenues. In 1994-96, livestock dominated the rangelands, representing about 84 percent of all the grazing animals in that area. The total population of large grazing wildlife species in the rangelands declined by 61 percent between 1977-78 and 1994-96. Competition for land and water from humans and their livestock, as well as illegal hunting, have been behind these declines. For example, maps of water sources, wildlife, and livestock distribution in the northern rangelands show that livestock near water points is "pushing" wildlife away from water.

Despite these overall and local declines of large grazing mammals, there was an increase in density in some areas between 1977-78 and 1994-96. Such gains were near the Masai Mara Game Reserve and Amboseli National Park, as well as in Lamu and Laikipia Districts. In the latter District, private and communal landowners have been a major contributor to this trend reversal, rather than initiatives based on new government protection policies.

CHAPTER 6: TOURISM

Tourism in Kenya is based primarily on the country's natural attractions, including wildlife in its native habitat as well as some of Africa's finest beaches. This natural endowment has turned Kenya's tourism industry into a leading economic sector, generating revenues of almost Ksh 49 billion (US$ 700 million) in 2005 and directly employing 176,000 people—about 10 percent of all jobs in the formal sector. This chapter shows that the tourism economy depends on a foundation of healthy ecosystems.

Viewing wildlife in its natural habitat is the primary objective for about 80 percent of the international visitors who come to Kenya for holidays. Wildlife is broadly distributed across Kenya, but particular species with high 'viewing value' exhibit specific patterns of spatial distribution. For example, the rangelands of Laikipia District as well as Amboseli, Marsabit, and Tsavo National Parks all have large elephant populations; the massive annual migration of wildebeest and zebra occurs in the plains of Kajiado District close to the Mara-Serengeti ecosystem. But declining wildlife numbers are undermining one of Kenya's principal tourist attractions. For instance, the wildebeest population in the Masai Mara ecosystem has fallen from 120,000 in 1977 to 31,000 in 2002.

In 2005, Kenya's protected areas welcomed 2.1 million visitors, the highest number ever registered. Of the country's 84 parks and reserves, Nairobi National Park (including the Animal Orphanage and Safari Walk), Lake Nakuru National Park, and Masai Mara National Reserve, together accounted for more than half of all visitors. International tourists accounted for more than 90 percent of revenues for all national parks where such revenue data are available. However, Kenyans are also heavy users of parks, particularly those near Nairobi, where they account for more than 70 percent of all revenues collected.

Beaches and coastal ecosystems also account for a large share of tourism earnings, including more than half of all nights spent by tourists in hotel accommodations in 2005. Coastal tourism includes both high-density beach tourism in and around Mombasa and tourism requiring lower visitor densities, such as snorkeling and diving.

To protect wildlife and ecosystems from serious damage caused by overly high visitor densities, tourism planners need to promote underutilized areas and spread visitor numbers more widely across destinations. This would also help to distribute tourism-related costs and benefits more evenly across the country. Improved spatial diversification of visitors will require increased and sustained investments in the transport system, safe water supplies, communications services, tourist accommodations, protected areas, and targeted marketing efforts. It will also require greater control and participation of local communities in wildlife management and tourism enterprises.

CHAPTER 7: WOOD

This chapter provides a brief overview of the ecosystems that provide Kenya with wood and how Kenyans use this resource. Estimates put Kenya's 1995 closed forest area at 984,000 hectares (1.7 percent of the land area). Other natural woody vegetation includes 2.1 million hectares of woodlands, 24.6 million hectares of bushlands, and 10.6 million hectares of wooded grasslands. Agricultural land can also have a high percentage of tree cover as reflected in the high tree density in the croplands of Central Province, for example. Woodlands, bushlands, and wooded grasslands contain most of Kenya's woody biomass. Closed canopy forests are only a minor contributor of wood fuel at the national level.

Kenyans use 80–90 percent of the wood from these ecosystems for energy (firewood and charcoal), and the remaining 10–20 percent for timber, posts, and poles. Biomass is Kenya's dominant fuel, accounting for over 80 percent of total energy consumption in 2000. Burning firewood and charcoal account for roughly equal percentages of total wood consumption.

About 89 percent of rural Kenyans rely on firewood for their energy needs. About 8 percent of the supply came from Trust Land, and another 8 percent from gazetted forests. The remaining 84 percent was supplied by agroforestry systems and on-farm sources. More than 80 percent of households obtain firewood within a 5-kilometer radius of their home.

Approximately 82 percent of urban households and 34 percent of rural households use charcoal regularly. Some 200,000 people are producers and another 300,000 transport and vend charcoal. Gross revenues from production are estimated at Ksh 17.5–32 billion per year (about US$ 250–457 million), putting them somewhere between revenues from horticulture exports and revenues from livestock products. Because the charcoal industry is not fully legalized, the government is foregoing tax revenues as high as Ksh 5.1 billion (US$ 72.9 million) per year.

The high-yield areas of theoretically harvestable biomass growth from natural vegetation closest to Nairobi would be the rangelands south of the city (in Narok and Kajiado Districts), but also in the southeast (in parts of Machakos District). For Mombasa, the closest areas would be the woodlands of Kwale and Kilifi Districts. These areas may be well suited for sustainable charcoal production once the industry becomes fully legalized and more transparently managed.

CHAPTER 8: THE UPPER TANA: PATTERNS OF ECOSYSTEM SERVICES AND POVERTY

This chapter examines maps of various ecosystem services and poverty patterns in a single region—the upper watersheds of the Tana River—to demonstrate how such maps can help to highlight the relationships among people, ecosystems, and poverty.

Home to 3.1 million people, this region represents an important supplier and consumer of ecosystem services. Smallholder agriculture is the dominant land use and is concentrated in the foothills of the Aberdare Range and Mount Kenya. The government has set aside a significant portion of the land for biodiversity and watershed protection, most of it in the mountainous areas. The area contains a broad cross-section of very poor and less poor communities, with the poorest communities located in drier plains downstream of the Aberdare Range and Mount Kenya.

After examining maps of selected ecosystem services (covering water-, food-, wood-, and biodiversity-related ecosystem services) in the upper Tana, Map 8.20 summarizes the poverty patterns in areas delineated by six indicators: high share of piped drinking water, presence of small-scale irrigation efforts, high share of food crops in cropland, high milk production, high number of crops grown, and high share of woodlots in cropland. These indicators reflect either investment areas for water infrastructure, or represent important supply areas of food-, wood-, and biodiversity-related ecosystem services. Such a side-by-side comparison of different ecosystem services is useful for describing poverty-ecosystem relationships and identifying locations where key supply areas and poverty patterns coincide. The maps show that for some of the selected indicators distinct spatial patterns emerge such as the poorest areas not benefiting in a major way from piped drinking water supplies, or high milk production being more prevalent in communities with lower poverty rates. However, they also show that for many of the selected indicators the key supply areas are not automatically associated with lower or higher poverty rates, suggesting determinants that are outside of the selected variables and not necessarily related to geography. This indicates the complexity of the poverty-environment relationship and the need for more detailed analysis that factors in the economic and social context in each subregion.

LESSONS LEARNED ON MAPPING ECOSYSTEM SERVICES AND POVERTY

1. By combining existing maps and data on ecosystem services and human well-being, analysts can create new ecosystem-development indicators.

2. Decision-makers can examine the spatial relationships among different ecosystem services to shed light on possible competition (i.e., tradeoffs) and synergies among various ecosystem services.

3. Decision-makers can examine the spatial relationships between poverty and combinations of ecosystem services.

4. In spite of the usefulness of overlaying maps of ecosystem services and poverty, there are limitations to this approach. These include: lack of data to map a comprehensive set of ecosystem services for all of Kenya; inherent limitations of spatial analyses (i.e. map overlays); limitations in the fundamental knowledge of ecosystems and their value; and the complexity of measuring and monitoring poverty and livelihoods.

5. There are important institutional barriers to measuring and mapping poverty-ecosystem relationships and using this information to inform national policies and decision-making. These include: lack of awareness about ecosystems and ecosystem processes; a sectoral mandate among government institutions that works against cross-cutting analyses involving multiple ecosystem services and poverty; and insufficient promotion of interdisciplinary analysis.

NEXT STEPS

Using the data and concepts demonstrated in this atlas, analysts and decision-makers in Kenyan institutions can initiate a comprehensive accounting of ecosystem services for the country. They can continue to develop new approaches to better integrate poverty-ecosystem relationships in national policies and decision-making. They can foster a better understanding among legislators of these poverty-ecosystem links. And they can apply ecosystem principles and the approach taken in the Millennium Ecosystem Assessment to national and local environmental reporting.

Accomplishing this would result in programs for poverty reduction that take into account where the poor live and what ecosystem services are available to them. It would improve the targeting of social expenditures and ecosystem interventions so that they reach areas of greatest need. And it would make available to decision-makers—in both the public and private sectors—an array of spatial information that could inform their decisions on a range of resource and social issues.

To achieve such outcomes will require leadership by the Ministry of Planning and National Development and the Ministry of Environment and Natural Resources, as well as creative contributions from actors outside of government. It will require actions in four areas:

1. Use and communicate the atlas.

- Make the underlying spatial data in this atlas publicly available. Encourage development of additional dissemination products.
- Incorporate maps and information on ecosystem services in Kenya's next state of the environment report and other environmental reporting efforts.
- Introduce poverty and ecosystem services maps into sectoral reporting.
- Inject maps and information on ecosystem services into future poverty analyses.
- Integrate maps and information on ecosystem services into higher education coursework.
- Prepare guidance and training materials to enable other countries to develop their own maps.

2. Build the knowledge base for mapping ecosystem services and for examining the relationships between poverty and ecosystem services.

- Expand mapping and spatial analyses to include more ecosystem services.
- Integrate ecological processes into future mapping of ecosystem services and use more sophisticated tools to analyze patterns and spatial relationships.

3. Use geospatial information to inform policy, planning, and implementation.

Efforts in three general areas would particularly benefit from the approach used in this atlas:

- Shaping national strategies and plans such as the *Economic Recovery Strategy* and the *Millennium Development Goals (MDGs)*.
- Formulating cross-sectoral policies.
- Improving local land use planning, zoning, and management plans.

4. Strengthen institutions to research and study poverty-ecosystem relationships.

- Continue to develop technical and analytical skills for spatial analyses within Kenyan institutions.
- Establish a technical working group to promote integrated spatial analyses for implementing the MDG needs assessment and the *Economic Recovery Strategy* (and its successor strategy).
- Establish a new technical unit that could spearhead more integrated and cross-cutting work involving multiple ecosystem services and poverty.
- Seek better integration of spatial information in monitoring and evaluation efforts.

Building Partnerships for Better Poverty-Environment Analyses

Nature's Benefits in Kenya: An Atlas of Ecosystems and Human Well-Being offers, for the first time in one volume, georeferenced information on poverty, water, food, biodiversity, wood, and tourism. It presents sectoral and intersectoral analyses in innovative ways and gives policymakers and decision-makers a quick national view of major spatial patterns in each sector. We are fully aware that Kenya needs a more holistic approach in planning and decision-making to address the complex interactions among different ecosystem processes and to achieve Kenya's multiple development targets. We therefore greatly appreciate the value of this publication and fully support future mapping and other analytical initiatives that take the complexity of nature and the important linkages between poverty and the environment into consideration.

Kenya has made significant investments in collecting environmental and poverty data over the years. This atlas demonstrates that information generated by institutions such as the Central Bureau of Statistics, the Department of Resource Surveys and Remote Sensing, and others can not only be used for better environmental reporting and poverty analyses but can also provide insights into linkages between poverty and the environment in specific locations. Such analyses can shed light on possible competition or synergies among various ecosystem services. Understanding such relationships can be extremely important as the country makes investment decisions and creates new economic opportunities.

The atlas also demonstrates that collecting census and household survey data and building technical skills to produce poverty maps within the Central Bureau of Statistics are useful investments and reach far beyond their more narrow application in the macroeconomic sector—such as disbursing development funds for Constituencies. We believe that investments to better integrate existing environmental and natural resources data and to fill important environmental data gaps will provide high returns and lead to more informed planning and decision-making at *both* national and local levels.

This report will allow decision-makers, both public and private, access to data and the ability to overlay high-quality, detailed maps of ecosystems and ecosystem services with maps of poverty. Integrating spatial information on human well-being and the environment in this way is relevant to many policy issues currently under discussion in Kenya. We see great opportunities to inject some of the ideas outlined in this atlas to help in land-use planning, prioritize livestock and tourism investments, enhance water management and food security planning, and improve environmental impact assessments. We encourage further use of the approaches set forth herein to guide policies under preparation (e.g., environment and geoinformation policy) and to assist in formulating new ones that cut across multiple sectors (e.g., wildlife and livestock policies). Making better use of maps and spatial information can certainly strengthen the implementation of the Millennium Development Goals and the *Economic Recovery Strategy* (and its successor strategy). It will certainly help the Government to formulate sound policies and implement realistic plans. It will help identify priority areas for interventions and assist in examining tradeoffs among different investment decisions.

Kenya needs to continue building partnerships within government institutions for better poverty-environment analyses. Only through such partnerships can the country build the necessary technical capacity to analyze and compile maps that document the extent of major ecosystems, the location of key supply areas of ecosystem services and their use, and the spatial distribution of poverty. We therefore support cross-sectoral units such as the Poverty Analysis and Research Unit at the Central Bureau of Statistics in the Ministry of Planning and National Development, the Geoinformation Section at the Department of Resource Surveys and Remote Sensing in the Ministry of Environment and Natural Resources, and the Arid Lands Resource Management Project in their current roles. In fact, we would like to see widening roles for these institutions, and, in terms of timeliness and countrywide coverage, expanded geospatial information. Such efforts will help target poverty reduction strategically and will help us to manage ecosystems in a more integrated way.

Nature's Benefits in Kenya required collaboration and contributions from national and international institutions covering various sectors and specialties. We believe that these working relationships and the experience gained in producing this atlas can become the foundation for developing more specific and more accurate tools and analyses, which we envision policymakers and other decision-makers in Kenya will request.

Professor James Ole Kiyiapi
Permanent Secretary
Ministry of Environment and Natural Resources

Dr. Edward Sambili, C.B.S.
Permanent Secretary
Ministry of Planning and National Development

Dr. Jacob Ole Miaron, C.B.S.
Permanent Secretary
Ministry of Livestock and Fisheries Development

Mr. Mahaboub Mohamed Maalim, O.G.W.
Permanent Secretary
Ministry of Water and Irrigation

Mrs. Rebecca Mwikali Nabutola, M.B.S.
Permanent Secretary
Ministry of Tourism and Wildlife

Preface

Nature's Benefits in Kenya: An Atlas of Ecosystems and Human Well-Being represents a step forward in the analysis of poverty in Kenya and its relation to the natural environment. It is the result of a partnership of national and international organizations, some of which were involved in preparing the first high-resolution poverty maps of Kenya. This publication springs from an effort among these partners to overlay the newly created poverty maps with environmental resource maps based on surveys and remote sensing data. The intent is to show the location and status of key environmental resources that are likely to have significant links with poverty. In creating this report, we worked with several purposes and audiences in mind.

One key purpose has been to build the information and analytical base for implementing Kenya's *Economic Recovery Strategy* and other national strategies. The maps highlight the benefits nature provides to people and the connections between poverty and ecosystem services. Our aim is to demonstrate how map-based analysis of poverty-ecosystem relationships can make a difference in policy development and implementation.

Secondly, we hope to encourage the private sector to give greater consideration to the role of environmental resources in alleviating poverty, with particular reference to the potential contribution of improved environmental management and investments in ecosystem restoration and enhancement. Likewise, we wish to assist environmental specialists in undertaking analyses that can shape anti-poverty policies.

The third purpose has been to conduct a multisectoral analysis of poverty-environment linkages. In Chapter 8, we analyze competing demands for diverse ecosystem services—including food crops, drinking water, irrigation water, and wood—across an entire region (the Upper Tana River watershed). We hope that this multidimensional geospatial analysis will inspire comparable studies involving additional environmental resources and other geographic regions of the country. Such an integrated look at poverty-environment relationships, we hope, will encourage increasing collaboration between institutions both inside and outside government.

We believe that now is the right time to put together an atlas that explores poverty through an ecosystem lens. There is a growing demand for integrated data and mapping of environmental resources, poverty, and the complex web of relationships between environment and livelihoods. The Kenyan Government has committed to several national plans, strategies, and international agreements requiring action toward achieving goals for development that are economically, socially, and environmentally sustainable.

Efforts are under way to include environment in poverty-reduction programs, such as the Poverty-Environment Initiative—a joint effort of the United Nations Development Programme, the United Nations Environment Programme, the Government of Kenya, and other national stakeholders. Various agencies, including the Kenyan Ministries of Finance and Planning as well as the Poverty Analysis and Research Unit at the Central Bureau of Statistics, have expressed interest in environmental profiles of high poverty areas. Following Kenya's *State of the Environment Report 2003* and *2004*, the National Environment Management Authority is exploring ways to use its environmental reporting data and expertise to inform national poverty-reduction efforts.

Another significant development is the growing interest of the media and the public in examining resource conflicts and competing demands for ecosystem services. Conflicts between wildlife conservation and cultivation of agricultural crops, competing demands for water resources by upstream and downstream users, and the conversion of public forests to other land uses are issues of particular concern.

We anticipate that the information presented in this atlas will be of value to various national and community-level groups. Kenya's policymakers form one core audience, encompassing national and District decision-makers and the analysts working with them in government, civil society, and the private sector. Other users include policymakers and analysts in international organizations who collaborate with Kenyan decision-makers. We hope that Kenya's students and teachers will use this study to enrich curricula in geography, environmental science, economics, and other disciplines and that the lessons learned in Kenya can be usefully applied to other countries and regions.

Jonathan Lash
President
World Resources Institute

Jaspat L. Agatsiva
Director
Department of Resource Surveys and Remote Sensing
Ministry of Environment and Natural Resources, Kenya

Anthony K.M. Kilele
Director of Statistics
Central Bureau of Statistics
Ministry of Planning and National Development, Kenya

Carlos Seré
Director General
International Livestock Research Institute

A Reader's Guide

This publication is not a traditional atlas. Rather, it combines traditional map elements with text-based policy analysis. In other words, it is an atlas with elements of a book. Readers can either use it as an atlas, paging through the maps and captions that interest them, or as a book, reading chapter by chapter.

For readers with specialized interests or limited time, we offer the following guide to this publication:

- *Local decision-makers interested in a specific region such as a District, large drainage area, or a Constituency.* A Member of Parliament or a District planner may be interested in looking at a set of maps for their respective areas and can use the map and the map captions to start their review. They may find the specific analysis for Chapter 8—looking at the upper Tana region—useful in helping them develop ideas on how to conduct a similar analysis for their location.
- *National or international decision-makers or analysts.* The box that links the maps to policymaking (Chapters 3 to 7) provides a natural entry point for these users. These boxes refer to broad national strategies and plans (such as the *Economic Recovery Strategy*) or address specific issues (such as wildlife management or charcoal production).
- *Planners at local and national levels and other decision-makers dealing with cross-cutting issues.* These users may want to familiarize themselves with the framework of ecosystem services (Chapter 1), which could be adapted to land use planning. Chapter 8 may provide ideas on how to examine various poverty and ecosystem indicators simultaneously. Finally, users could draw important conclusions based on the limitations in the presented data and identify priorities for future data collection: Are all relevant regions of Kenya mapped? Do District and national planners need to invest in more up-to-date data? Does the country need to collect information on other ecosystem services (e.g., hydrological flows or use of ecosystems for food security) because they are important for the economy and livelihoods? We intend to make the underlying data behind these maps available. They can then be used to create online tools and other decision-support products.
- *Specialists working on issues related to water, food, biodiversity, tourism, and wood.* These specialists will most likely turn first to the chapter dealing with their topic. In most cases, they will have a much more thorough understanding of the issues than provided by the introductory text. However, even these experts will find some new material. For example, the following maps are being published here for the first time: predominant drinking water sources for small administrative areas in Chapter 3; predominant livelihood systems, share of food crops, and number of crops in Chapter 4; wildlife density numbers in the 1970s and 1990s in Chapter 5; spatial distribution of selected charismatic species and coastal ecosystem assets in Chapter 6; woodlots in croplands, and importance of firewood collection and charcoal making for cash income in Chapter 7.
- *Information specialists and policymakers responsible for strengthening Kenya's data infrastructure and capacity for improved poverty-environment analysis.* The conclusions and recommendations would be the starting point for these users.
- *Journalists, speechwriters, students, and analysts in search of facts, maps, and other reference material.* Scanning the list of maps at the beginning of each chapter, the boxes with the poverty and demographic profiles, and the bullets in the 'Summing Up' section can provide a quick overview of what topics and indicators are covered. We plan to release a separate online product that will include all the maps and associated map captions in presentation format.
- *Educators.* They may use the publication to identify specific maps, concepts, or ideas that can enrich curricula or teaching materials. The underlying spatial data should be useful for GIS training and student projects.

All readers should be aware that Chapters 3–7 conclude with two text boxes of particular note:

- *Linking the maps to policymaking.* This box—highlighted in beige—illustrates how the presented maps could be used for more specific policy analysis or targeting of programs. In some chapters, the box uses broad national strategies and plans as an entry point (chapters on water, food, and tourism). Other chapters address important issues such as wildlife management, preservation of biodiversity, or the charcoal industry.
- *Creating a demographic and poverty profile for new geographic units.* This box—highlighted in green—emphasizes that the underlying spatial data behind the maps can be used to create demographic and poverty indicators for new units of analysis. For example, we calculated the number of people and the number of poor for the upper watersheds of Kenya's 'water towers,' the communities within 25 kilometers of the most visited national parks, and croplands with high shares of food crops or woodlots in five Provinces. These boxes also examine—in a first rough analysis—certain relationships between poverty and the environment.

Introduction

Nature's Benefits in Kenya: An Atlas of Ecosystems and Human Well-Being integrates spatial data on poverty and the environment in Kenya, providing a new approach to examining the links between ecosystem services (the benefits derived from nature) and the poor. This publication focuses on the environmental resources most Kenyans rely on to earn their livelihoods, such as soil, water, forest, rangeland, livestock, and wildlife. The atlas overlays georeferenced statistical information on population and household expenditures with spatial data on ecosystems and their services (water availability, wood supply, wildlife populations, and the like) to yield a picture of how land, people, and prosperity are related in Kenya.

RATIONALE

Maps—and the geographic information systems (GIS) that underlie them—are powerful tools for integrating data from various sources and are becoming increasingly important for investigating poverty-environment interactions. Policymakers need spatial information to help them identify areas where development lags and environmental resources are at risk of degradation. Spatial information is also essential to help target areas where investment in physical infrastructure, improved health and education services, and better ecosystem management could have the greatest impact. Maps are also powerful tools for communicating information and findings to experts in multiple disciplines as well as to the public. Both specialists and non-specialists can examine mapped data to identify patterns, trends, and clusters.

Analyses that integrate geospatial data on poverty and the environment can shed light on many important questions: How does the location of poverty compare to the distribution of key environmental resources and services? Which areas provide critically important ecosystem services? How do the supply areas for various services overlap? Who has access to environmental resources and benefits from their use? Who bears the cost of alterations to ecosystems that affect their capacity to supply services?

Moreover, better and more detailed spatial analyses of poverty-ecosystem relationships can be used to put government priorities in perspective: Do current policies target the crucial issues and localities? Are these policies based on sustainable use of environmental resources and services?

Access to improved spatial information can help empower the public to question government priorities, advocate for alternative policies, and exert pressure for better decision-making. Over time, public access to policy-relevant information and analysis will tend to increase the transparency and accountability of government decision-making related to poverty and the environment. This will enhance the likelihood that pro-poor policies and interventions that target and fully integrate the environment's contribution to poverty reduction can germinate and take root.

However, a map-based approach such as that used here does have some limitations. Not all ecosystem services and social processes relevant to poverty are easily mapped. In addition, the ability to show spatial relationships between ecosystem services and poverty depends greatly on the availability of high-resolution georeferenced data. Even when the required data are available, the analysis may reveal little about the causes of poverty, or changes in the underlying processes and functions of natural environmental systems. Nonetheless, such a visual and geographic approach may let policymakers "see" Kenya's natural systems in a new light, helping them to visualize ways to use those systems to alleviate poverty.

SEIZING THE MOMENT

The advent of new datasets (and the growing popularity of web-based geospatial communication tools) makes this an opportune time to create a specialized atlas linking poverty and environment. An extensive supply of geospatial data and expertise on Kenya's environmental resources has been assembled in various national and international agencies in recent years. Examples include aerial surveys of wildlife, livestock, crops, and forests; maps of coastal resources and irrigation infrastructure; and a new high-resolution land cover map.

At the same time, high-resolution poverty maps for Kenya have recently come into use in several national agencies. Within the Central Bureau of Statistics, the Poverty Research and Analysis Unit is now producing and distributing an array of tools, analyses, data, and publications on poverty.

For the most part, this new trove of environmental data has yet to be integrated across different environmental sectors (such as agriculture, wildlife, water, forestry, energy, climate change, etc.), or to be integrated with spatial data on poverty. Encouraging such integration is one of the main goals of this atlas.

ABOUT THE ATLAS

The atlas begins with a brief overview of key concepts related to ecosystems, their contributions to human well-being, and their potential to contribute to poverty reduction and economic development. Chapter 2 presents the most comprehensive, up-to-date maps and other spatial information on the extent and location of poverty in Kenya. Chapters 3 through 7 present maps and analyses on specific environmental resources and ecosystem services, including water, food, biodiversity, tourism, and wood.

Chapter 8 takes a more cross-cutting look at poverty-environment relationships. This chapter examines competing demands for ecosystem services in a single region—the area surrounding the headwaters of the Tana River—and compares these with spatial patterns of poverty in this area.

The final section provides general findings about the use of the introduced maps for sociogeographic analysis. It concludes with four recommendations that are expected to advance a more comprehensive accounting of ecosystem services and to improve the understanding of poverty-environment relationships in Kenya.

In Chapter 1

CONTENTS

LIST OF MAPS

AUTHORS AND CONTRIBUTORS

Norbert Henninger (WRI)
Dan Tunstall (WRI)
Mohammed Said (ILRI)
Greg Mock (consultant)
Karen Holmes (consultant)
Florence Landsberg (WRI)
Janet Nackoney (WRI)
Carolina de Rosas (WRI)
Jo Tunstall (consultant)
Hyacinth Billings (WRI)
Jamie Worms (WRI)

WHAT THIS CHAPTER SHOWS

After highlighting the relevance of nature to people's livelihoods and the importance of natural resources in Kenya's economy, this chapter introduces the concept of ecosystems and ecosystem services. It describes various categories of *ecosystem services*, briefly explains some underlying *ecosystem processes*, and emphasizes that understanding the processes that affect the flow of ecosystem services is essential for effective *ecosystem management*. Six maps showing the physical geography; rainfall patterns; major ecosystem types; and densities of wildlife, livestock, and people provide a synoptic view of Kenya as a context for the subsequent chapters on poverty and selected ecosystem services.

Ecosystems and Ecosystem Services

Kenyans—like all people on Earth—depend on nature to sustain their lives and livelihoods. Not only do they obtain from nature the basic goods needed for survival—such as water, food, and fiber—they also rely on nature to purify air and water, produce healthy soils, cycle nutrients, and regulate climate. Collectively, these benefits derived from nature's systems are known as *ecosystem services*. They fuel the Kenyan economy and, if wisely used and invested, build the nation's wealth.

The lives of ordinary Kenyans confirm the importance of ecosystems. Indeed, their incomes often directly reflect their access to ecosystem services and their exposure to the risks of nature's cycles. For farmers in western Kenya, for instance, the family harvest—of maize, beans, tea, or other typical crops—will reflect the level of rainfall or access to irrigation; the fertility of the soil and access to fertilizer; the genetic profile of their seed; and the crop's exposure to damaging insects, weeds, or hailstorms. Nor will their link with ecosystems end there. Other critical elements of their lives—such as drinking water, firewood, or building materials—will come from nearby ecosystems. From distant ecosystems they may obtain their clothes, tools, or medicines—a testament to the international market for ecosystem services facilitated by global trade.

The life of a herder in the north will show a similar dependence on access to nature's services, including good pastures, healthy animals of good genetic stock, and reliable watering points for livestock. Likewise, fisherfolk from Lake Victoria will depend on the health of fish stocks and the availability of firewood to smoke their catch.

Even an office worker in Nairobi has many links to the environment. When she turns on her computer in the morning, she relies on a hydropower plant on the Tana River, whose turbines are powered by water from Mount Kenya and the Aberdare Range. She may enjoy some *nyama choma* with meat from Kenya's rangelands, seasoned with sea salt from Malindi, and roasted over charcoal from acacia trees in Kajiado. Or perhaps a *chapati* made with wheat planted in Narok District.

The dependence of all Kenyans—urban and rural—on ecosystem services demonstrates the importance of managing natural systems wisely. For example, to ensure an adequate and safe supply of drinking water, Kenyans must take care with how they use the land upstream from drinking water reservoirs—whether they build roads, remove vegetation, establish industrial areas, add fertilizer, or spray pesticides—all these are activities that affect water quantity or quality. Similarly, the continued supply of forest, range, and ocean resources depends on how sustainably these resources are harvested. To be sustainable, fish, timber, woodfuel, and fodder must be harvested below the rate at which the resources are replenished. Otherwise, the natural capital on which future health and prosperity depends will erode. Likewise, crop yields can rise only if soils are maintained and their fertility increased. Revenues from nature-based tourism will benefit future generations only if wildlife is plentiful and diverse, and oceans and coral reefs are healthy.

WHAT IS AN ECOSYSTEM?

An ecosystem is a dynamic complex of plant, animal (including human), and microorganism communities interacting with their physical environment (including soil, water, climate, and atmosphere) as a functional unit (Biggs et al. 2004; MA 2005). The physical boundaries of ecosystems are not fixed and sharp; scientists and planners change the size of ecosystems for different purposes. The entire world—with its landmasses, oceans, and shared atmosphere—can be thought of as an ecosystem; so can the plants, animals, and humans living in Kenya's arid and semi-arid lands (its rangelands and desert ecosystems) or the species interacting in a small tidal pool (a tidal pool ecosystem).

THE IMPORTANCE OF NATURAL RESOURCES IN THE ECONOMY OF KENYA

In 2004, the agriculture sector alone contributed 26 percent of gross domestic product (53 percent, if indirect links to other economic sectors are counted), 60 percent of total export earnings, 45 percent of government revenue, and 62 percent of jobs in the formal economy. Accounting for employment in the informal sector, the share of Kenyans depending on agricultural resources for their livelihoods rises to almost 80 percent (RoK 2006; CBS 2004, 2005). Other environmental income contributions to the economy come from tourism based on Kenya's natural endowment of wildlife, mountains, rangelands, beaches, and coral reefs, as well as timber production from forests and fish catches from lakes, rivers, and the Indian Ocean.

Improving the health and prosperity of Kenya's people, while also safeguarding the natural environment and the many important economic and spiritual benefits it provides, are identified as top priorities in national development plans (GoK 2003; MoPND et al. 2005). Attaining these multiple development goals means that policymakers and civil society groups need to have access to information and analysis that will make clear the numerous interconnections among environmental resources, human well-being, and economic expansion.

UNDERSTANDING ECOSYSTEM SERVICES AND PROCESSES

The array of ecosystem services enjoyed by humans can be divided into four main categories (MA 2003):

- *Provisioning services*, which include the production of basic goods such as crops, livestock, water for drinking and irrigation, fodder, timber, biomass fuels, fibers such as cotton and wool; and wild plants and animals used as sources of foods, hides, building materials, and medicines;
- *Regulating services*, which encompass the benefits obtained as ecosystem processes affect the physical and biological world around them; these include flood protection, coastal protection, regulation of air and water quality, absorption of wastes, control of disease vectors, and regulation of climate;
- *Cultural services*, which are the nonmaterial benefits that people derive from ecosystems through spiritual enrichment, recreation, tourism, education, and aesthetic enjoyment; and
- *Supporting services*, such as nutrient cycling, production of atmospheric oxygen, soil formation, and primary production of biomass through plant photosynthesis; these services are necessary for the production and maintenance of the three other categories of ecosystem services.

Effective ecosystem management requires an understanding of the processes that affect the flow of ecosystem services. Ecosystem processes are sequences of interactions among ecosystem components, and are governed by the feedbacks among these components. For example, soil erosion is an ecosystem process resulting in the loss of soil due to the interaction of soil types, landscape characteristics, animal and human factors, and weather. This, in turn, can reduce crop yields in a farmer's field and lower the harvest of food crops (a provisioning service).

So-called "drivers" of ecosystem processes can directly speed things up or slow them down. In many countries, a principal driver of ecosystem change is changes in land use, such as conversion of forests to cropland and pasture, or the draining of wetlands for crops and infrastructure. Indirect drivers of change encompass demographic, economic, and socio-political factors, including population growth, changes in technology, economic growth, trade, economic globalization, violent conflicts, and legal and governance reform (Biggs et al. 2004).

Ecosystem processes take place all the time, everywhere. Some processes, such as changes in the composition of species occupying a given rangeland, proceed slowly. Others, such as floods, fires, and animal migration, can occur much more rapidly. When processes operating at different speeds or different spatial scales interact, unforeseen consequences can result (Biggs et al. 2004).

Moreover, ecosystem services and their drivers are distributed unevenly across the landscape. For a given ecosystem service, the supply is often concentrated in key resource areas that are characterized by a large number of ecosystem processes. Thus, understanding where key resource areas are located, the ecosystem processes operating to create and maintain these areas, and the services produced and valued by the community is essential for managing resources for improved livelihoods and sustained use (Biggs et al. 2004).

Figure 1.1 Ecosystem Services: The Benefits People Derive from Ecosystems

PROVISIONING SERVICES
Products obtained from ecosystems
- Food
- Fresh water
- Fuelwood
- Fiber
- Biochemicals
- Genetic resources

REGULATING SERVICES
Benefits obtained from regulation of ecosystem processes
- Climate regulation
- Disease regulation
- Water regulation
- Water purification

CULTURAL SERVICES
Nonmaterial benefits obtained from ecosystems
- Spiritual and religious
- Recreation and ecotourism
- Aesthetic
- Inspirational
- Educational
- Sense of place
- Cultural heritage

SUPPORTING SERVICES
Services necessary for the production of all other ecosystem services
- Soil formation
- Nutrient cycling
- Primary production

Source: MA 2003.

Ecosystem productivity is broadly determined by the availability of water, nutrients, and energy, but human actions can lead to positive or negative changes in productivity levels. The concept of resilience is the degree to which an ecosystem can be disturbed before it crosses a threshold to a different state. When ecosystems are subjected to sufficient stresses, particularly of a kind or degree that the system has never before experienced, the resulting changes in ecosystem state and functioning can significantly reduce their ability to support human existence and livelihoods.

In many instances, people degrade ecosystems because of a delay between their uses of ecosystems and the impact of these uses. For example, herders who are grazing more cattle than a given range can support may not be aware of the degradation they are causing until it is too late.

Another problem is that the *effects* of ecosystem damage may occur far from the cause, resulting in damage that may unfairly burden some people, while the benefits of ecosystem use may unfairly accrue to others. For example, deforestation in the upper part of a river system can produce changes in water quality and flow that are felt primarily by people living downstream, while the benefits of timber cutting or conversion to agriculture are reaped locally. Similarly, urban shareholders may benefit from the profits of a mining operation along a rural river, but local fisherfolk, whose livelihoods are affected by polluted runoff from the mine, usually reap no financial benefits (Biggs et al. 2004).

In any case, it is usually far more difficult to reverse a change in an ecosystem than to cause it in the first place. For instance, in a few hours a severe storm can wash away soils that took centuries to form. Particular processes, once set in motion, acquire so much momentum that they can be difficult or impossible to slow down, such as runaway cycles of soil erosion or invasions of alien plant or animal species. Thus, from a human perspective, some ecosystem change is functionally irreversible, and the system is stuck in its new state (Biggs et al. 2004).

MAJOR ECOSYSTEM COMPONENTS: A NATIONAL VIEW OF KENYA

Ecosystems provide humans with water, food, fiber, building materials, and spiritual enrichment. They may be relatively undisturbed systems, such as a natural forest or nature reserve, or they may be extensively modified by human activities, such as agricultural land and urban areas. Different ecosystem processes can determine the distribution of floods, pollutants, and disease vectors, such as mosquitoes or cholera organisms. People's livelihoods and economies therefore depend on a reliable flow of multiple ecosystem services, all of which are the result of complex interactions among the physical, biological, and chemical environments. To describe and analyze these relationships, scientists and planners first delineate different ecosystem types and then inventory their components and processes. This publication translates some of this information into a spatial representation.

Maps can show the location of major ecosystem elements such as rivers and lakes, mountains and plains, the clustering of certain plant communities, the home areas of wild and domesticated animals, or the densities of human populations. Moreover, maps can display where people are obtaining certain ecosystem services, for example, important production and harvest areas for food, fiber, or animal products. They can pinpoint locations affected by the construction of roads, canals, pipelines, or dams; by the expansion of settlements and croplands; or by the introduction of new species—each of these activities can influence the availability and flow of multiple ecosystem services. Maps can also highlight important areas that supply other ecosystem services, such as flood protection provided by mangrove forests, or sediment and pollutant removal provided by certain wetlands.

The following six maps, each representing major ecosystem components, give a brief national overview of Kenya. They also provide some general context for the subsequent chapters on poverty and selected ecosystem services.

Map 1.1 outlines Kenya's physical geography, water bodies, and major drainage areas. The country covers 582,650 square kilometers—about

Map 1.1 **Physical Geography, Water Bodies, and Major Drainage Areas**

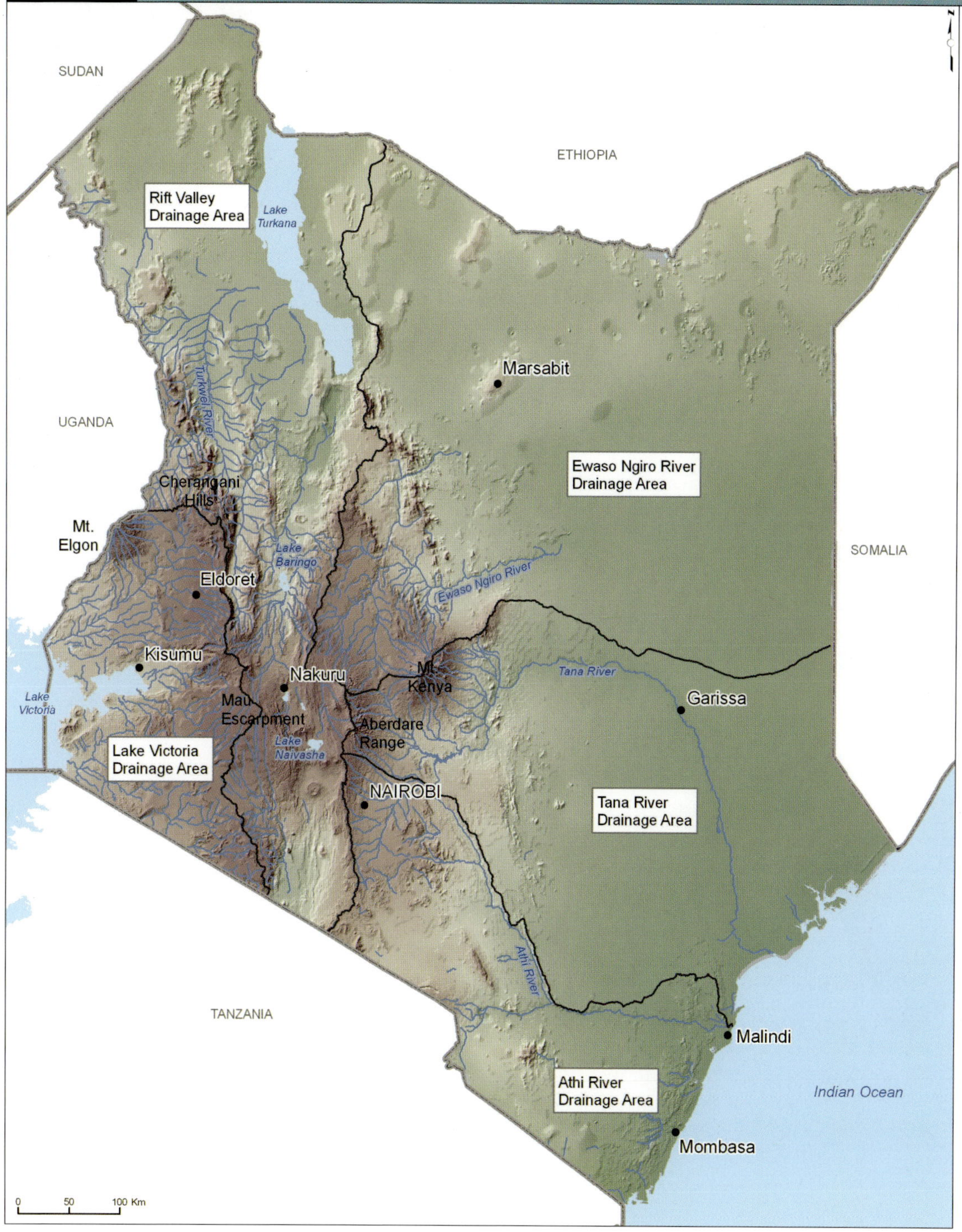

Sources: Cities (SoK and ILRI 2000), water bodies (FAO 2000), permanent rivers (NIMA 1997), 250-meter Digital Elevation Model (SoK, JICA, and ILRI 1996), and major drainage areas (MoWD and JICA 1992).

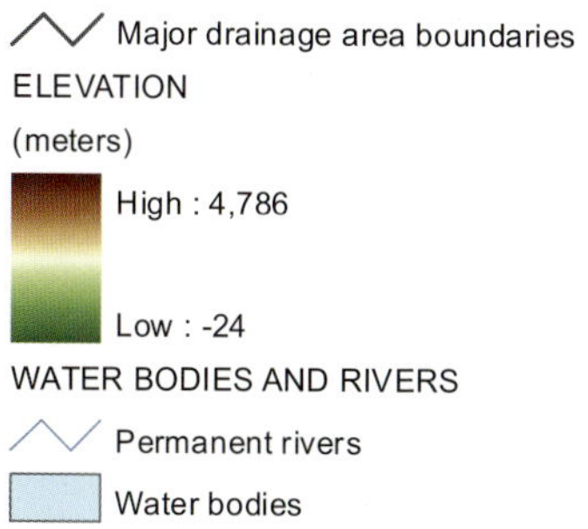

Map 1.2 **Average Annual Rainfall**

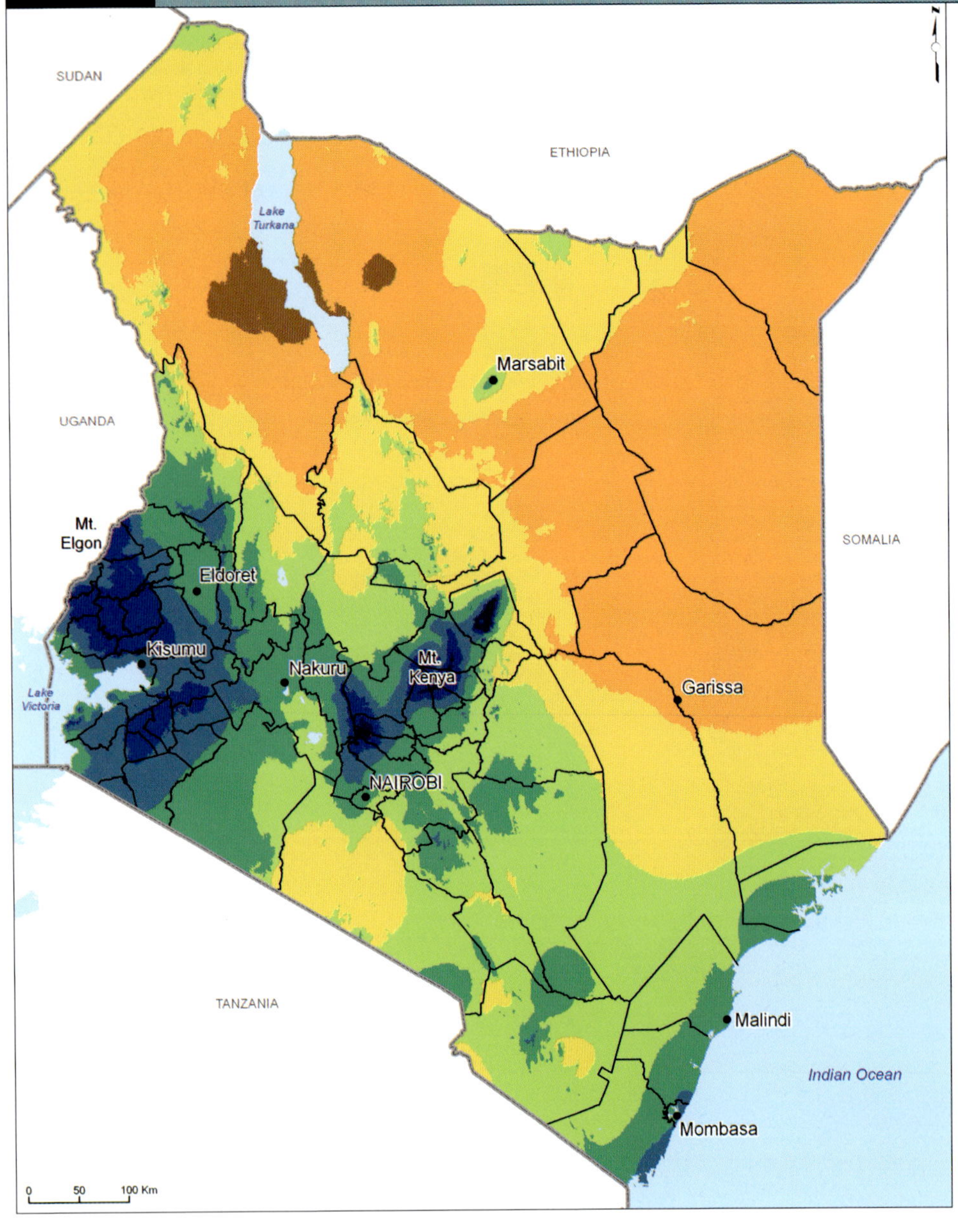

Sources: Administrative boundaries (CBS 2003), cities (SoK and ILRI 2000), water bodies (FAO 2000), and average annual rainfall (Hijmans et al. 2005).

AVERAGE ANNUAL RAINFALL (millimeters)

- > 2,000
- 1,600 - 2,000
- 1,200 - 1,600
- 800 - 1,200
- 600 - 800
- 400 - 600
- 200 - 400
- <= 200

OTHER FEATURES

- District boundaries
- Water bodies

twice the size of the state of Arizona or slightly larger than France or Thailand. About 67 percent of Kenya lies at an elevation below 900 meters, shown in the map as green lowlands. The areas above 1,200 meters—the highlands—are depicted in darker browns, covering about 23 percent of the country. They include five major mountain ranges (Mount Kenya, Mount Elgon, Aberdare Range, Mau Escarpement, and Cherangani Hills), which are surrounded by high-elevation plateaus and foothills shown in lighter browns. The Great Rift Valley, stretching north-south from the Ethiopian border at Lake Turkana, to Lake Baringo and Lake Naivasha, and then to the Tanzanian border, splits the highlands into a western and eastern part.

About 1.9 percent (SoK 2003) of Kenya is covered by water with Lake Victoria, Lake Turkana, Lake Naivasha, and Lake Baringo being the four largest inland water bodies. The highlands are the source of the major permanent rivers traversing the drier lowlands, such as the Tana River—Kenya's longest river. To facilitate water management, planners have grouped the drainage pattern of surface water into five major drainage areas (MoWD and JICA 1992). The rivers draining into Lake Victoria (extending over 8 percent of Kenya's land area) provide the largest share of internal renewable surface water supply (65 percent), while the Athi River drainage area (11 percent of Kenya's land area) provides 7 percent, the lowest share among the five major drainage areas (SoK 2003).

Nairobi and Mombasa are Kenya's two largest urban areas with a projected 2006 population of 2.8 million and 0.8 million, respectively. For the same year, Nakuru's inhabitants are projected at 260,000, and Eldoret's and Kisumu's at 220,000 each (World Gazetteer 2006).

Kenya's average annual rainfall is approximately 630 millimeters per year (FAO 2005). However, the annual rainfall amount varies significantly across the country (Map 1.2). It ranges from 200–400 millimeters in northern and eastern Kenya to more than 1,600 millimeters in western Kenya, bordering Lake Victoria, and in central Kenya, close to its high mountain ranges. More than 80 percent of Kenya is arid and semi-arid (SoK 2003).

The annual rainfall amounts in Map 1.2 hide the distinct patterns of rainy and dry seasons (see Maps 3.2 to 3.5 in Chapter 3) in a year, as well as the great variance of rainfall amounts between years. The high variability of rainfall throughout the seasons, between years, and across space has influenced the distribution of plants, animals, and humans. It is posing significant challenges for Kenya's natural-resource based economy and the livelihood of its citizens.

For most of Kenya, rainfall alone is not sufficient to grow crops without irrigation. In unirrigated areas a patchwork of grasses, shrubs, and trees dominate the landscape, with water availability and soil types determining the exact spatial patterns of plant communities. They are shown in Map 1.3 either as savanna and grassland ecosystems (39 percent of Kenya) or as bushland and woodland ecosystems (36 percent of Kenya). The map also includes a small percentage of areas naturally devoid of vegetation (bare areas).

Croplands stretch from the higher rainfall areas in the highlands to more marginal cropping areas—often classified as agropastoral—where the major land use is some cropping mixed with livestock raising, due to scant and erratic rainfall. Croplands and the associated agroecosystems cover about 19 percent of Kenya in Map 1.3.

Areas with the highest rainfall amounts support a denser tree cover. These usually occur in Kenya's mountain ranges and within a belt along the Indian Ocean. Densely forested areas (closed forests) make up about 1.7 percent of Kenya's land area (UNEP 2001).

Urban ecosystems (large urban areas on the map) cover only about 0.2 percent of the country. These are areas where buildings and streets are the dominant features. Urban ecosystems capture a large proportion of Kenya's human population and economic output. Here, dense human populations, with their domesticated animals and plants, cohabit with wild animal and plant species that are well-adapted to these highly modified habitats.

Map 1.3 **Major Ecosystem Types, 2000**

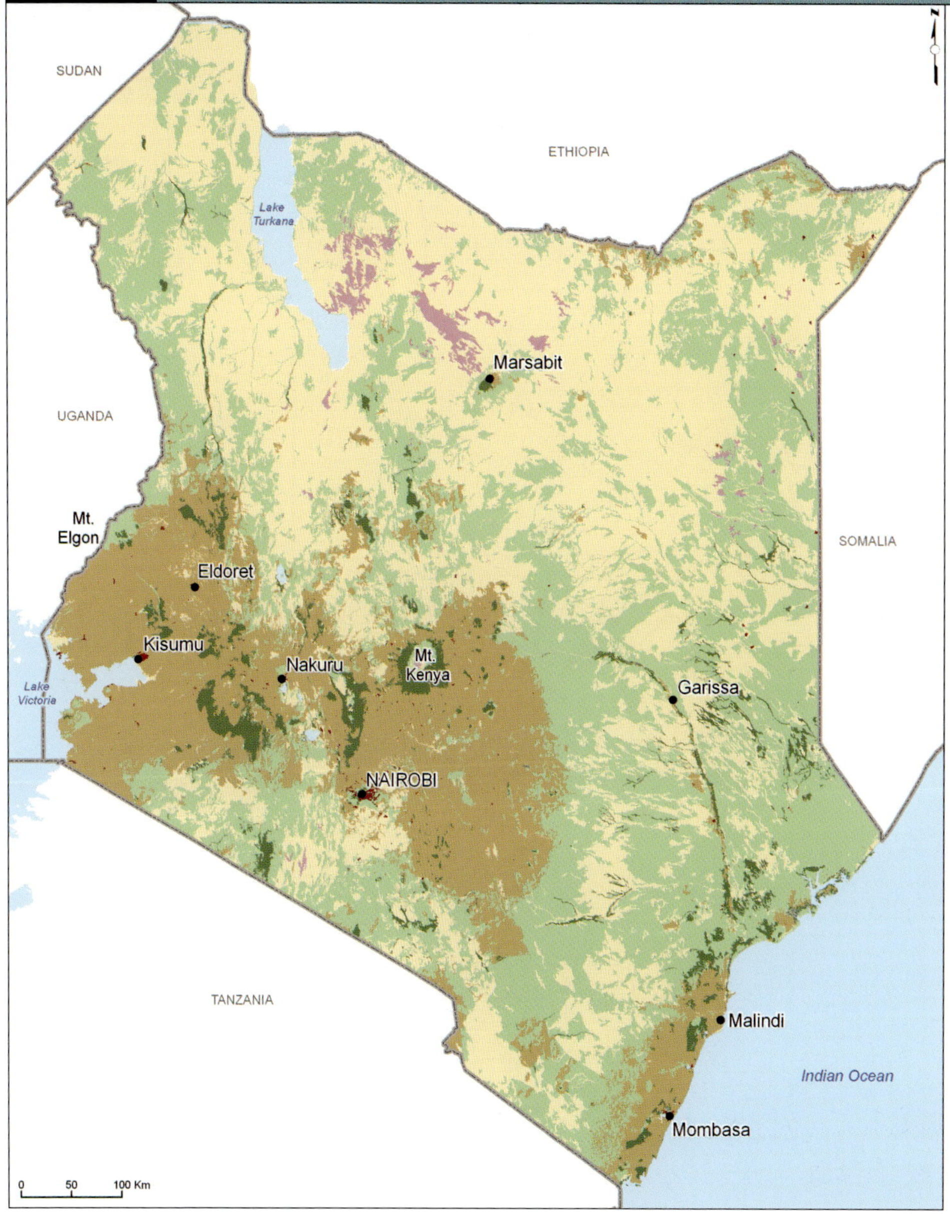

Sources: Cities (SoK and ILRI 2000), water bodies (FAO 2000), and major ecosystem types (FAO 2000).

Note: Forest is the aggregate of two categories in the *Africover* legend (closed and multilayered trees). Bushland and woodland combine various *Africover* classes such as open trees, thickets, and shrublands. Savannas are grasslands with shrubs or sparse trees. Bare areas include areas covered by rocks or rock fragments.

MAJOR ECOSYSTEM TYPES
- Forest
- Bush- and woodland
- Cropland
- Savanna and grassland
- Bare areas
- Urban areas
- Water bodies

Large shares of Kenya's wildlife have their home in the grassland, savanna, shrubland, and woodland ecosystems. The densities of wild grazing animals in Kenya's rangeland Districts vary across the country (Map 1.4). Highest wildlife densities are in Narok and Kajiado Districts, close to the Tanzania border, and in Laikipia District, just north of Mount Kenya. Wild grazing animals can be found throughout the northeastern rangelands as well, but at lower densities. Large parts of Makueni, Kitui, and Mwingi Districts, as well as some coastal and northern Districts, had no observation of wild grazing animals in 1994-96. (Large grazing animals are also present, at much lower numbers, in the densely settled and cropped Districts shown in grey.)

Map 1.5 depicts the spatial distribution of livestock in Kenya's rangeland Districts. (In the Districts shown in grey, livestock is also plentiful but not easily observed by aerial surveys because it is more integrated within croplands and human settlements.) Along with wild grazing animals, most of Kenya's rangelands contain livestock. Higher livestock numbers can be found in the wetter part of the rangelands (e.g., Trans Mara District) and closer to permanent water sources (both natural and human-made). In Districts covered mostly by low livestock densities (shown in light purple), livestock raising is either combined with cropping (e.g., Kitui, Machakos, Makueni, Baringo, and West Pokot Districts) or can rely on a relatively dense network of boreholes or other more permanent water sources (e.g., Turkana District). Areas with more patchy livestock distribution in the rest of the country generally reflect pastoral production systems where herders move livestock periodically to follow the seasonal supply of water and feed.

Map 1.4 Wildlife Density in the Rangelands, 1994-96

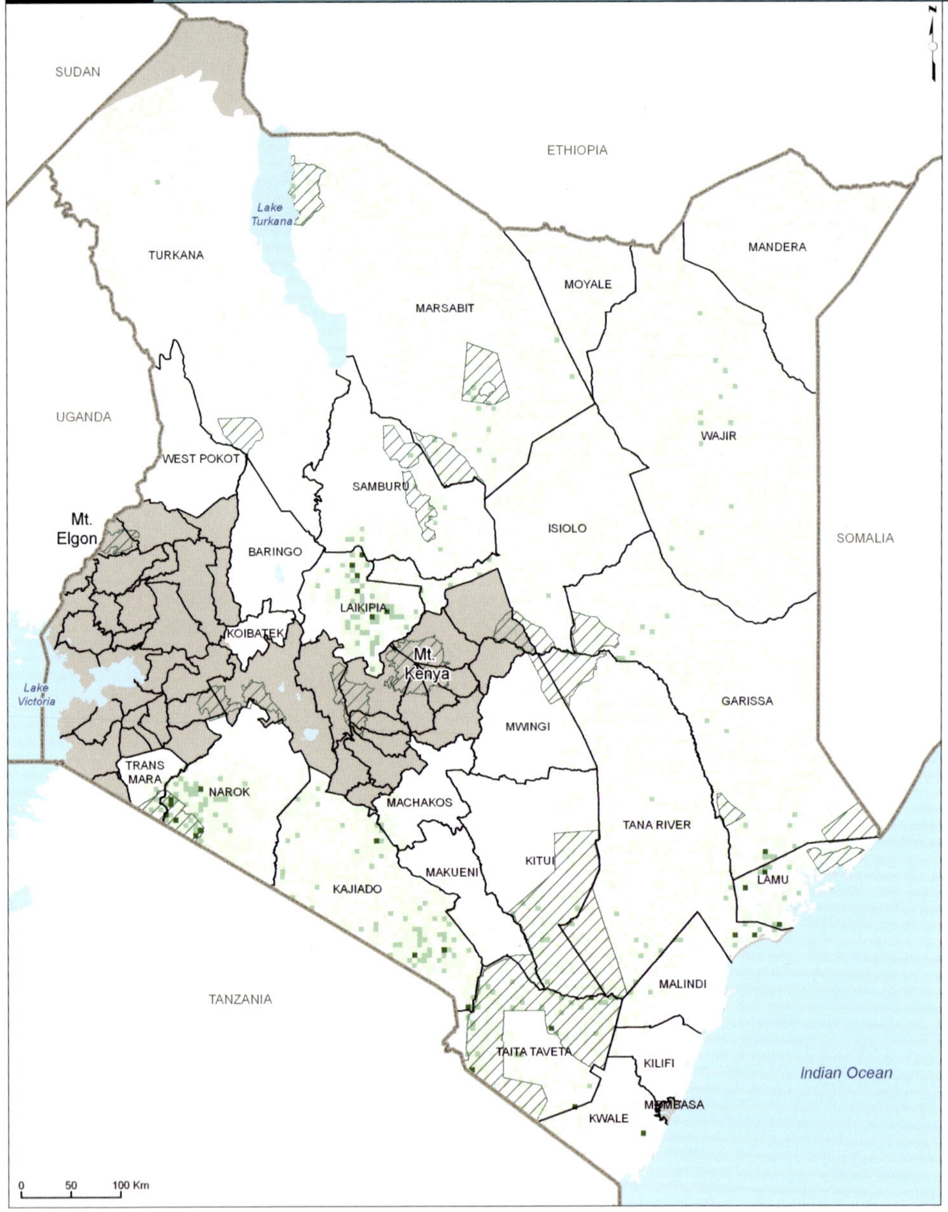

Sources: Administrative boundaries (CBS 2003), water bodies (FAO 2000), parks and reserves (IUCN and UNEP/WCMC 2006), and 1994-96 wildlife density (ILRI calculation based on DRSRS 2003; Grunblatt et al. 1995; Grunblatt et al. 1996).

Note: Species numbers are aggregated (using the *tropical livestock unit* which is equivalent to an animal weight of 250 kilograms) to squares of 5 kilometers by 5 kilometers and then averaged by square kilometer. The wildlife counts include 21 different large grazing animals that can be observed during low-altitude flights.

WILDLIFE DENSITY
(tropical livestock unit per sq. km)
- > 100
- 20 - 100
- 0 - 20
- No observation of animals
- Non-rangeland areas or no data

OTHER FEATURES
- District boundaries
- Major national parks and reserves (over 5,000 ha)
- Water bodies

Map 1.5 Livestock Density in the Rangelands, 1994-96

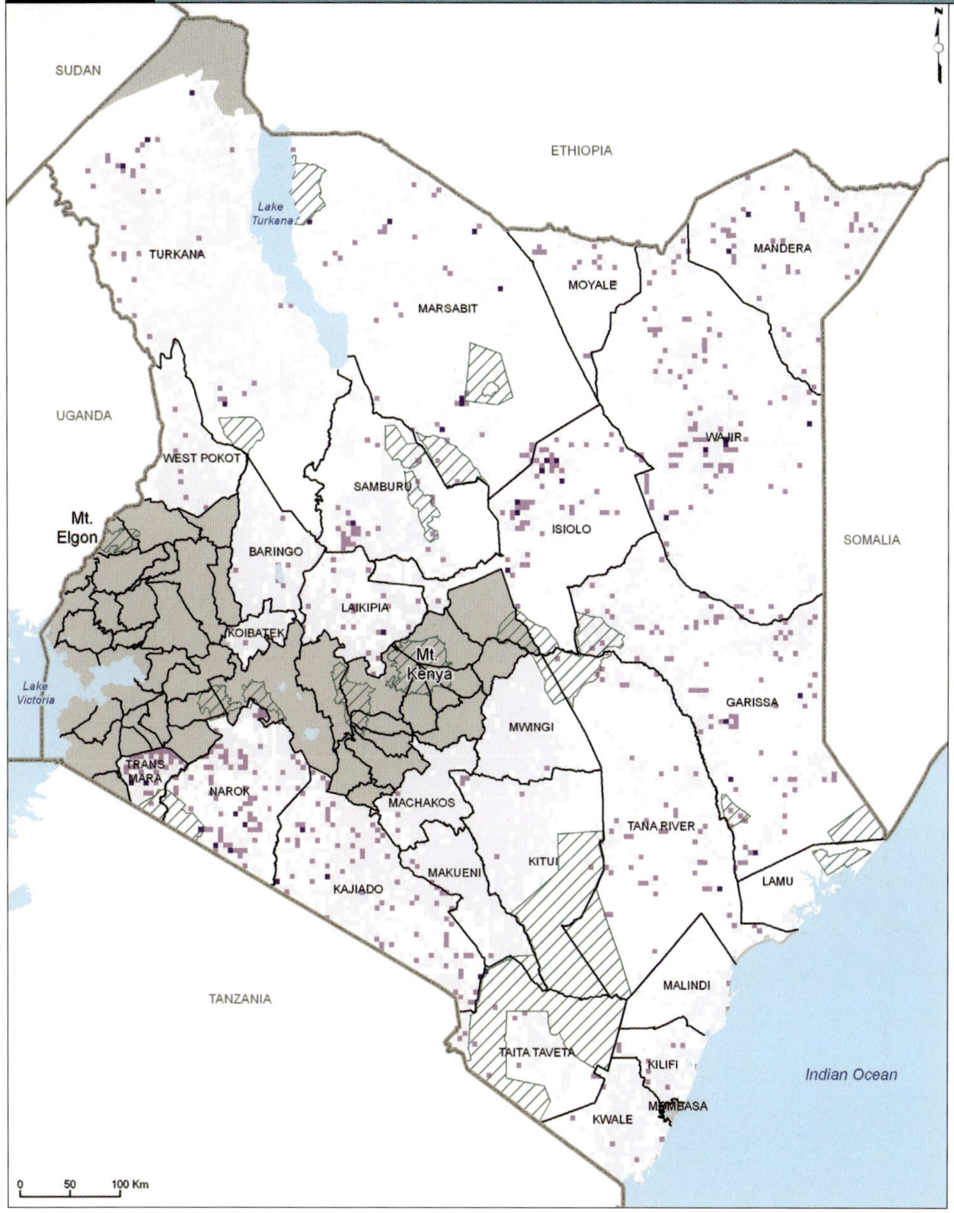

Sources: Administrative boundaries (CBS 2003), water bodies (FAO 2000), parks and reserves (IUCN and UNEP/WCMC 2006), and livestock density (ILRI calculation based on DRSRS 2003; Grunblatt et al. 1995; Grunblatt et al. 1996).

Note: Species numbers are aggregated (using the *tropical livestock unit* which is equivalent to an animal weight of 250 kilograms) to squares of 5 kilometers by 5 kilometers and then averaged by square kilometer. Livestock includes cattle, sheep, goats, camels, and donkeys that were observed during low-altitude flights.

LIVESTOCK DENSITY
(tropical livestock unit per sq. km)
- > 175
- 50 - 175
- 0 - 50
- No observation of animals
- Non-rangeland areas or no data

OTHER FEATURES
- District boundaries
- Major national parks and reserves (over 5,000 ha)
- Water bodies

Livestock raising coupled with higher population densities, cropping, and infrastructure development, usually displaces wildlife. However, in a number of locations in Narok, Kajiado, and Laikipia Districts, high wildlife and livestock densities coincide.

Map 1.6 portrays the population distribution in Kenya (humans are an important component of ecosystems) as of 1999, when the nation's last official census was conducted. At that time, the Kenyan population totaled 28.7 million (CBS 2001). More recent estimates suggest a population of 32.8 million in 2004 (CBS 2006).

As Map 1.6 clearly shows, most Kenyans inhabit the most productive agricultural lands or live along the coast of Lake Victoria and the Indian Ocean. The areas in and around Nairobi as well as in the central highlands support the highest population densities (dark purple-shaded map areas), with more than 600 people per square kilometer. Similar high densities occur in the western part of the country, mainly northwest of Kisumu and in the three Districts slightly inland from the southern shores of Lake Victoria. Pockets of high population density can also be seen along the Indian Ocean coast, primarily around Mombasa.

Only 24 percent of Kenyans live in the rangeland Districts shown in Maps 1.4 and 1.5. Population densities in these arid and semi-arid lands are generally low, with people clustering more densely around towns, market centers, and "temporary" refugee settlements (close to the border with Sudan and Somalia).

Map 1.6 Human Population Density, 1999

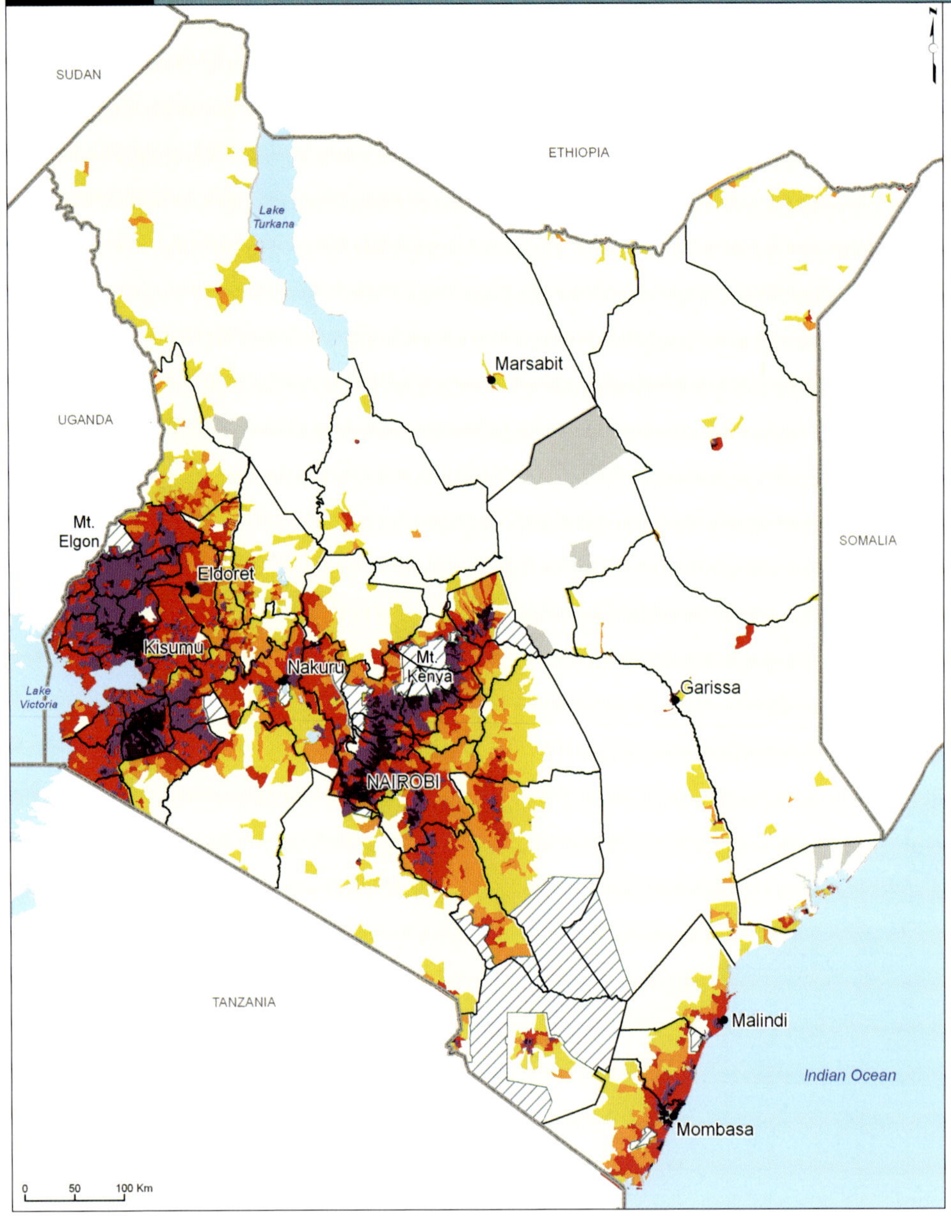

Sources: Administrative boundaries (CBS 2003), cities (SoK and ILRI 2000), water bodies (FAO 2000), parks and reserves (IUCN and UNEP/WCMC 2006), and 1999 population density (CBS 2002).

POPULATION DENSITY
(number of people per sq. km)
- > 600
- 300 - 600
- 100 - 300
- 50 - 100
- 20 - 50
- <= 20
- No data

OTHER FEATURES
- District boundaries
- Selected national parks and reserves
- Water bodies

SUMMING UP

- An ecosystem is a dynamic complex of plant, animal (including human), and microorganism communities interacting with their physical environment (including soil, water, climate, and atmosphere) as a functional unit.
- The benefits derived from nature's systems are known as *ecosystem services*.
- The array of ecosystem services enjoyed by humans can be divided into four main categories: *provisioning services* (food, water, timber, biomass fuels, fibers, medicines, etc.), *regulating services* (flood protection, coastal protection, regulation of air and water quality, etc.), *cultural services* (tourism, education, aesthetic enjoyment, etc.), and *supporting services* (nutrient cycling, soil formation, etc.).
- Ecosystems and ecosystem services play a significant role in Kenya's economy and people's livelihoods: About 80 percent of Kenyans derive their livelihoods from agricultural activities; agriculture contributes, directly and indirectly, about 53 percent to the economy; and nature-based tourism, fishing, and timber production are other important sources of environmental income.
- For a given ecosystem service, the supply is often concentrated in key resource areas. Understanding where key resource areas are located, the ecosystem processes operating to create and maintain these areas, and the services produced and valued by the community is essential for managing resources for improved livelihoods and sustained use.
- The areas above 1,200 meters—the highlands—cover about 23 percent of the country.
- The annual rainfall amount varies significantly across the country. It ranges from 200–400 millimeters in northern and eastern Kenya to more than 1,600 millimeters in western and central Kenya. More than 80 percent of Kenya is arid and semi-arid.
- The high variability of rainfall throughout the seasons, between years, and across space has influenced the distribution of plants, animals, and humans. It poses significant challenges for Kenya's natural-resource based economy and the livelihood of its citizens.
- Savanna and grassland ecosystems and bushland and woodland ecosystems cover 39 and 36 percent of Kenya, respectively. Agroecosystems extend over another 19 percent and closed forests make up about 1.7 percent of Kenya's land area. Urban ecosystems cover only about 0.2 percent of the country.
- The highest densities of wild grazing animals are in Narok, Kajiado, and Laikipia Districts.
- Most of Kenya's rangelands contain livestock. Higher livestock numbers can be found in the wetter part of the rangelands and closer to permanent water sources. Livestock densities in large parts of the northeastern rangelands reflect pastoral production systems where herders move livestock periodically to follow the seasonal supply of water and feed. In a number of locations in Narok, Kajiado, and Laikipia Districts, high wildlife and livestock densities coincide.
- The areas close to Nairobi in the central highlands support the highest population densities, with more than 600 people per square kilometer. Similar high densities occur in the western part of the country, slightly inland from Lake Victoria. Only 24 percent of Kenyans live in Kenya's rangeland Districts.

In Chapter 2

CONTENTS

LIST OF MAPS

AUTHORS AND CONTRIBUTORS

Paul Okwi (ILRI)
Godfrey Ndeng'e (CBS)
Russ Kruska (ILRI)
Patti Kristjanson (ILRI)
Norbert Henninger (WRI)
Janet Nackoney (WRI)
Florence Landsberg (WRI)
Greg Mock (consultant)
Karen Holmes (consultant)
Eunice Kariuki (ILRI)
Pamela Ochungo (ILRI)
Dan Tunstall (WRI)
Hyacinth Billings (WRI)
Carolina de Rosas (WRI)
Jamie Worms (WRI)

WHAT THIS CHAPTER SHOWS

This chapter provides a brief overview of the major elements of human well-being, the definition of poverty in official Kenyan statistics, and various indicators related to well-being or lack of well-being. National maps familiarize the reader with changes in Kenya's population distribution between 1989 and 1999. Subsequent maps of selected indicators of poverty, inequity, and housing quality exhibit varying geographic patterns across Kenya. The maps indicate some convergence in spatial patterns showing that a great number of administrative areas in the central part of the country are among those considered to be better off. The maps also highlight the exceptions to this trend: some areas with low poverty rates nonetheless retain a significant density of poor people. At the same time, not all areas with high poverty rates and high poverty densities have high levels of poor housing or high inequity. A careful analysis of the spatial patterns of multiple indicators of well-being is therefore needed to better describe the poverty situation, and design and implement poverty reduction efforts.

2

Spatial Patterns of Poverty and Human Well-Being

This chapter presents a geospatial profile of poverty and human well-being in Kenya. Although poverty and human well-being are familiar concepts, these seemingly simple terms tend to defy precise, universally agreed definition.

Most modern experts agree that poverty is a multi-dimensional phenomenon, involving not only a lack of financial means, but also various kinds of non-monetary deprivation, such as lack of access to social services and lack of ability to participate in political, social, and cultural institutions and decision-making. As Nobel laureate economist Amartya Sen (1999) has observed, "Policy debates have indeed been distorted by overemphasis on income poverty and income inequality, to the neglect of deprivation that relates to other variables, such as unemployment, ill health, lack of education, and social exclusion."

The maps and figures represent an attempt to capture diverse dimensions of poverty and human well-being in Kenya. Different geospatial indicators paint different pictures of poverty and human well-being; thus, it is crucially important that analysts choose indicators that are appropriate to illuminate the issue or policy choice under consideration.

The mapped indicators presented here build on the results of a 2003 poverty mapping analysis conducted by the Kenyan Central Bureau of Statistics with several partner organizations (CBS 2003). The chapter also draws on a 2005 successor analysis to fill data gaps for North Eastern Province and to obtain information on inequity (CBS 2005). In addition, this chapter relies on detailed information from Kenya's 1999 *Population and Housing Census* to show the spatial distribution of Kenya's population and to construct an index of housing quality (CBS 2002).

The maps look at both rural and urban poverty, two distinctly different phenomena in Kenya. For example, the expenditure-based poverty measures from the Central Bureau of Statistics reflect cost-of-living differences for rural and urban areas. In rural areas, expenditure poverty is defined as spending less than Ksh 1,239 per month (about US$ 0.59 per day), whereas in urban areas, the poverty line is defined as spending less than Ksh 2,648 per month (about US$ 1.26 per day).

In addition, this chapter presents information on poverty and human well-being that is locally specific—that is, information based on data aggregated separately for each of Kenya's local administrative units (see Boxes 2.1 and 2.2). Depending on the chosen indicator, this information may either represent a Constituency area (there are a total of 210 Constituencies in the country), or a Location (the maps show 2,070 rural Locations and 496 urban Locations), or a Sublocation (there are 6,622 Sublocations in the country).

The first cluster of maps deals with conventional economic measures of human welfare based on expenditures (so-called money-metric indicators). The rest of the chapter explores other measures of well-being, such as the Gini coefficient, which measures economic inequality; and housing quality, which reflects the overall wealth of a household.

THE DIMENSIONS OF WELL-BEING

Human well-being has many elements. Sufficient income to obtain adequate food and shelter is certainly important, but other dimensions of well-being are crucial as well. These include security, good health, social acceptance, access to opportunities, and freedom of choice. Poverty is defined as the lack of these elements of well-being (MA 2005).

Figure 2.1 Constituents of Well-Being

SECURITY
- PERSONAL SAFETY
- SECURE RESOURCE ACCESS
- SECURITY FROM DISASTERS

BASIC MATERIAL FOR GOOD LIFE
- ADEQUATE LIVELIHOODS
- SUFFICIENT NUTRITIOUS FOOD
- SHELTER
- ACCESS TO GOODS

HEALTH
- STRENGTH
- FEELING WELL
- ACCESS TO CLEAN AIR AND WATER

GOOD SOCIAL RELATIONS
- SOCIAL COHESION
- MUTUAL RESPECT
- ABILITY TO HELP OTHERS

FREEDOM OF CHOICE AND ACTION
OPPORTUNITY TO BE ABLE TO ACHIEVE WHAT AN INDIVIDUAL VALUES DOING AND BEING

Readers should note that these maps give only a snapshot for a single period (all well-being indicators are for 1999). Spatial poverty analyses could greatly benefit from regularly updated poverty maps, especially since rapid changes in economic, environmental, and household conditions can throw people into or help people exit from poverty.

It is also important to bear in mind that all poverty indicators have shortcomings. For instance, data on poverty are often collected and recorded at the level of the household, masking important differences among family members with respect to nutritional status, access to education, and other important dimensions of well-being. In addition, there are inherent limitations in the ability to aggregate locally derived data to give meaningful results at the national level.

A central tenet of this atlas is that human well-being relies fundamentally on the ability to access a wide variety of ecosystem services. Because many of these services do not flow through markets and do not have a market price attached to their use, they are not accounted for in conventional money-metric measures of welfare, such as income or expenditures. A careful reading of this chapter should be continually informed by the awareness that, for poor people in Kenya, as elsewhere, great gains in well-being can be obtained through more equitable and secure access to local ecosystem services that are central to environmentally sustainable livelihoods.

Box 2.1 Kenya's Administrative Units

For administrative purposes, Kenya is divided into a hierarchical system of Provinces, Districts, Divisions, and smaller local administrative units known as Locations and Sublocations. This atlas presents data at the following administrative levels:

8	Provinces (including Nairobi)
69	Districts
210	Constituencies
2,566	Locations separated into 2,070 rural Locations and 496 urban Locations (covering the whole country except North East Province)
6,622	Sublocations (covering the whole country)

Map 2.1 shows Kenya's Provinces and Districts.

To organize elections and national parliamentary representation, the Electoral Commission of Kenya divides the country into 210 Constituencies. The voters in each Constituency area select one elected representative (i.e., Member of Parliament) to the national Parliament (CBS 2005).

The number of administrative units and their exact boundaries has varied over the years (especially for the smaller administrative areas) due to changes in administrative or political priorities. Administrative areas shown in this publication reflect the 2003 boundaries provided by Kenya's Central Bureau of Statistics.

Sources: Administrative boundaries (CBS 2003) and water bodies (FAO 2000).

Map 2.1 Provinces and Districts, 2003

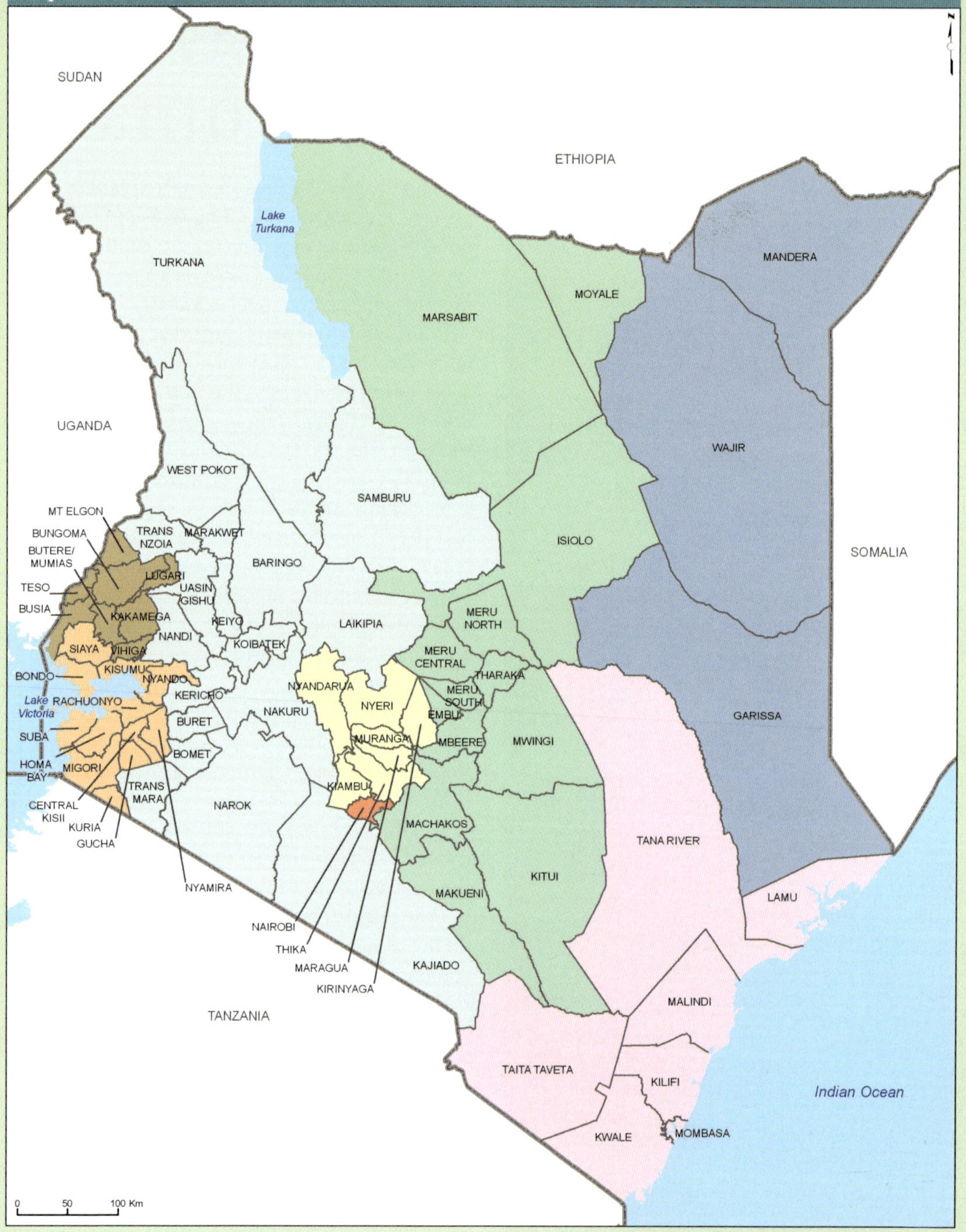

PROVINCE NAMES

- Central
- Coast
- Eastern
- Nairobi
- North Eastern
- Nyanza
- Rift Valley
- Western
- District boundaries

Box 2.2 **The Power of Information: Disaggregated Data Makes the Difference**

For spatially complex phenomena such as poverty and human well-being, data that are averaged at the Provincial and District levels can mask important local variation. By contrast, data disaggregated to more local levels can often reveal such variations.

For instance, as shown in the maps of Nyanza Province, the incidence of poverty (that is, the percentage of the population living below the poverty line) at the District level is quite high across the Province's 12 Districts (Map 2.2). However, a higher-resolution map (Map 2.3) showing the incidence of poverty for the 420 Locations indicates much wider variation. It can now be seen that Locations positioned next to each other often have very different poverty rates, and that Nyanza Province, a relatively poor area, contains several pockets of relatively low poverty (shades of green in the map).

Map 2.2 Nyanza Province: Poverty Rate at District Level, 1999

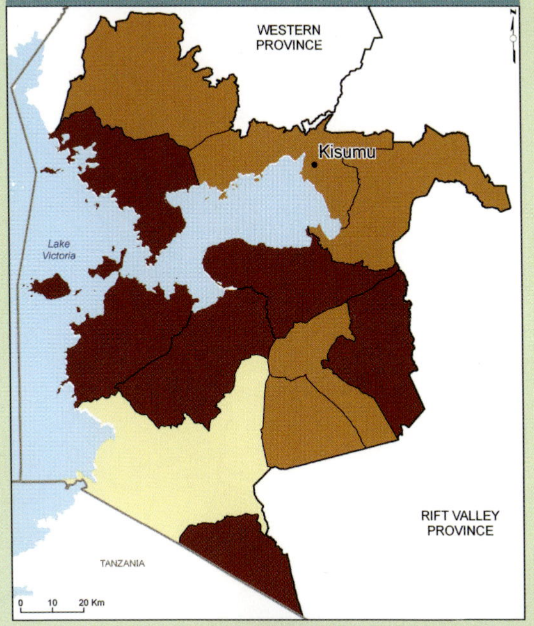

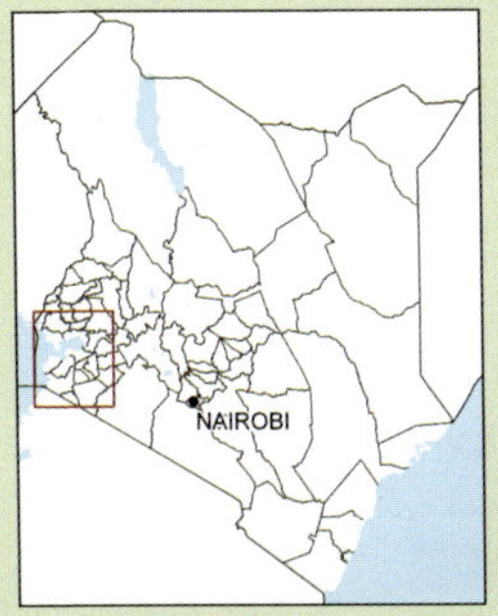

POVERTY RATE
(percent of the population below the poverty line)

- > 65
- 55 - 65
- 45 - 55
- 35 - 45
- <= 35

OTHER FEATURES

- Province boundaries
- District boundaries
- Location boundaries
- Water bodies

Map 2.3 Nyanza Province: Poverty Rate at Location Level, 1999

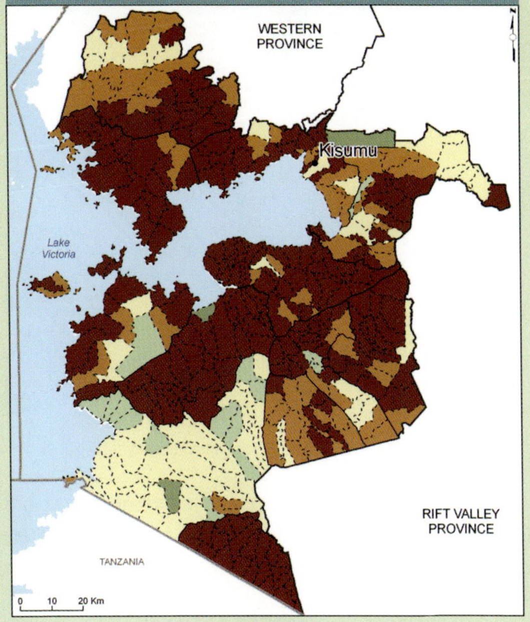

Sources: Administrative boundaries and 1999 poverty rate (CBS 2003), cities (SoK and ILRI 2000), and water bodies (FAO 2000).

POPULATION DENSITY, POVERTY RATE, AND POVERTY DENSITY

Kenya's population has grown rapidly in recent years, rising to 28.7 million at the last census in 1999 (CBS 2001), a 34-percent increase over the 1989 census (CBS 1994). The country is also becoming more densely settled; population density reached 49 people per square kilometer in 1999, versus only 37 per square kilometer a decade earlier (see Maps 2.4 and 2.5). The latest estimate puts Kenya's 2004 population at 32.8 million increasing the average population density to 56 persons per square kilometer (CBS 2006).

Official Kenyan poverty statistics are based on detailed information about household expenditures on food and other items such as health and education. A poverty line—the level below which a household is considered poor—is estimated based on the minimum amount needed to purchase a basket of food providing 2,250 calories per day, along with a basic set of non-food requirements. Using survey data from 1997, Kenya's poverty line was estimated to be Ksh 1,239 per month (about US$ 0.59 per day) for rural households and Ksh 2,648 per month (about US$ 1.26 per day) for urban households. Given these poverty lines, about 53 percent of the rural population and 50 percent of the urban population were poor in 1997 (CBS 2003).[1]

A nation's *poverty rate* is the percentage of the population below the nation's poverty line (this is also known as the "headcount ratio"). Map 2.6 shows that Kenya's spatial pattern of poverty rates varies widely across the nation. Map 2.7 depicts another way to look at the spatial distribution of poverty using *poverty density*, which is defined as the number of poor people living in a given area.

Maps of poverty density exhibit geospatial patterns that are quite different from those of poverty rates. Administrative areas in arid and semi-arid regions generally have high poverty rates but overall very low densities of poor persons per square kilometer. The spatial patterns of these two indicators for large parts of the more densely settled areas are inversely related. For example, highly productive agricultural areas in Central Province have generally low poverty rates but still fairly large concentrations of poor people. Exceptions to this inverse relationship occur in western Kenya, some isolated areas in central Kenya, and along the coast, where both poverty rates and poverty densities are high (shown as dark brown in Maps 2.6 and 2.7).

Understanding the relationships between the poverty rate and the poverty density is important for designing and implementing poverty reduction interventions. Using either the poverty rate or the poverty density alone to identify areas to focus poverty programs will likely be ineffective, either missing many poor people or wasting resources on families that are not poor. For example, targeting only areas with the highest poverty rates will not reach all or most of Kenya's poor, leading to "undercoverage" of people in need, most of them in the densely settled areas of central Kenya. On the other hand, providing resources only to areas with the highest poverty densities will bypass the poor in the arid and semi-arid areas and increase the likelihood of "leakage" of poverty aid to the non-poor in areas with low poverty rates, such as the area between Nairobi and Mount Kenya.

[1] Note that the maps in this chapter rely on a statistical estimation technique that combines information from the 1997 *Welfare Monitoring Survey* and the 1999 *Population and Housing Census*. Given the statistical estimation technique applied, the final estimates of these indicators refer to 1999, and the maps are labeled as such (CBS 2003 and CBS 2005).

Map 2.4 Human Population Density, 1989

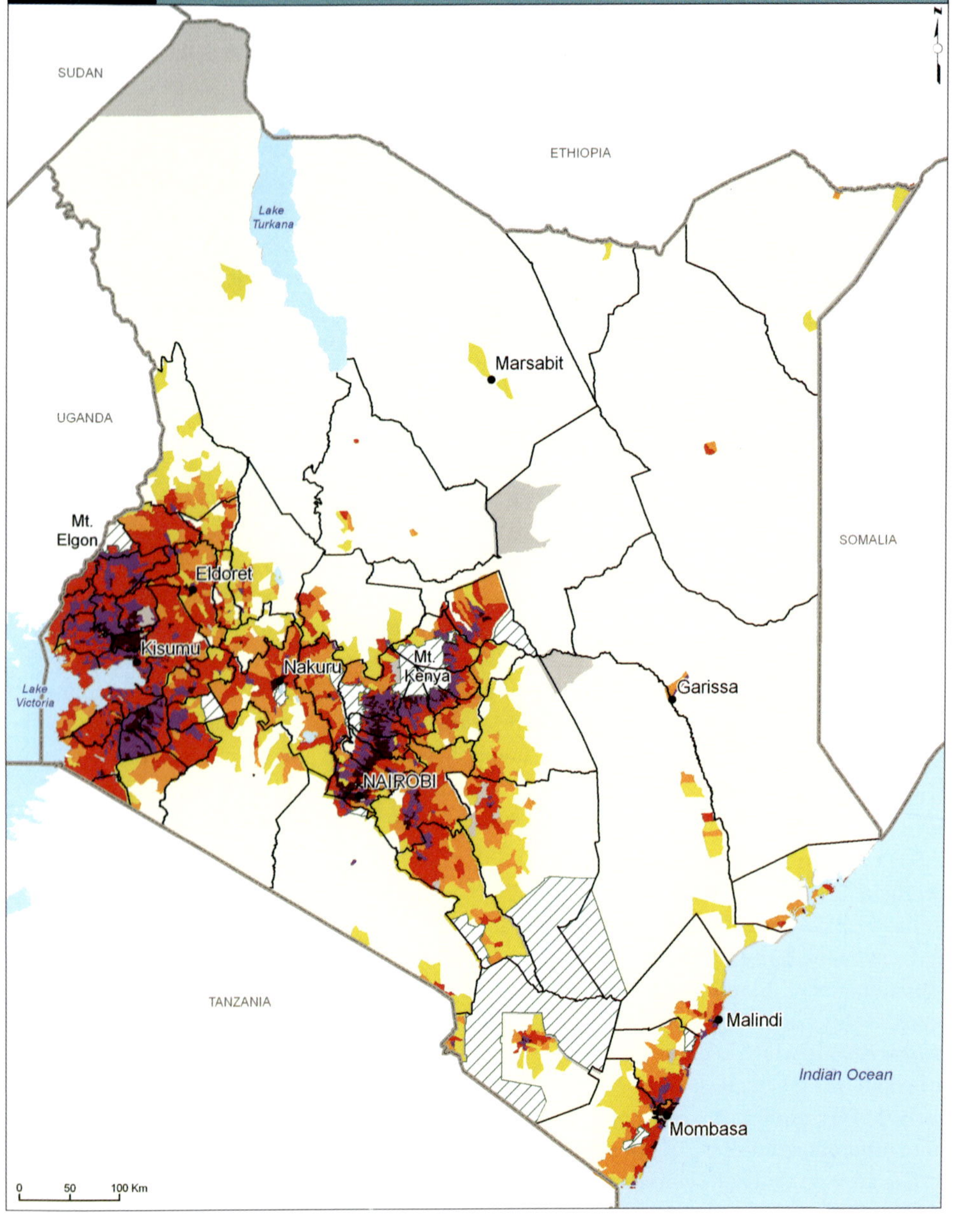

Map 2.5 Human Population Density, 1999

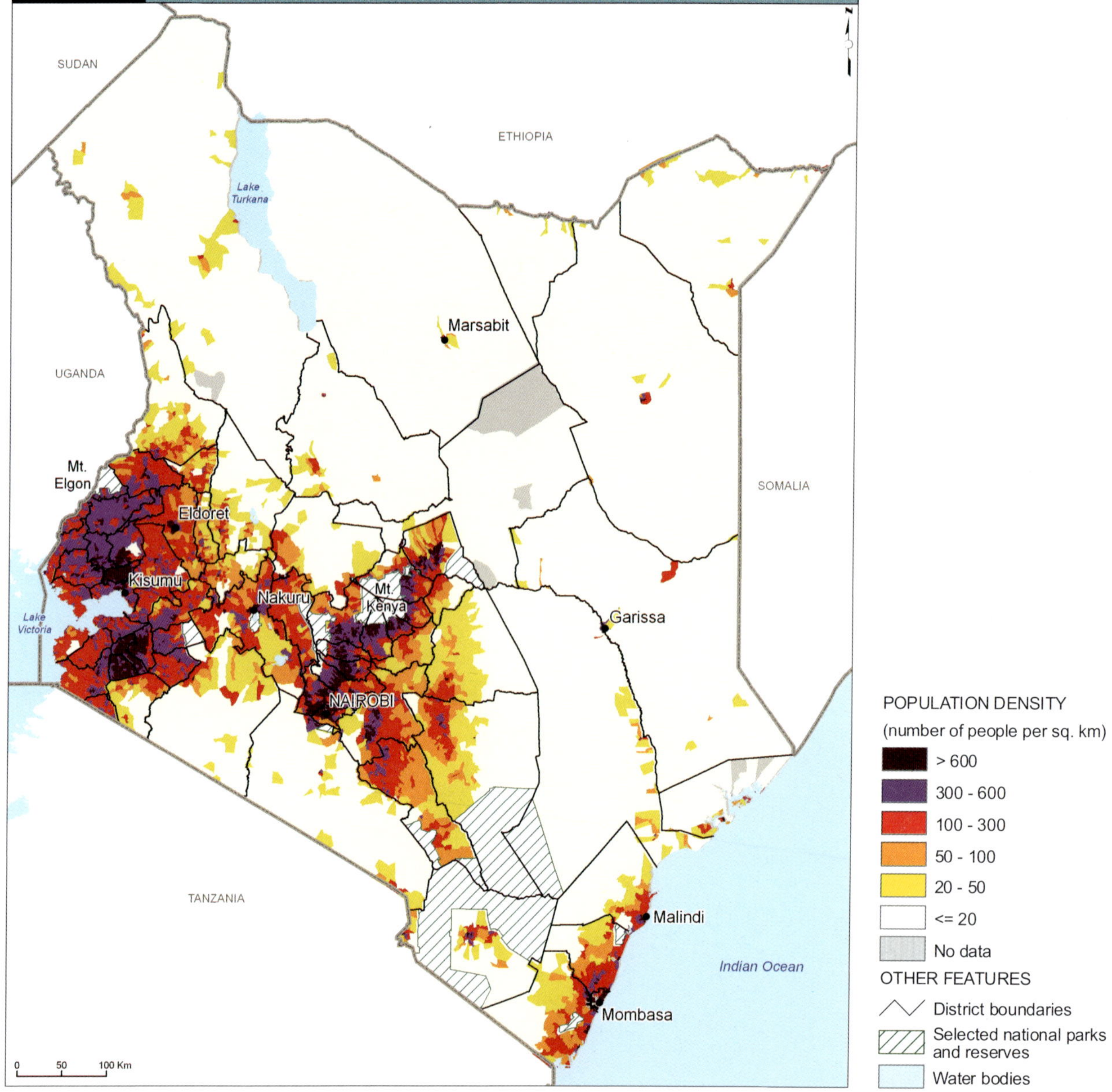

Sources: Administrative boundaries (CBS 2003), cities (SoK and ILRI 2000), water bodies (FAO 2000), parks and reserves (IUCN and UNEP/WCMC 2006), 1989 population density (CBS 1995), and 1999 population density (CBS 2002).

Kenya's population is concentrated in the most productive agricultural lands, near major fisheries in Lake Victoria and along the Indian Ocean coast, and around important market and economic centers. Areas with the highest population density (dark purple-shaded map areas, with more than 600 people per square kilometer) are found in the central highlands between Nairobi and Mount Kenya. In western Kenya, the number of very densely populated areas has risen sharply over the past decade, with such areas found mainly northwest of Kisumu town (Western Province) and in the three Districts slightly inland from the southern shores of Lake Victoria (Nyanza Province).

Map 2.6 Poverty Rate: Percentage of Population Below the Poverty Line, 1999

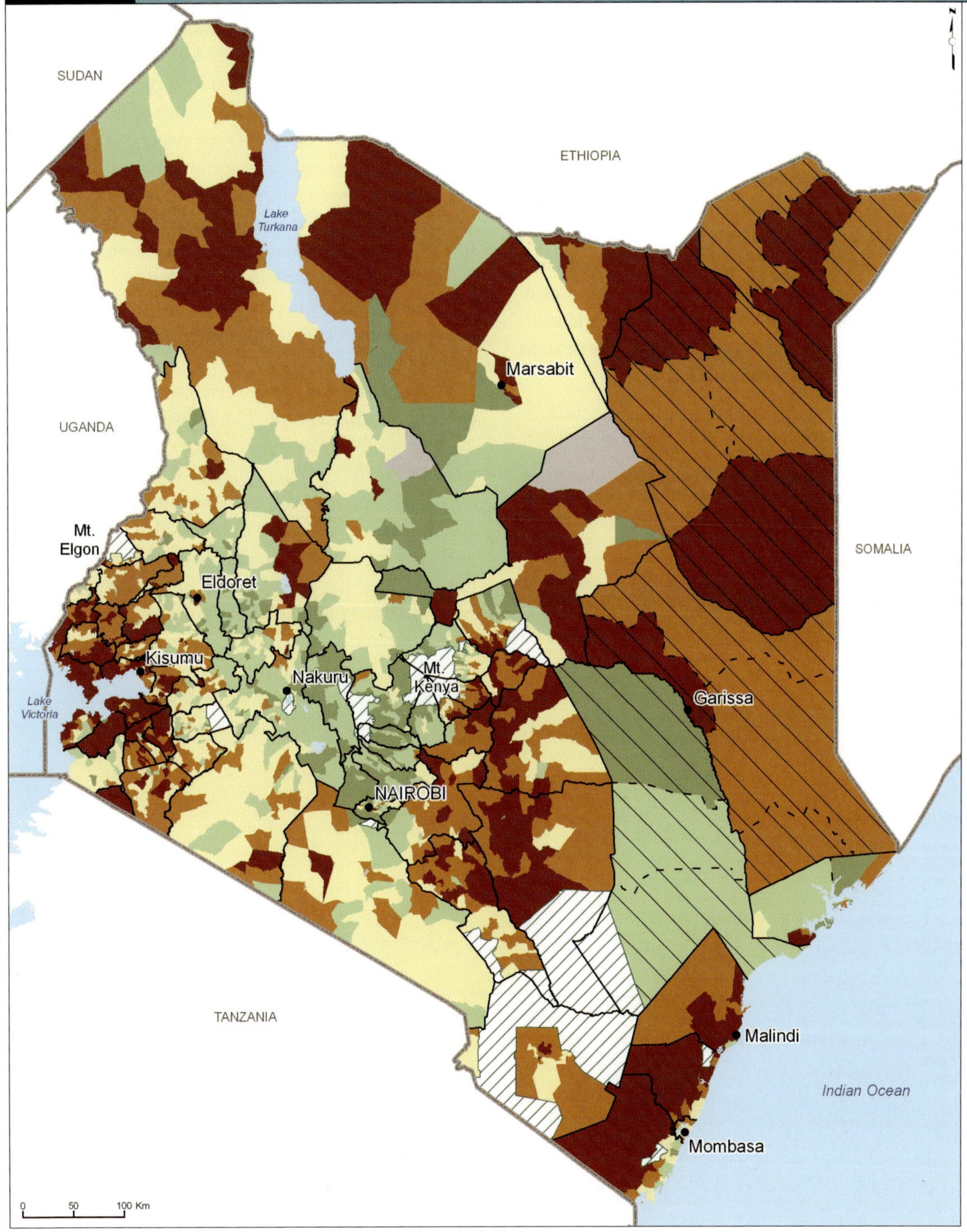

Sources: Administrative boundaries (CBS 2003), cities (SoK and ILRI 2000), water bodies (FAO 2000), parks and reserves (IUCN and UNEP/WCMC 2006), 1999 poverty rate for rural Locations and urban Sublocations (CBS 2003), and 1999 poverty rate for Constituencies (CBS 2005).

The spatial distribution of poverty rates varies markedly across Kenya. Mapping poverty incidence at the local level reveals variation that is masked by maps based on average values across entire Provinces and Districts. Less poor Districts, such as those to the north and east of Mount Kenya, tend to be more spatially heterogeneous with respect to local poverty rates; here, pockets of relatively high poverty rates frequently adjoin more prosperous administrative areas. Two thirds of the 69 Districts shown contain at least one administrative area with poverty rates in excess of 65 percent.

Note: This map shows poverty rates for the smallest administrative areas available, combining estimates at three different scales: 2,056 rural Locations (covering most of Kenya), 80 urban Sublocations (Nairobi, Mombasa, Nakuru, Kisumu, and Eldoret), and 14 Constituencies (covering the northeastern part of the country). The urban estimates are based on a poverty line of Ksh 2,648 per month while the rest of the country is based on the rural poverty line of Ksh 1,239 per month. The poverty estimates for the 14 Constituencies are generally associated with a higher standard error than the other administrative units, a result of the statistical estimation technique (CBS 2005).

POVERTY RATE
(percent of the population below the poverty line)
- > 65
- 55 - 65
- 45 - 55
- 35 - 45
- <= 35
- No data
- Data calculated by Constituency

OTHER FEATURES
- District boundaries
- Constituency boundaries
- Selected national parks and reserves
- Water bodies

Map 2.7 Poverty Density: Number of Poor People Per Square Kilometer, 1999

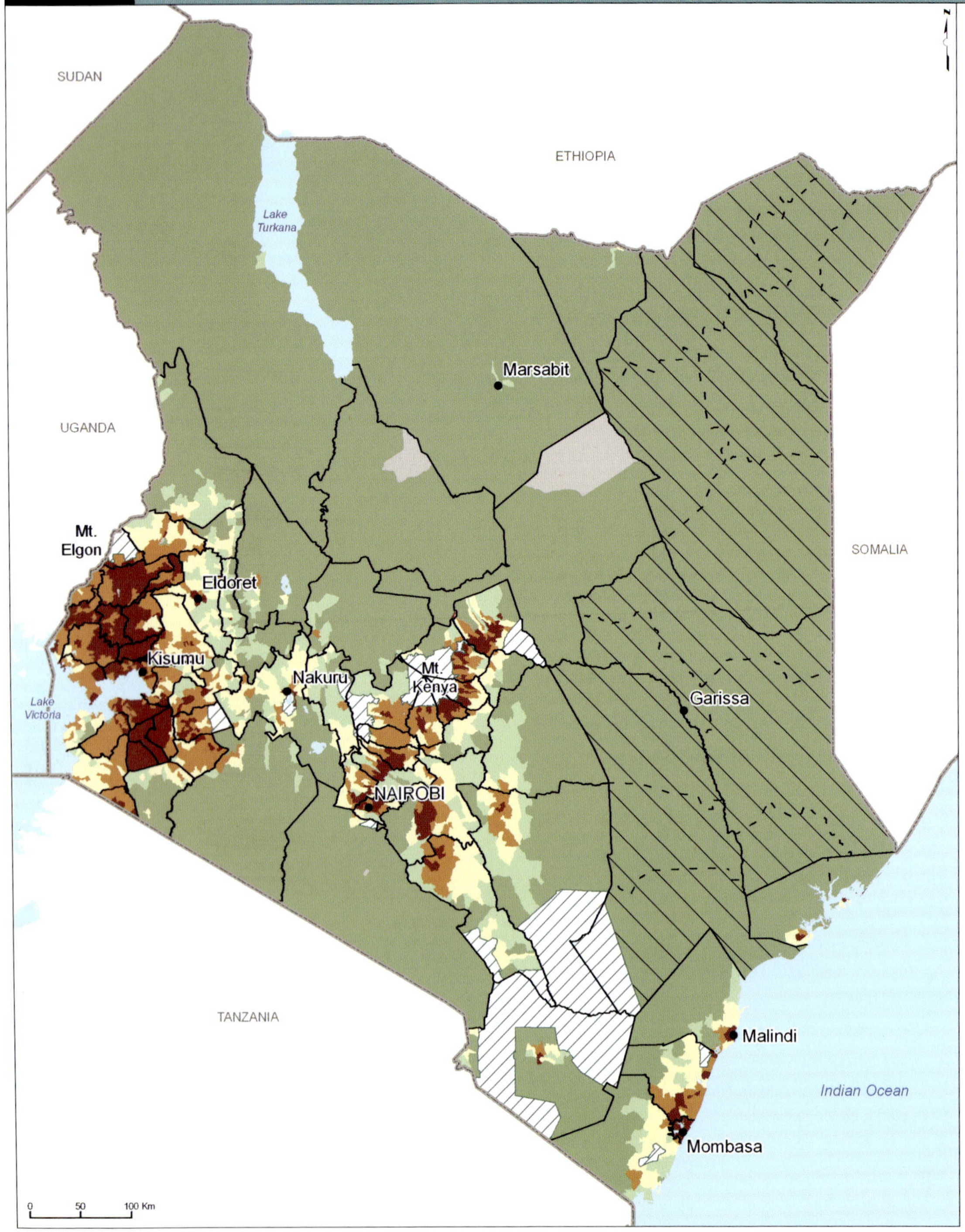

Sources: Administrative boundaries (CBS 2003), cities (SoK and ILRI 2000), water bodies (FAO 2000), parks and reserves (IUCN and UNEP/WCMC 2006), and 1999 poverty density (WRI/ILRI calculation based on 1999 poverty rates from Map 2.6).

***Poverty density* follows spatial patterns that are quite distinct from the distribution of *poverty rates*. Poverty density generally reflects patterns of overall population density. As a result, more than 60 percent of Kenya's rural poor live in just 31 percent of the 2,056 rural Locations shown on the map. Conversely, poverty density is lowest in remote, sparsely populated areas (mostly in arid and semi-arid ecosystems), even though many of these areas exhibit extremely high poverty rates.**

Note: This map shows poverty densities for the smallest administrative areas available, combining estimates at three different scales: 2,056 rural Locations (covering most of Kenya), 80 urban Sublocations (Nairobi, Mombasa, Nakuru, Kisumu, and Eldoret), and 14 Constituencies (covering the northeastern part of the country). The urban estimates are based on a poverty line of Ksh 2,648 per month while the rest of the country is based on the rural poverty line of Ksh 1,239 per month. The poverty estimates for the 14 Constituencies are generally associated with a higher standard error than the other administrative units, a result of the statistical estimation technique (CBS 2005).

POVERTY GAP AND RESOURCES NEEDED TO CLOSE THE GAP

For some policy analysis and decision-making, it is important to know not only *how many* people are poor, but also *how poor* they are, on average. The *poverty gap* (also known as the depth of poverty) is an indicator that captures this aspect of poverty. It measures the average expenditure shortfall (gap) for the poor in a given administrative area relative to the poverty line, that is, how far below the poverty line the poor in a given area are. For example, a poverty gap of 10 percent means that, on average, the household expenditures of the poor are 10 percent below the expenditure level that defines the poverty line.

Map 2.8 depicts the poverty gap for administrative areas within Kenya. The household survey data underlying this map shows a poverty gap of 19.3 percent for the rural population of Kenya. This means that, on average, each poor person in a rural area would require an additional Ksh 239 (US$ 3.41) per month to move out of poverty (i.e., 19.3 percent times the rural poverty line of Ksh 1,239). Note that this national average masks considerable spatial variation, with poverty gaps ranging from less than 10 percent in wealthier areas of Central and Nairobi Provinces to more than 30 percent in the poorest areas of Nyanza and Coast Provinces.

Map 2.8 Poverty Gap as a Percent of Poverty Line, 1999

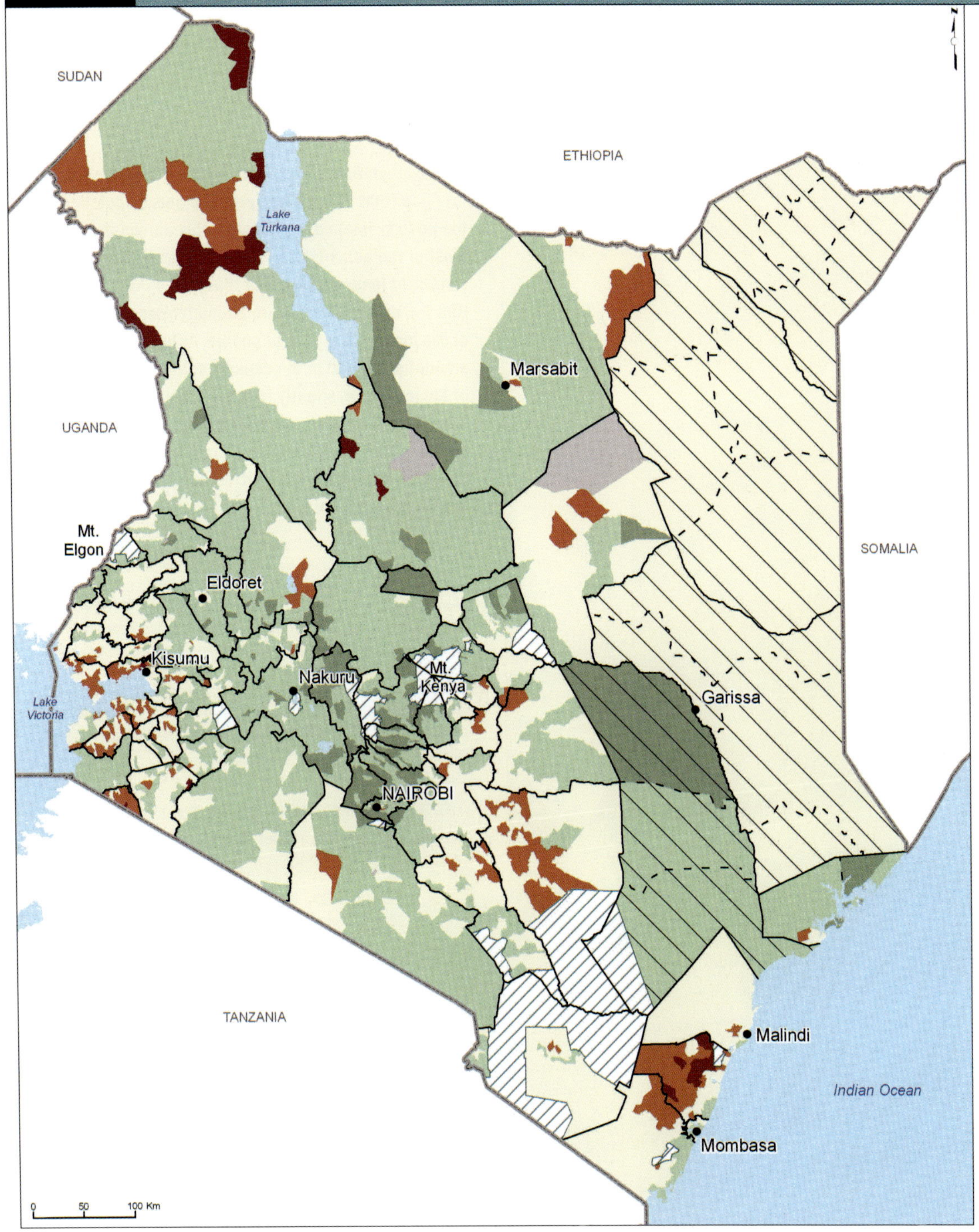

Sources: Administrative boundaries (CBS 2003), cities (SoK and ILRI 2000), water bodies (FAO 2000), parks and reserves (IUCN and UNEP 2006), 1999 poverty gap for rural Locations and urban Sublocations (CBS 2003), and 1999 poverty gap for Constituencies (CBS 2005).

The poverty gap—a measure capturing not only the proportion of the population that is poor but also how poor they are—varies across Provinces and Districts. In many Locations within Central and Nairobi Provinces, poverty gaps are on the order of less than 10 percent of the poverty line (dark green-shaded map areas), meaning that the average poor person needs only about Ksh 124 (US$ 1.77) per month to move out of poverty. In contrast, areas with large poverty gaps (brown-shaded map areas showing rates greater than 30 percent) occur in many parts of Nyanza and Coast Provinces. In these areas, more than Ksh 350 (US$ 5.00) per month may be needed to lift the average poor person above the poverty line.

Note: This map shows poverty gaps for the smallest administrative areas available, combining estimates at three different scales: 2,056 rural Locations (covering most of Kenya), 80 urban Sublocations (Nairobi, Mombasa, Nakuru, Kisumu, and Eldoret), and 14 Constituencies (covering the north-eastern part of the country). The urban estimates are based on a poverty line of Ksh 2,648 per month while the rest of the country is based on the rural poverty line of Ksh 1,239 per month. The poverty gap estimates for the 14 Constituencies are generally associated with a higher standard error than the other administrative units, a result of the statistical estimation technique (CBS 2005).

POVERTY GAP
(percent of poverty line)

- > 40
- 30 - 40
- 20 - 30
- 10 - 20
- <= 10
- No data
- Data calculated by Constituency

OTHER FEATURES

- District boundaries
- Constituency boundaries
- Selected national parks and reserves
- Water bodies

POVERTY METRICS

Poverty rate (head count index), poverty gap (depth of poverty), and squared poverty gap (poverty intensity) are the three most widely used metrics to gauge a country's poverty situation. The most easily understood measure is the *poverty rate*, which reflects how widespread poverty is in a given area (it is typically shown as the percentage of people falling below the poverty line). This measure captures changes in poverty as soon as a family moves above or below the poverty line. However, it does not reflect any changes in household expenditures or incomes that those who remain below the poverty line may make. That is, it does not probe the depth of poverty.

The *poverty gap* can capture such changes. It measures how far below the poverty line the poor in a given area are. Economists calculate the poverty gap by adding up all the shortfalls of the poor (ignoring the non-poor) and dividing it by the total population. It is possible for an area to experience a decline in the poverty gap, but no change in the poverty rate (i.e., slight increases in household expenditures or incomes that do not allow families to cross the poverty line). One of the caveats of the poverty gap is that the average used in its calculation conceals that some poor households in an area might only be a few shillings below the poverty line, while others in the same area might be much farther below the poverty line.

The *squared poverty gap* incorporates the inequality among poor people into its calculation. This measure accounts for the number of poor, the depth of poverty, and the inequality among the poor. It gives the strongest weighting to the poorest of the poor and lower weighting to less poor households (i.e., households that have higher expenditures or incomes but still fall below the poverty line).

(Ravallion 1992 and CBS 2005).

The poverty gap is a crude estimate of the minimum amount of resources needed to eradicate poverty. By multiplying the poverty gap with the poverty line and the number of poor in an administrative area, analysts can determine the amount of shillings needed to lift out of poverty all of the poor in a given area. For example, given a nationwide average rural poverty gap of 19.3 percent, a rural poverty line of Ksh 1,239 per household per month, and a total population of rural poor of 11.4 million, at least Ksh 2.74 billion (US$ 39.1 million at US$ 1 = Ksh 70) per month would be needed to eliminate poverty for all rural Kenyan families. This is a minimum estimate based on assumptions of perfect targeting, no corruption, and no program costs. In practice more resources and different approaches will be required because perfectly targeted cash transfers are neither feasible nor the best intervention to eradicate poverty.

Map 2.9 converts the percentage figures of Map 2.8 into Kenyan shillings. It presents a standardized measure dividing the total shillings needed in each administrative area by its size in square kilometers. These standardized expenditure shortfalls are not distributed evenly across the country, but unlike in Map 2.8 the greatest amount of total resources (shown in dark brown) are now needed in areas with high poverty densities (as shown in Map 2.7).

Map 2.9 **Minimum Amount of Kenyan Shillings Needed Per Square Kilometer Per Month to Close the Poverty Gap, 1999**

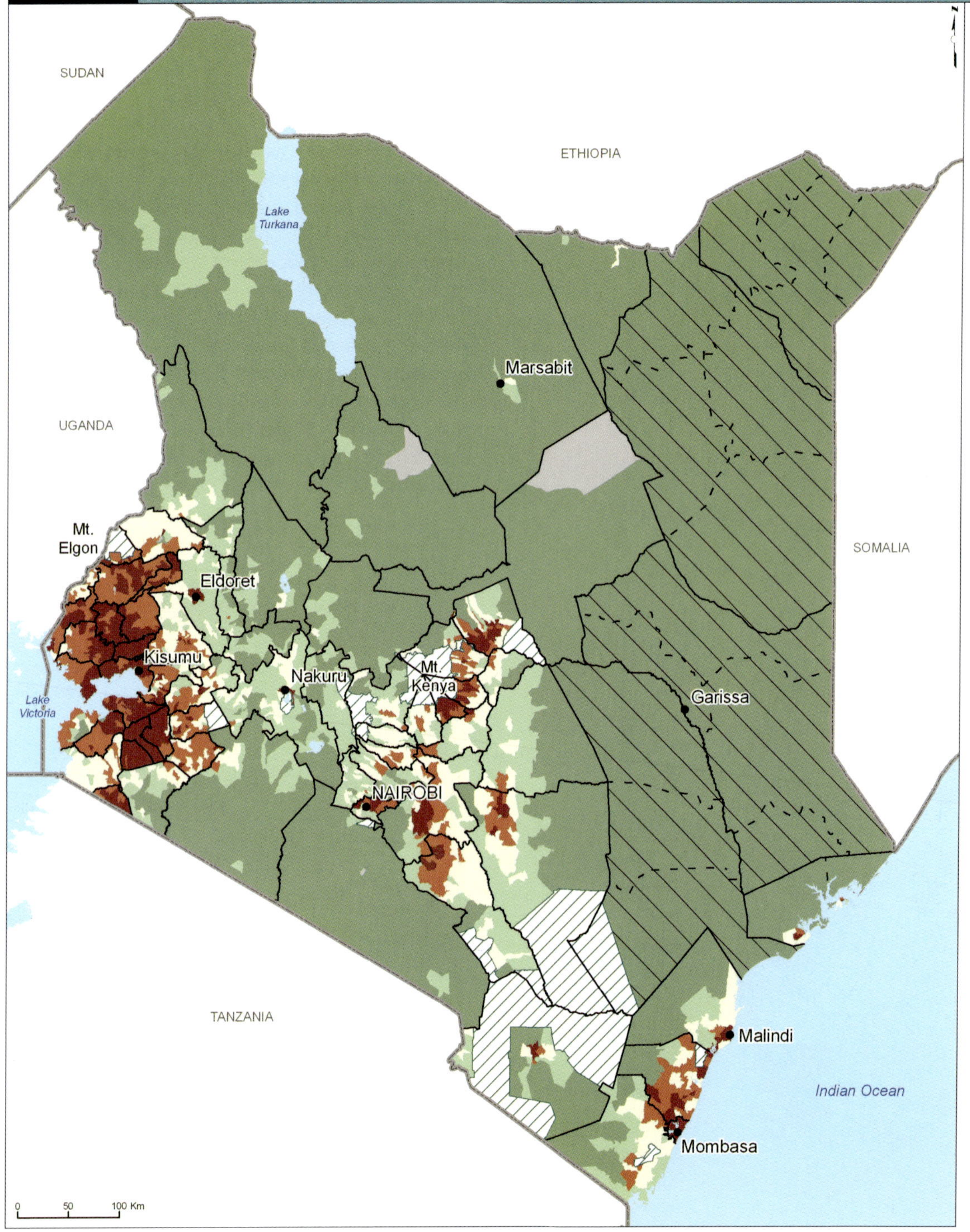

Sources: Administrative boundaries (CBS 2003), cities (SoK and ILRI 2000), water bodies (FAO 2000), parks and reserves (IUCN and UNEP 2006), and 1999 Kenyan shillings per square kilometer (WRI/ILRI calculation for Locations, Sublocations, and Constituencies based on poverty gaps in Map 2.8).

The magnitude of resources needed to close the poverty gap (that is, to raise the entire poor population above the poverty line) varies considerably across the country. Most of the administrative areas in Kenya's arid and semi-arid lands require less than Ksh 4,000 (US$ 57 at US$ 1 = Ksh 70) per square kilometer per month, a result of the low density of poor people. At least 15 times that amount is needed in the densely settled areas northwest of Kisumu town, slightly inland from the southern shore of Lake Victoria, and in parts of Nairobi and Coastal Provinces. (Note that these estimates are minimum investments, based on assumptions of perfect targeting, no corruption, and no program costs.)

Note: This map shows data for the smallest administrative areas available, combining estimates at three different scales: 2,056 rural Locations (covering most of Kenya), 80 urban Sublocations (Nairobi, Mombasa, Nakuru, Kisumu, and Eldoret), and 14 Constituencies (covering the northeastern part of the country). The urban estimates are based on a poverty line of Ksh 2,648 per month while the rest of the country is based on the rural poverty line of Ksh 1,239 per month. The estimates for the 14 Constituencies are generally associated with a higher standard error than the other administrative units, a result of the statistical estimation technique (CBS 2005).

THEORETICAL INVESTMENT TO FILL THE POVERTY GAP
(Kenyan shillings per month per sq. km)

- > 60,000
- 28,000 - 60,000
- 12,000 - 28,000
- 4,000 - 12,000
- <= 4,000
- No data
- Data calculated by Constituency

OTHER FEATURES

- District boundaries
- Constituency boundaries
- Selected national parks and reserves
- Water bodies

MEASURES OF INEQUALITY AND HOUSING QUALITY

In addition to looking at spatial patterns of poverty incidence and the poverty gap, examining the spatial dimensions of economic inequality can also provide important insights for policy analysis and decision-making. Measures of inequality look at the distribution of economic welfare across the entire population (both poor and non-poor), rather than just considering the income or consumption shortfalls of the poor. Such measures can be useful indicators of a society's well-being, since high levels of economic inequality can strain the fabric of society, eroding social capital and diminishing social cohesion.

One of the principal indicators used by economists to measure inequality is the Gini coefficient, which varies between 0 (total equality, where each individual or household has the same income or expenditure) and 1 (total inequality, where one person has everything). As indicated in Map 2.10, inequality as measured by the Gini coefficient varies considerably within and between administrative areas.

As discussed earlier in this chapter, not all dimensions of human well-being can be captured by money-metric indicators of poverty or inequality. To capture one such non-monetary dimension of well-being, we constructed an index of *housing quality* using data collected from each household in Kenya's 1999 census. Map 2.11 depicts this housing quality index throughout Kenya's Districts. The index combines measures of the quality of the materials used to provide roofing, flooring, and walls in Kenyan homes. A dwelling was considered to be "poor quality housing" if it was rated as "poor" in all three categories, that is, having a "poor quality" roof, floor, and walls (see box beside Map 2.11 for further details).

Map 2.10 **Average Inequality of Per Capita Expenditures, 1999**

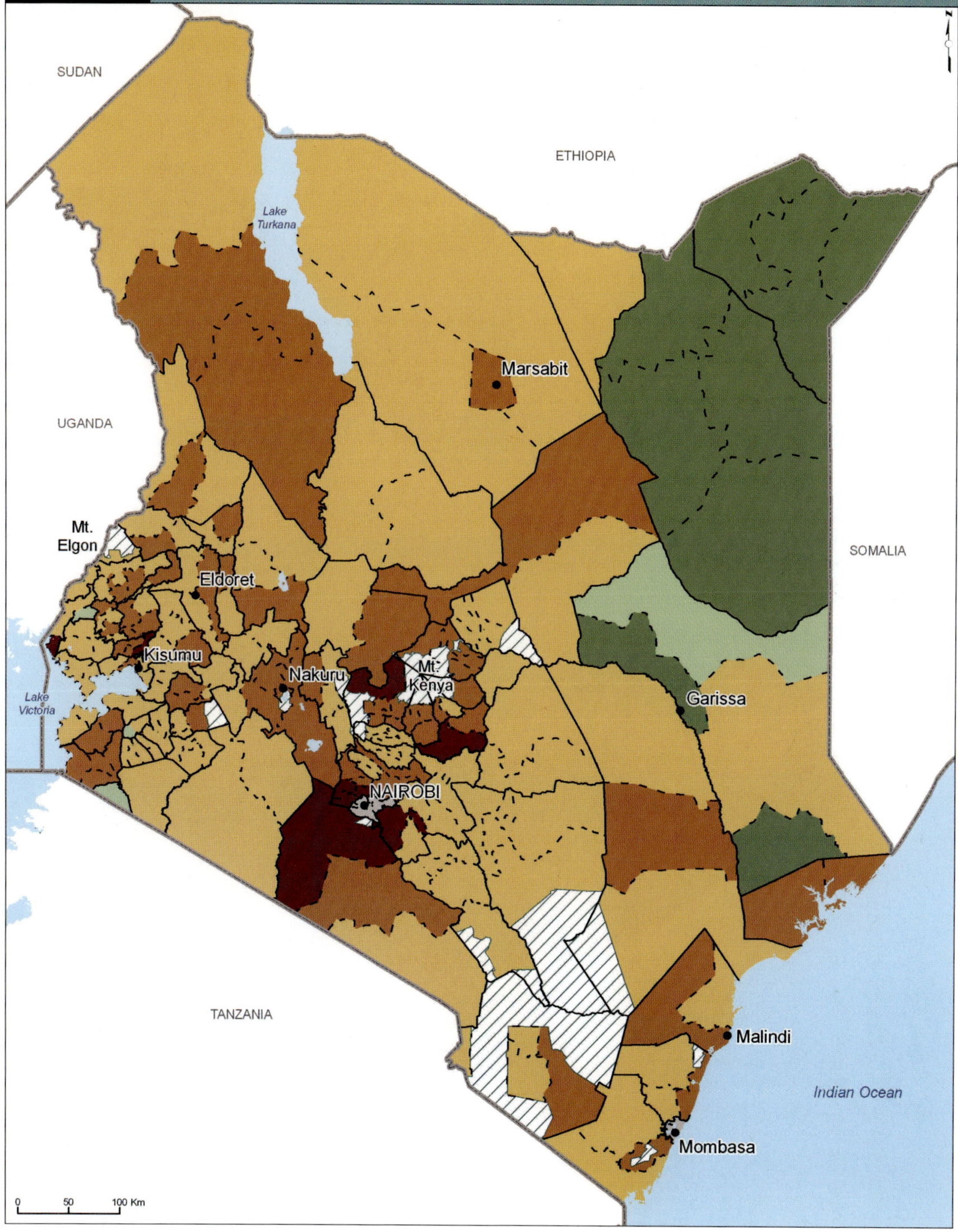

Sources: Administrative boundaries (CBS 2003), cities (SoK and ILRI 2000), water bodies (FAO 2000), parks and reserves (IUCN and UNEP/WCMC 2006), and 1999 Gini coefficient for 210 Constituencies (CBS 2005).

Spatial patterns of inequality in per capita expenditure (as measured by the Gini coefficient) differ at the sub-provincial level. Areas of highest inequality (shaded dark brown) are found near urban areas, including Nairobi and large towns such as Kisumu. Inequality also is quite high in some less poor areas of the central highlands and Rift Valley, perhaps due to very poor subsistence farmers living side by side with more prosperous households earning higher incomes from commercial agriculture. Not surprisingly, inequality is lower in areas of the north, Western Province, North East Province, and Coastal Province where the populace is more uniformly poor.

GINI COEFFICIENT
- 0.41 - 0.43
- 0.36 - 0.40
- 0.31 - 0.35
- 0.26 - 0.30
- 0.22 - 0.25
- No data

OTHER FEATURES
- District boundaries
- Constituency boundaries
- Selected national parks and reserves
- Water bodies

CONSTRUCTING AN INDEX OF HOUSING QUALITY

To move beyond the money-metric indicators of welfare and inequality, we calculated an index of housing quality. Housing quality captures the "shelter" dimension of well-being and may also represent a proxy of wealth.

Kenya's 1999 *Population and Housing Census* (an often under-used source of well-being data) collected data on housing quality from each household (CBS 2002). To calculate the index, we categorized the building materials used for the roof, walls, and floors of a household's dwelling. The census groups the roofing materials into eight classes: corrugated iron sheets, tiles, concrete, asbestos sheets, grass, *makuti* (thatched roofing material made from dried coconut palm leaves), tin, and "others." If a household uses grass, *makuti*, tin, or "others," then we classify it as having a "poor quality roof." We disregard the fact that some households prefer grass-thatched houses to others. For the wall type, we use the same approach. We consider nine types of wall: stone, brick/block, mud/wood, mud/cement, wood only, corrugated iron sheet, grass/reeds, tin, and others. Households that have mud/wood, wood only, grass/reeds, tin, and "others" are classified as having "poor quality walls." Accordingly households with "earth" and "other" floor types are classified as having "poor quality floors' compared to those that have "cement," "tiles," or "wood only" floor types. We assumed that no household prefers a "poor floor type" to a "non-poor" one for any reason other than the inability to afford it.

To derive an index reflecting the quality of shelter, we combined these three measures. The proportion of households in a Location with "poor quality housing" is defined as those families that rate "poor" on all three dimensions, that is, having a "poor quality roof," "poor quality floors," and "poor quality walls."

Map 2.11 Percentage of Households With Poor Quality Housing, 1999

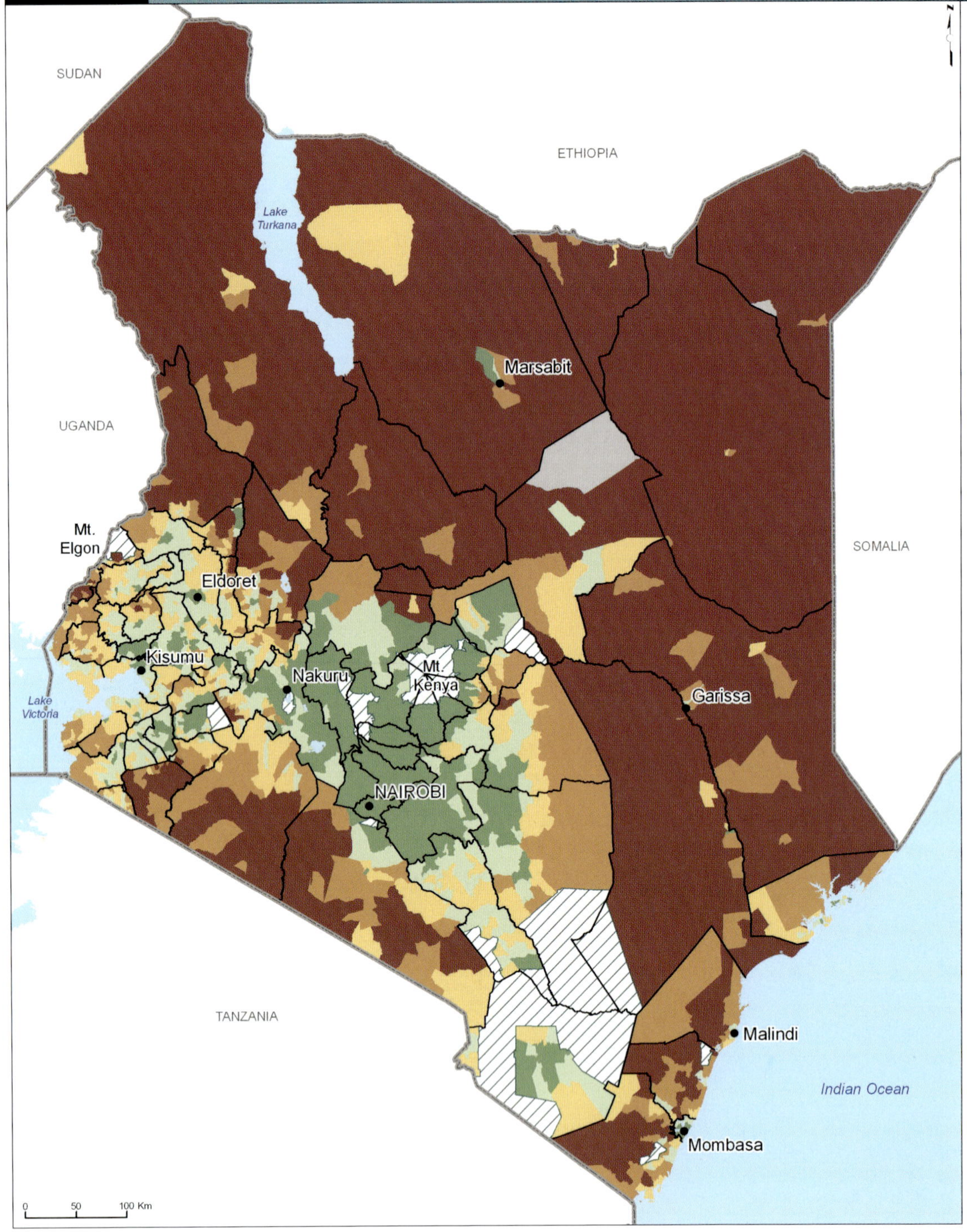

Sources: Administrative boundaries (CBS 2003), cities (SoK and ILRI 2000), water bodies (FAO 2000), parks and reserves (IUCN and UNEP/WCMC 2006), and 1999 Housing Quality for Locations (CBS/ILRI/WRI calculation based on 1999 *Population and Housing Census*).

In most parts of Kenya (dark brown-shaded map areas), the majority of households live in "poor quality" homes made of inferior materials for roofing, flooring, and walls (see Box on the left). Housing quality is higher in the central regions of the country (green-shaded map areas). This echoes the spatial pattern of poverty rates. One exception is the administrative areas in the Locations northwest of Kisumu and slightly inland from the southern shores of Lake Victoria. These Locations show a higher share of better quality housing, but are very poor in terms of per capita expenditure indicators (i.e., poverty rate and poverty density) as shown in Maps 2.4 and 2.5. (Note that Map 2.11 hides high concentrations of very poor housing in small areas such as the informal settlements of Nairobi. It is a result of the scale of administrative areas, the percentage thresholds, and the index components selected for this national view.)

SUMMING UP

- Human well-being has many elements. Sufficient income to obtain adequate food and shelter are important dimensions as are security, good health, social acceptance, access to opportunities, and freedom of choice.
- Poverty is defined as a lack of these elements of well-being.
- Human well-being relies fundamentally on the ability to access a wide variety of ecosystem services.
- The majority of the 32.8 million Kenyans (2004) live in the most productive agricultural lands, near major fisheries in Lake Victoria, and along the Indian Ocean coast. Here, rural population densities of greater than 600 persons per square kilometer are not uncommon. Most of Kenya's arid and semi-arid lands show population densities of less than 20 persons per square kilometer.
- Official Kenyan poverty statistics are based on detailed information about household expenditures. They use a rural poverty line of Ksh 1,239 per month (about US$ 0.59 per day) and an urban poverty line of Ksh 2,648 per month (about US$ 1.26 per day). Given these poverty lines, about 53 percent of rural and 50 percent of urban Kenyans were poor in 1997.
- *Poverty rate* (i.e., the percentage of the population below the poverty line) and *poverty density* (the number of poor in a given area) provide two distinct ways to depict the spatial distribution of poverty. Maps of these two indicators often show quite different patterns. Understanding the relationship between *poverty rate* and *poverty density* is important for designing and implementing poverty reduction interventions.
- The *poverty gap* measures how far below the poverty line the poor in a given area are. The poor in Kenya's rural areas have household expenditures that are on average 19.3 percent below the rural poverty line. On average, each rural Kenyan would require an additional Ksh 239 (US$ 3.41 at US$ 1 = Ksh 70) per month to move out of poverty.
- Most of the administrative areas in Kenya's arid and semi-arid lands require less than Ksh 4,000 (US$ 57 at US$ 1 = Ksh 70) per square kilometer per month to close the poverty gap, that is, to raise all families above the poverty line. At least 15 times that amount is needed in the more densely settled parts of the country.
- Measures of *inequality* look at the distribution of economic welfare across the entire population (both poor and non-poor). Areas of highest inequality are found near urban areas and large towns. Inequality is low in rural areas with the highest poverty rates.
- *Housing quality*, a measure reflecting overall wealth of a household, is higher in the central regions of the country.

In Chapter 3

CONTENTS

LIST OF MAPS

AUTHORS AND CONTRIBUTORS

Janet Nackoney (WRI)
Bancy Mati (ICRISAT)
Greg Mock (consultant)
Linda Cotton (consultant)
Norbert Henninger (WRI)
Florence Landsberg (WRI)
An Notenbaert (ILRI)
Russ Kruska (ILRI)
Eunice Kariuki (ILRI)
Pamela Ochungo (ILRI)
Dan Tunstall (WRI)
Carolina de Rosas (WRI)
Hyacinth Billings (WRI)
Jamie Worms (WRI)

WHAT THIS CHAPTER SHOWS

Water is unique from an ecosystem perspective because water and the associated freshwater systems are linked to all four categories of ecosystem services. This chapter provides an overview of water availability and demand, and describes where in Kenya specific water uses are concentrated. A first set of maps shows the uneven availability of surface water as exemplified by annual and seasonal rainfall, as well as the network of permanent and intermittent rivers. The next map compares water availability to projected demand from households, industry, and agriculture and highlights where demand is projected to exceed local surface and groundwater supplies. The following section presents a series of maps reflecting the main uses and users of water in Kenya: sources of drinking water supply across the country, water transfers to Kenya's two largest urban areas, subdrainage areas important for electricity generation, location of irrigated crop production, and water demand from livestock and wildlife in the rangelands. Two maps showing the occurrence of floods conclude this chapter. They serve as a reminder that impacts from ecosystem processes are not always benign: what constitutes a service for one group or area may be very detrimental to another group or area.

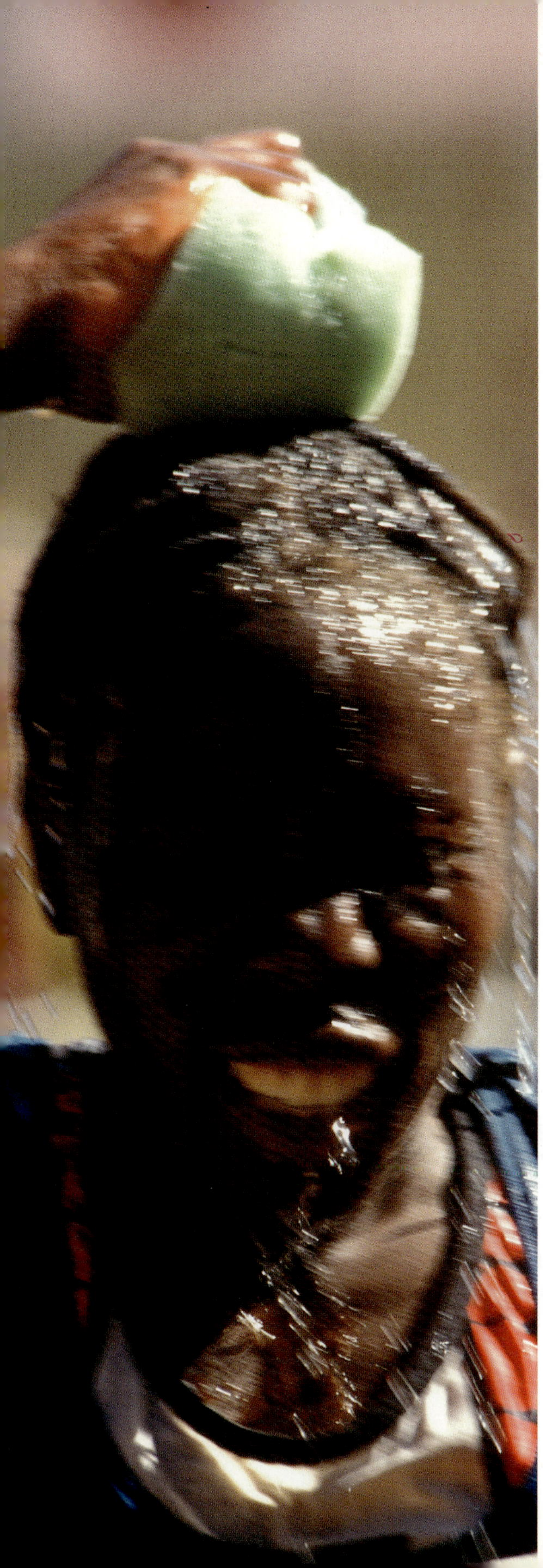

3

Water

Water in sufficient quantity and quality is essential for human well-being. Kenyans use water for drinking, energy generation, livestock production, agriculture, tourism, industry, and many other livelihoods. Lack of adequate, good-quality water is therefore a significant obstacle to development. Exposure to unsafe water, for example, is a major contributor to child mortality and disease in Kenya. Reduced access to water increases collection time—a burden that falls disproportionately on women and children—taking time away from other productive tasks, such as going to school.

Water is also the lifeblood of Kenya's ecosystems. The hydrological cycle sustains life: all organisms need water to survive. Water enters the terrestrial environment as precipitation and then turns into surface flows and groundwater. In the process, aquatic systems such as rivers, lakes, wetlands, and other freshwater habitats are created. Ecological processes such as the cycling of nutrients also depend on water. Unfortunately, water is not always plentiful in Kenya, and the country has been characterized as water scarce. This poses challenges for water management now and in the future.

From an ecosystem standpoint, water is unique in that it is linked to all four categories of ecosystem services (MA 2005):

- *Provisioning services* of freshwater systems include the storage and retention of water (in lakes, rivers, and as groundwater) for domestic, agricultural, and industrial use. Water is a vital input for the production of food (e.g., fish, irrigated crops, and livestock), timber, fiber, and fuel. Of course, freshwater itself is a product for consumption.
- *Regulating services* of freshwater systems and important freshwater habitats such as wetlands include modifying water flows (hydrological flows), recharging and discharging groundwater resources, and diluting or removing pollutants. The ability of freshwater systems to provide these services is strongly linked to the type of vegetation cover and to land cover changes, such as conversion of wetlands or expansion of urban areas.
- *Supporting services* of the hydrological cycle are important for soil formation and soil loss (erosion) and nutrient cycling. Freshwater systems also provide habitat for a great number of species, promoting biodiversity, which underlies the resilience and productivity of ecosystems.
- *Cultural services* include the important recreational benefits provided by lakes and rivers, as well as their spiritual and inspirational roles in different cultures.

Service provision from water often leads to conflicting benefits and costs, depending on the perspective of the different users. A service for one group may be a "disservice" for another. For example, damming rivers for hydroelectric power generation may benefit urban electricity users but harm local fishers. Floods can have both positive and negative impacts depending on the context. While floods can destroy homes, crops, and kill people and animals, they often serve as an important supplier of nutrients to floodplains and are an important factor in maintaining biodiversity and freshwater systems.

This chapter provides an overview of water availability in Kenya as reflected by its annual and seasonal rainfall patterns and networks of permanent and seasonal rivers. It also compares water supply and demand and examines the different uses of water in Kenya's economy. The chapter also highlights floods, one of the potentially hazardous characteristics of water.

The chapter addresses the following questions:

- What is the geographic distribution of water resources in the country?
- How is drinking water obtained in rural and urban areas?
- What is the water demand from livestock and wildlife, and how does it vary across the country?
- How do water and freshwater ecosystems contribute to the economy?

Although this chapter does not specifically examine the topic of wetlands, they deserve a brief note because of their ecological importance. Wetlands cover only 2-3 percent (640,000 ha) of Kenya's surface area (SoK 2003) but play a critical role in Kenya's ecosystems. They provide groundwater recharge and discharge, water storage, filtering of nutrients and pollutants, shoreline stabilization, microclimate stabilization, and habitat for biodiversity. Kenyans raising livestock or growing crops depend on wetlands as a refuge from drought, especially in arid and semi-arid areas (Emerton and Vorhies 1998). Given their high diversity of bird species, wetlands also support tourism activities. It has been a common practice for wetlands in Kenya to be converted to cropland, (e.g., Campbell et al. 2003) undermining their supply of other ecosystem services.

WATER SUPPLY AND DEMAND

Kenya is characterized as having marginal rainfall over most of the country. More than 80 percent of its land area, including much of the northern and eastern regions, is arid or semi-arid and receives very little rain each year (SoK 2003). The area in southwestern Kenya that gets the most rain, known as the highlands, supports 75 percent of the nation's population and generates a significant percentage of Kenya's Gross Domestic Product (SoK 2003).

About 32 percent of Kenyan households rely on groundwater for their drinking water supply (CBS et al. 2004). It is also important for industrial use and for crop and livestock production. People living in arid and semi-arid areas rely heavily on groundwater, as it is often the only reliable source of water. Rainfall permeating the soil provides most of Kenya's groundwater resources (Nyaoro 1999).

Rainfall

Rainfall in Kenya is closely linked to the livelihoods of its citizens and the health of the nation's economy. For example, the La Niña drought of 1998-2000 caused damages (loss of hydropower and industrial production, crop and livestock loss, and health impacts) estimated at 16 percent of GDP in each of the following two years (World Bank 2004). Even this number underestimates the full costs of the drought, because it does not reflect costs associated with famine and malnutrition, including loss of lives and livelihoods. The costs of the El Niño floods of 1997-98 are estimated to be of similar magnitude (11 percent of annual GDP).

For a country straddling the equator, Kenya's annual rainfall is relatively low and varies significantly between seasons and from year to year. The average annual rainfall is 630 millimeters per year (FAO 2005), but it is unevenly distributed across

Map 3.1 Average Annual Rainfall

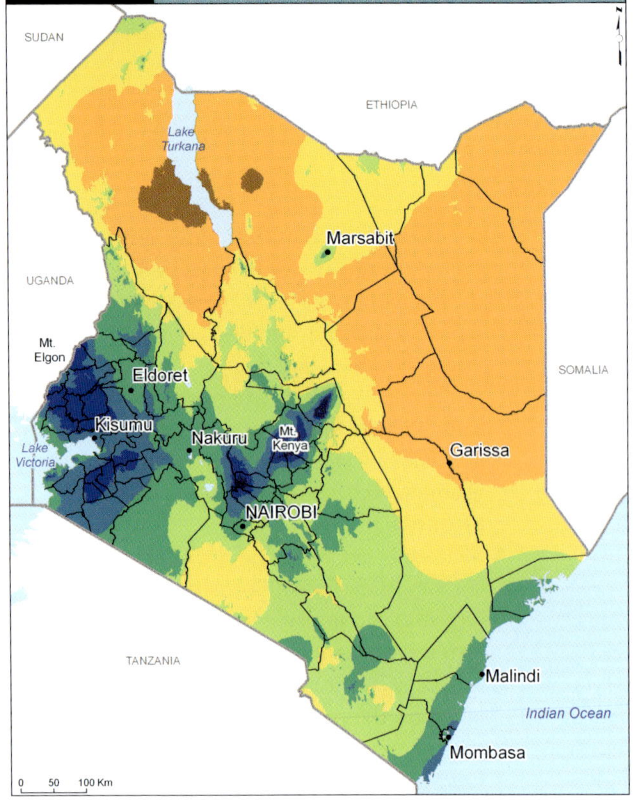

Sources: Administrative boundaries (CBS 2003), cities (SoK and ILRI 2000), water bodies (FAO 2000a), and average annual rainfall (Hijmans et al. 2005).

AVERAGE ANNUAL RAINFALL (millimeters)
- > 2,000
- 1,600 - 2,000
- 1,200 - 1,600
- 800 - 1,200
- 600 - 800
- 400 - 600
- 200 - 400
- <= 200

OTHER FEATURES
- District boundaries
- Water bodies

Areas along the Indian Ocean, in central Kenya close to Nairobi, and in western Kenya bordering Lake Victoria have annual rainfall totals of more than 800 millimeters (a rough benchmark for growing maize). The peaks of high mountain ranges are also associated with elevated rainfall. Rainfall amounts of less than 400 millimeters (which are common in the northern and eastern parts of the country) and of 400–600 millimeters roughly demark Kenya's arid and semi-arid regions.

Map 3.2 Average Monthly Rainfall, February

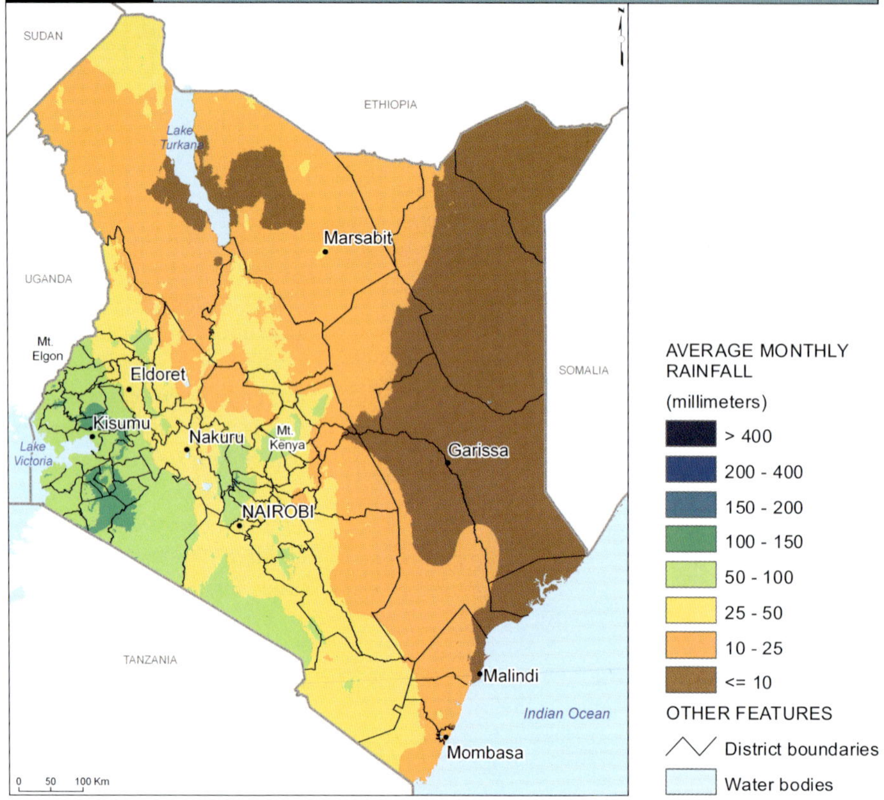

Sources: Administrative boundaries (CBS 2003), cities (SoK and ILRI 2000), water bodies (FAO 2000a), and average monthly rainfall (Hijmans et al. 2005).

Rainfall in Kenya is highly variable throughout the year. These four maps represent the seasonal variability of the country's rainfall, with average statistics for the months of February, April, July, and November. East of the Rift Valley, two distinct rainy seasons occur. The "long" rains, shown in Map 3.3 (approximated by the monthly rainfall in April), fall from March to May, and the "short" rains (approximated in Map 3.5 with the monthly data in November) fall from October to November. However, the areas in the western part of the country bordering Lake Victoria generally experience one long rainy season from March to September (SoK 2003). During the rest of the year, most of Kenya remains relatively dry (Maps 3.2 and 3.4).

In most parts of the country, the "long" rains account for much of the annual rainfall, but the "short" rains nevertheless play a critical role in many areas. The "short" rains (Map 3.5) are essential for crops to mature in the Districts between Mombasa and Nairobi (Makueni, Kitui, Mwingi, and eastern Machakos), all areas with more marginal annual rainfall amounts of 600–800 millimeters (Map 3.1).

Map 3.3 Average Monthly Rainfall, April

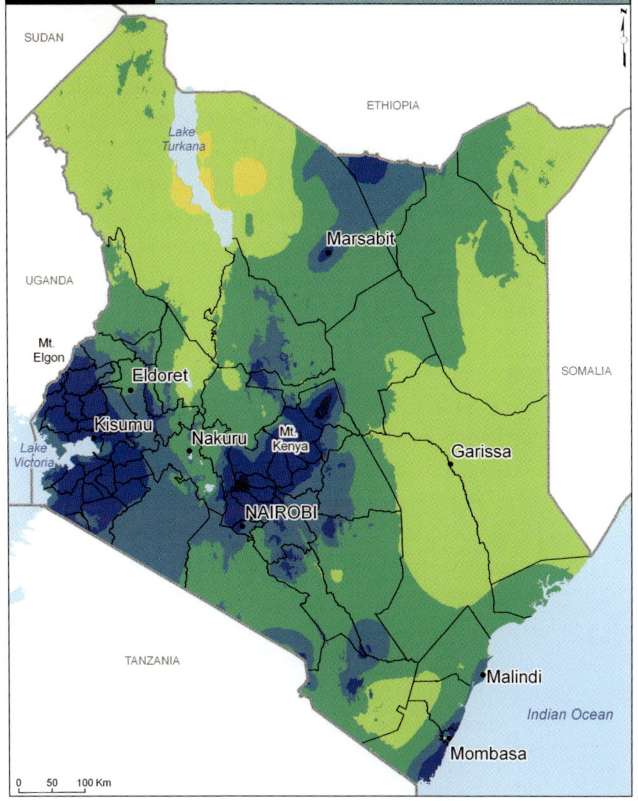

Map 3.4 Average Monthly Rainfall, July

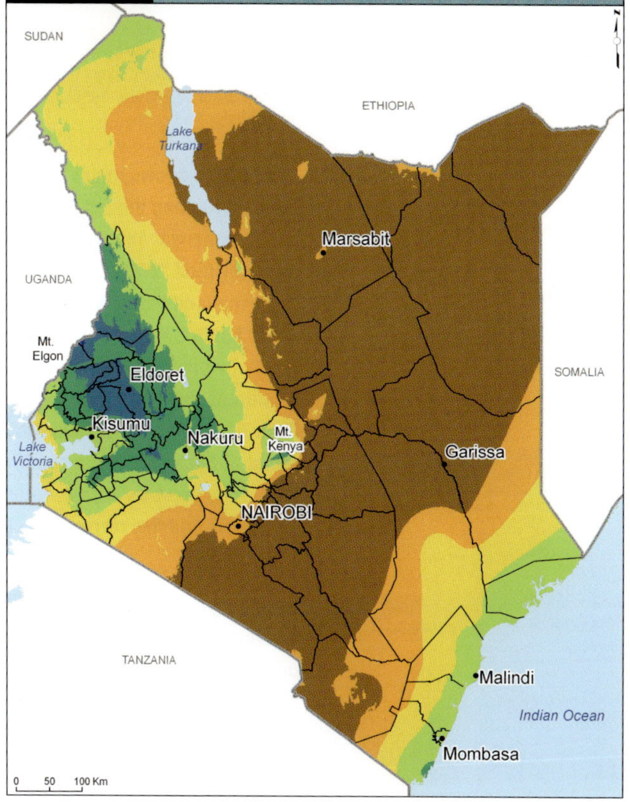

Map 3.5 Average Monthly Rainfall, November

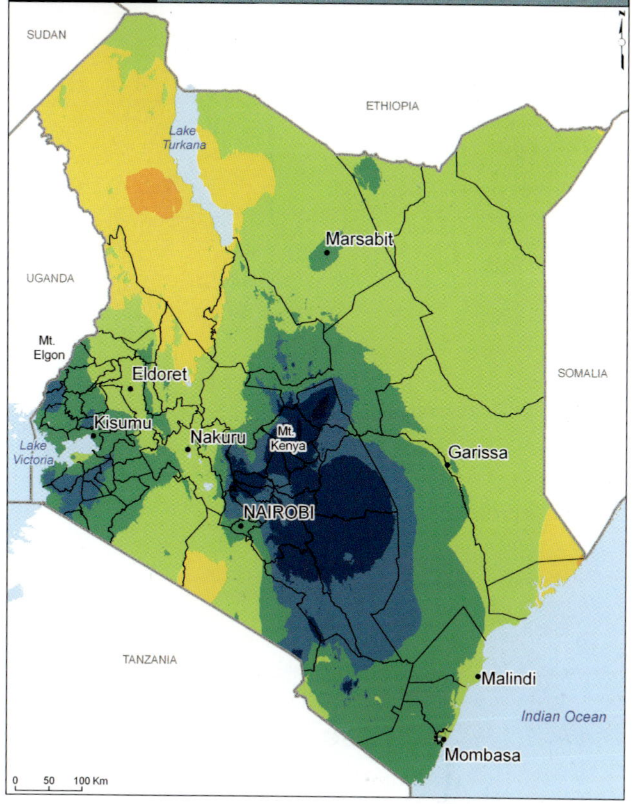

the country (Map 3.1). About 15 percent of Kenya receives sufficient rain to grow maize and other non-drought-resistant crops. Another 13 percent are classified as having more marginal rainfall that is sufficient only to grow selected drought-resistant crops. The remaining 72 percent has no agronomically useful growing season (SoK 2003).

This annual rainfall amount hides the distinct pattern of dry and wet seasons, which vary across Kenya as well (Maps 3.2 – 3.5). In the western and Lake Victoria areas, rainfall is high from March to September, with lower rainfall in January and February (SoK 2003). Areas east of the Rift Valley essentially have two main rainy seasons, referred to as "short" and "long" rains. Kenya is unique in that more of its land area is under two rainy seasons than any other country (Jones and Thornton 1999). This seasonal variation in water availability is reflected in Kenya's great diversity of wild plant and animal communities, which have adapted to these seasonal changes (Oindo and Skidmore 2002). But the unique rainfall pattern also creates a special challenge for growing crops: none of the two rainy seasons is quite long enough to allow very high yields.

Rainfall amounts and distribution also vary a great deal from year to year. Over the past three decades, eastern Africa has experienced at least one major drought in each decade and floods have occurred frequently (UNEP 2006). Periods of below- and above-average rainfall are somewhat linked (Amissah-Arthur et al. 2002) to sea surface temperature, ocean currents, and atmospheric winds in the southern hemisphere (popularly known as La Niña and El Niño events).

Rivers and Drainage Networks

Surface water from rivers, streams, and lakes provides Kenyans with an important source of water and food. Kenya's major rivers originate in five mountain ranges or 'water towers,' as they are known: Mount Elgon, the Aberdare Range, the Mau Escarpment, Cherangani Hills, and Mount Kenya (See Box 3.1, Map 3.16).

Kenya's network of perennial rivers is most dense in the central and western parts of the country, leading to uneven supplies of surface water. Water resource managers have divided Kenya's surface waters into five large drainage areas: Ewaso Ngiro, Tana River, Rift Valley, Athi River, and Lake Victoria (Map 3.6). A look at the annual renewable water supplies for each of these major drainage areas echoes the patterns shown by a map of Kenya's perennial and intermittent rivers: The Lake Victoria drainage area with its dense network of perennial rivers provides 65 percent of Kenya's internal renewable surface water supply per year. The Athi River drainage area provides the lowest share—7 percent (SoK 2003).

Map 3.6 Major Drainage Areas and Rivers

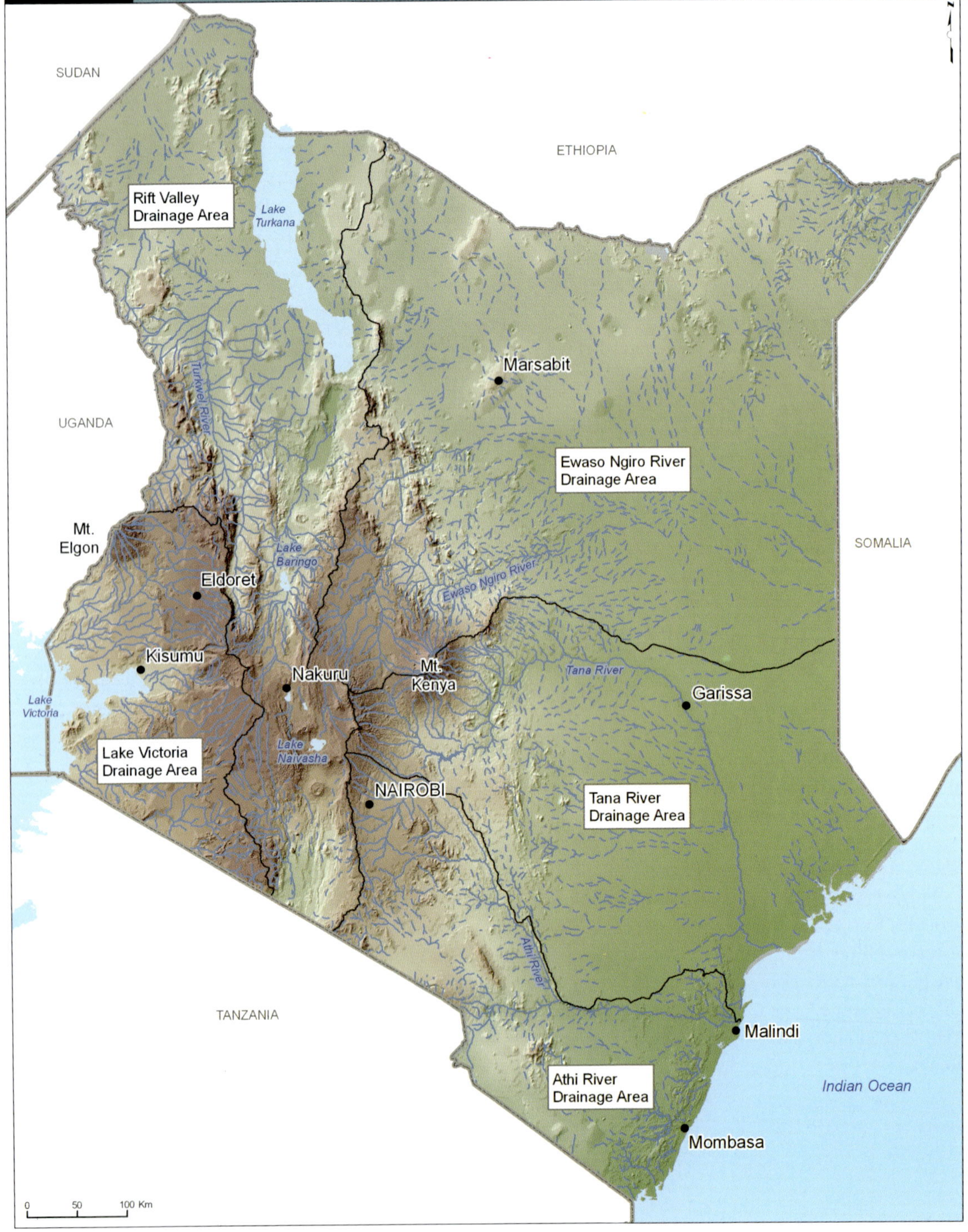

Sources: Cities (SoK and ILRI 2000), water bodies (FAO 2000a), permanent and intermittent rivers (NIMA 1997), 250-meter Digital Elevation Model (SoK, JICA, and ILRI 1996), and major drainage areas (MoWD and JICA 1992a).

This map shows major water bodies and drainage areas, reflecting the spatial distribution of water availability in Kenya. About 1.9 percent of Kenya is covered by water (SoK 2003). Lake Victoria, Lake Turkana, Lake Naivasha, and Lake Baringo are the four largest inland water bodies. Also displayed are the permanent rivers, most of which are found in the highlands, while the intermittent rivers are located in the rangelands. The Tana River (Kenya's longest) and the Athi River flow year-round and travel through significant stretches of dry land. They serve as a vital water and energy resource for people and the surrounding ecosystems.

Major drainage area boundaries

ELEVATION (meters)

High : 4,786

Low : -24

WATER BODIES AND RIVERS

Permanent rivers

Intermittent rivers

Water bodies

Demand Versus Supply of Water

Kenya's total annual renewable water resource is estimated at 30.7 billion cubic meters per year, with 20.2 billion cubic meters coming from internal renewable surface water, and the remainder supplied by groundwater and incoming flows from transboundary rivers (MoWD and JICA 1992b). Using a 2004 population of 32.8 million (CBS 2006), the total renewable water resource available per year is 936 cubic meters per person. Population growth alone will continue to reduce per capita water availability.

Average water availability of less than 1,000 cubic meters per capita per year designates Kenya as *water scarce*. This signifies that policymakers must pay particular attention to managing water resources so as to avoid hampering food production or impeding economic development. While this national average highlights the challenge posed by water availability to Kenya's development, it masks the great spatial and temporal variability of water supplies. A more detailed analysis of water demand and supply by subdrainage area can reveal where water is scarce and where it is plentiful.

The 1992 *Study on the National Water Master Plan* (MoWD and JICA 1992b) compared potential annual water supply (based on long-term average annual rainfall and maximum exploitable groundwater yield) to annual water demand (for households, agriculture, and industry) for 214 different subdrainage areas, each representing an aggregation of smaller watersheds. Map 3.7 aggregates both average annual surface water and maximum exploitable groundwater resources from the 1992 study and highlights where the projected annual water demand for 2000 and 2010 would exceed supplies. The map shows that subdrainages with densely settled urban populations such as Nairobi and Mombasa cannot cover water needs from their local supplies. In fact both cities have relied on long-distance water transfers for decades (see Maps 3.9 and 3.10). In the Rift Valley subdrainages north and south of Nakuru, local water resources are not sufficient to meet demand. Without continued water transfers from other areas that will keep pace with growing demand, these subdrainages will experience water shortages. Map 3.7 also highlights that even in areas with perennial rivers, demand can outstrip local supplies when a high number of people settle within a subdrainage. For example, some subdrainages in the upper Ewaso Ngiro, Tana River, and in western Kenya are projected to experience a local water deficit.

While such an analysis can pinpoint more location-specific problems, the projections are still based on historic patterns of water use and assumptions about future demographic and economic changes, as well as consumption patterns and investments in water resource development. These assumptions can easily change. For example, the level of rural-urban migration could increase or decrease from the projected rate. Studies show that increased human migration from rural to urban areas multiplies water demand (Thompson et al. 2002; Katui-Katua 2004) and creates a challenge for cities to provide residents and businesses with adequate amounts of clean, piped water for household, commercial, and industrial use. Urban dwellers tend to use about twice as much water as rural residents, and households with piped connections (mostly in urban areas) use, on average, three times more water than those without connections (Katui-Katua 2004).

Moreover, analysis at the scale of a subdrainage still hides issues of water scarcity within smaller watersheds and within communities. Respondents to a 1994 survey (Nakagawa et al. 1994) of Kenyans living in both wet and dry areas found that access to and quality of water was a constant preoccupation. The population living in the wetter areas of Kakamega and Bungoma Districts experienced water shortages only during the three driest months of the year. Residents of the drier areas in Kitui District faced a water shortage almost every month. In the wet areas, each person used on average about 40 liters per day, while in the drier areas it was about half of this amount. Interestingly, individual conceptions of a "severe water shortage" in the wet areas were classified as "average or above average water conditions" in the dry areas.

Map 3.7 Annual Projected Water Balance by Subdrainage Area, 2000 and 2010

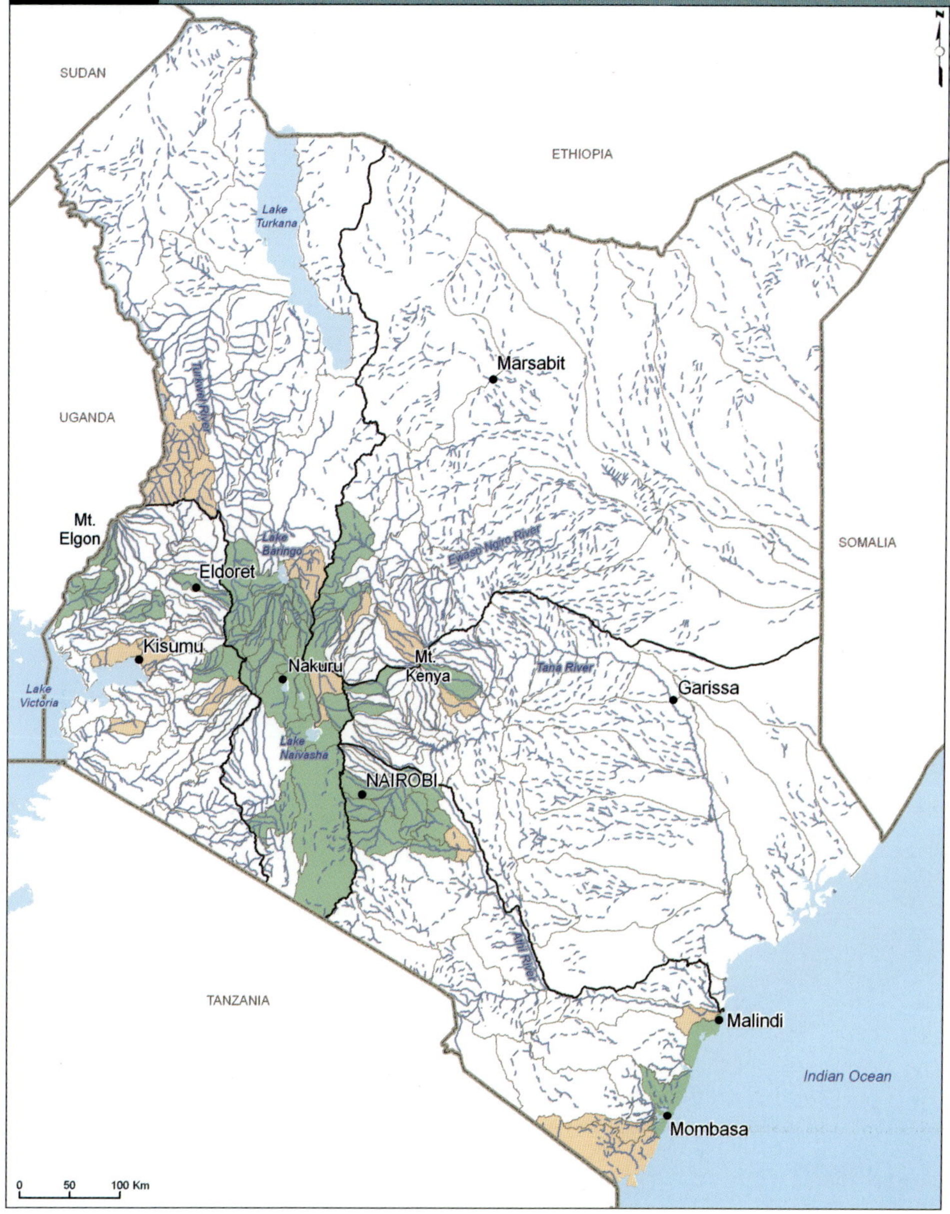

Sources: Cities (SoK and ILRI 2000), water bodies (FAO 2000a), permanent and intermittent rivers (NIMA 1997), 250-meter Digital Elevation Model (SoK, JICA, and ILRI 1996), subdrainage and major drainage areas (MoWD and JICA 1992a), and annual projected water balance by subdrainage area (MoWD and JICA 1992b).

This map compares potential annual water supply (which includes both surface water based on long-term average annual rainfall, and groundwater based on maximum exploitable groundwater yield) to projected annual water demand from households, agriculture, and industry. Areas with a water deficit in 2000 (in light green) cannot currently meet their annual water needs from supplies within their subdrainage area and the situation is not expected to change by 2010. These areas either require water transfers from other subdrainages to meet growing demand or they experience water shortages. The subdrainage areas marked in light orange do not currently have shortages but are projected to experience water deficits by 2010.

For almost all arid and semi-arid subdrainage areas showing no deficit on this map, current surface water availability alone is not sufficient to meet demand. These areas have to tap into their groundwater supplies to meet current and future demand.

The map tends to overestimate the positive balance between annual water supply and demand for a large number of subdrainage areas, due to the fact that water shortages often occur more locally in smaller watersheds within the subdrainage areas. In addition, the map is limited in that it does not show seasonal or annual variation in water availability. In many of the arid and semi-arid subdrainages, lower-than-average rainfall or droughts are frequent, leading to serious water shortages.

WATER BALANCE
- Deficit in both 2000 and 2010
- Deficit in 2010
- No deficit foreseen

DRAINAGE BOUNDARIES
- Major drainage area boundaries
- Subdrainage area boundaries

WATER BODIES AND RIVERS
- Permanent rivers
- Intermittent rivers
- Water bodies

Land areas with negative water balances (where water supply is outstripped by demand) will require investment in water resource infrastructure to cover their needs. In addition to increasing water supply, resource managers need to boost the efficiency of water use as well. This includes monitoring water use, especially groundwater uptake. It also requires technologies and policies for regulating water use and for promoting conservation and reuse of water. Such techniques include capturing and storing more of the annual rainfall or runoff (water harvesting), planting crops that are more water efficient, using more efficient technology for irrigation, and using more efficient methods of transporting water (e.g., avoiding leakage).

WATER-BASED ECOSYSTEM SERVICES

The maps in this section show the main uses and users of water in Kenya: drinking, industrial, and other uses in urban areas; energy generation; crop production; livestock production; and wildlife demand.

Drinking Water

Accessibility to water remains a major problem for rural people in Kenya, as well as for the urban poor. Connection to piped water is often considered a privilege of the more affluent in urban areas. But supply problems can arise with piped water too because of inadequate infrastructure, such as stalled water projects, delays in repairing leaks or damaged equipment at key supply areas, clogged water supplies, or vandalism. It is not uncommon for Kenyans with piped water to experience lengthy water shortages (Njuguna-Githinji 2001; Katui-Katua 2004). Women and girls are generally responsible for collecting water for household use when water is not piped directly to the home—a task requiring heavy physical labor and a great deal of time (Were et al. 2004). Map 3.8 shows the main sources of drinking water for households in Kenya.

In 2003, open surface water (lakes, ponds, rivers, and streams) was the major source of drinking water for 29 percent of Kenyan households, almost all of them in rural areas (CBS et al. 2004). These house-

Map 3.8 Dependence on Ecosystem for Drinking Water, 1999

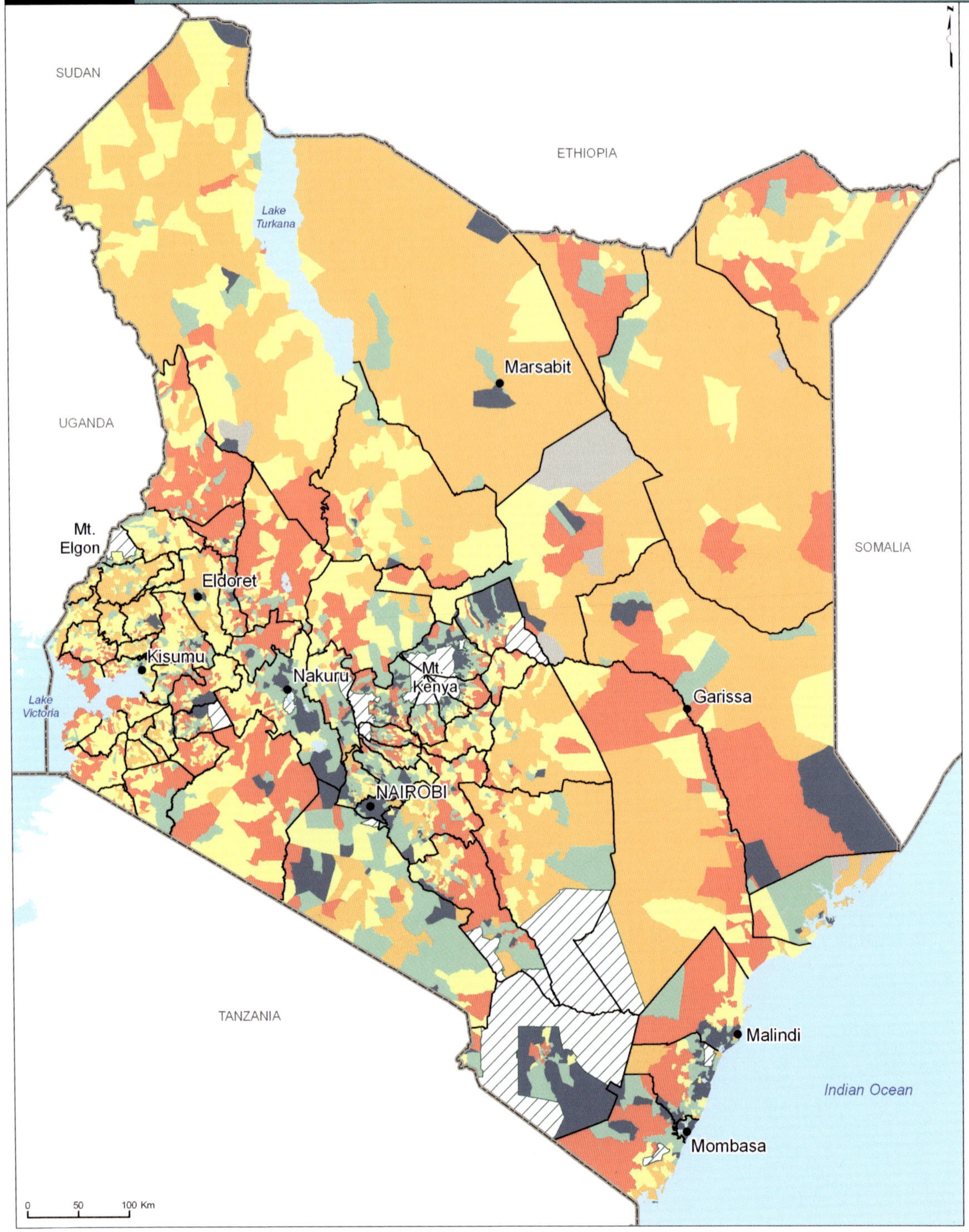

Sources: Administrative boundaries (CBS 2003), cities (SoK and ILRI 2000), water bodies (FAO 2000a), parks and reserves (IUCN and UNEP/WCMC 2006), and 1999 drinking water sources (CBS/ILRI/WRI calculation based on CBS 2002).

In most rural parts of Kenya, people obtain their drinking water from untreated surface water, groundwater, or a combination of surface and groundwater (depicted in red, orange, and yellow, respectively). Dependence on surface water (shown in red areas, where more than 75 percent of households rely on surface water) is most prevalent along permanent streams and other freshwater bodies in the highlands, along Lake Victoria, and close to permanent rivers crossing arid and semi-arid areas (e.g. north of Eldoret and close to Garissa).

Areas in which more than 75 percent of households depend solely on groundwater for drinking water are shown in orange. They are in the arid and semi-arid areas and in a few communities along the Indian Ocean. Here, households obtain their water from wells and boreholes. Groundwater, in this case from springs, is also a dominant source in selected Districts in western Kenya.

Areas where more than 75 percent of households receive piped drinking water are shown in blue. Such areas are clustered around Mombasa, Nairobi, Nakuru, and other more densely populated areas.

SINGLE DOMINANT DRINKING WATER SOURCES
- More than 75% of households rely on surface water
- More than 75% of households rely on groundwater
- More than 75% of households rely on piped water

MIX OF DRINKING WATER SOURCES
- More than 75% of households rely on surface and groundwater
- Mix of piped, surface, and groundwater with no dominant source
- No data

OTHER FEATURES
- District boundaries
- Selected national parks and reserves
- Water bodies

holds are particularly vulnerable since the quantity of water available at any given time depends directly on natural flows of water and the rainfall patterns that generate them. Use of surface waters also implies direct reliance on ecosystems for their natural waste removal capacity, such as filtering by wetlands and the dilution capacity of freshwater systems.

About 32 percent of Kenyan households (CBS et al. 2004) relied on groundwater sources (wells and springs) for their drinking water in 2003. Reliable supplies require sufficient and regular recharge from surface sources. Communities that obtain drinking water from groundwater are generally less vulnerable to water quality issues because of the natural filtering of groundwater supplies. However, high salinity and fluoride levels can make this source of water unsuitable for drinking, especially in coastal areas, as well as some areas in eastern and northeastern Kenya. In these cases, groundwater may still be used for irrigation, livestock, and industrial purposes (Nyaoro 1999).

By 2003, 32 percent of Kenyan households had benefited from piped water—either directly to their homes or through public taps (CBS et al. 2004). However, the differences between urban and rural areas remain great, with 71 percent of urban households and only 19 percent of rural households having piped water. Households with piped water are more indirectly linked to nature. They are relying on water management planning and water delivery systems to ensure adequate supplies and on municipal water treatment to protect them from water contamination.

Map 3.9 Water Sources: Nairobi

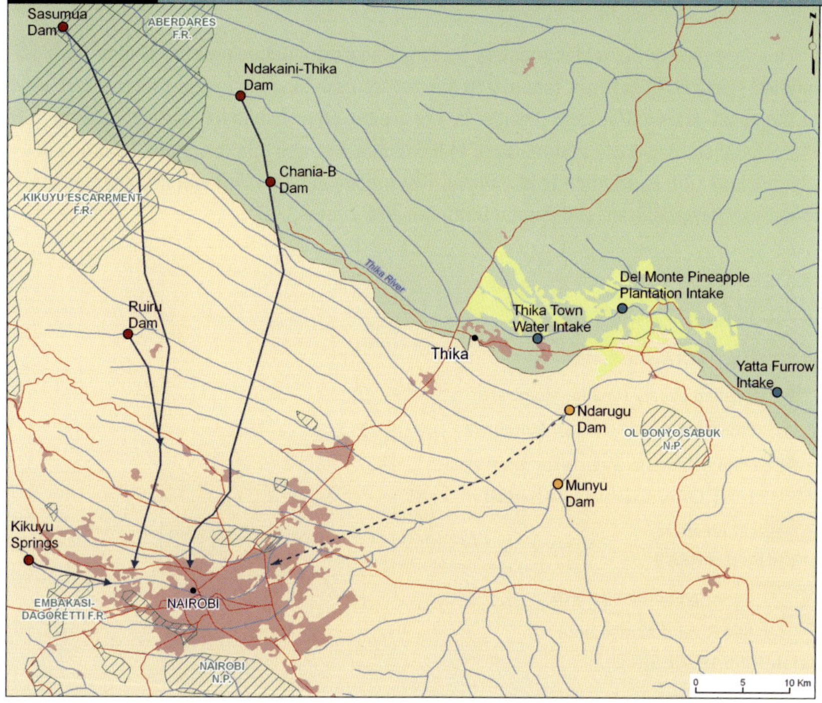

This map shows the water systems and dams on which Nairobi depends. These are located in the Athi River and Tana River basins. The map also presents water supply points for Thika town and the DelMonte water intake northeast of Nairobi (Nyaoro 1999). The DelMonte plantation extracts water from the Thika River for irrigation of its crops. Many conflicts have arisen between the DelMonte company and the Nairobi City Council, as well as with the Thika Town Council, over the use of this water (Nyaoro 1999). The Yatta Furrow Intake, located further downstream of the DelMonte Intake, supplies water for domestic, livestock, and irrigation use to the North and South Yatta areas of Machakos and Kitui (Wambua 2003). The map also shows the sites of the proposed Ndarugu Dam and Munyu Dam, planned drinking water sources for Nairobi.

WATER PIPELINES
- Existing pipeline
- Proposed pipeline

WATER WITHDRAWAL SITES
- Existing supply sites for Nairobi
- Proposed supply sites for Nairobi
- Other important water withdrawal sites

IMPORTANT LAND COVER FEATURES
- Urban areas
- Pineapple plantations

DRAINAGE AREAS
- Tana River drainage area
- Athi River drainage area

OTHER FEATURES
- Major roads
- Permanent rivers
- National parks and reserves

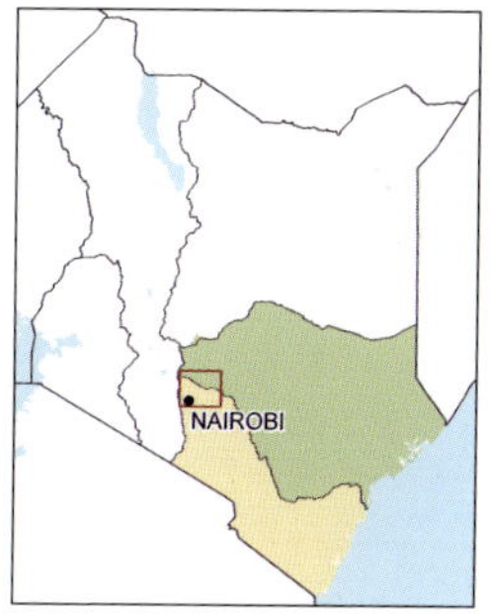

Sources: Cities (SoK and ILRI 2000), water bodies, urban areas and pineapple plantations (FAO 2000a), parks and reserves (IUCN and UNEP/WCMC 2006), major roads (SoK and ILRI 1997), permanent rivers and Sasumua pipeline (NIMA 1997), Athi and Tana River major drainage areas (MoWD and JICA 1992a), and location of dams, withdrawal points, and other pipelines (approximately placed by WRI based on SoK 1971, and on information from MoWD and JICA 1992c, 1992d, 1992e, and from Nyaoro 1999, Wambua 2003).

Water Supply in Urban Areas

Population and economic activities are highly concentrated in urban areas. Water is used not only for drinking but also for industrial production and urban agricultural activities (see Box 4.1 in Chapter 4). Water for Kenya's two largest cities, Nairobi and Mombasa, is transported over significant distances because supplies in the immediate vicinity are not sufficient.

Industrial use of public water is relatively minor in the country as a whole, consuming only about 4 percent of the total public water supply. In urban areas, the manufacturing industry utilizes a greater percentage of the public water supply, ranging from 13 percent to close to 40 percent (Onjala 2002). However, industrial water use is likely underestimated since it only accounts for withdrawals from public water supply; many companies extract additional water from rivers and private boreholes as well (Onjala 2002).

Nairobi draws its water from five different sources (Map 3.9) with a total capacity of approximately 460,000 cubic meters per day (Owore 2004). Over the past 100 years (Nairobi City Council 2006), Nairobi's sources of water have expanded outwards from nearby springs (Kikuyu Springs) to sources in the Athi River drainage area (Ruiru River Dam) and finally to reservoirs in the Tana River drainage area (Sasumua, Chania-B, and Ndakaini-Thika reservoirs). Despite recent investments in water delivery infrastructure, supplies have difficulty keeping pace with demand. Moreover, uneven distribution, waste through leakage, and illegal connections exacerbate supply shortages in certain areas of Nairobi (Owore 2004).

On the coast, the majority of Kenyans rely heavily on sources further inland for piped-in drinking water. Mombasa District's main sources of water (Munga et al. 2004) are Mzima Springs (through

Map 3.10 Water Sources: Mombasa

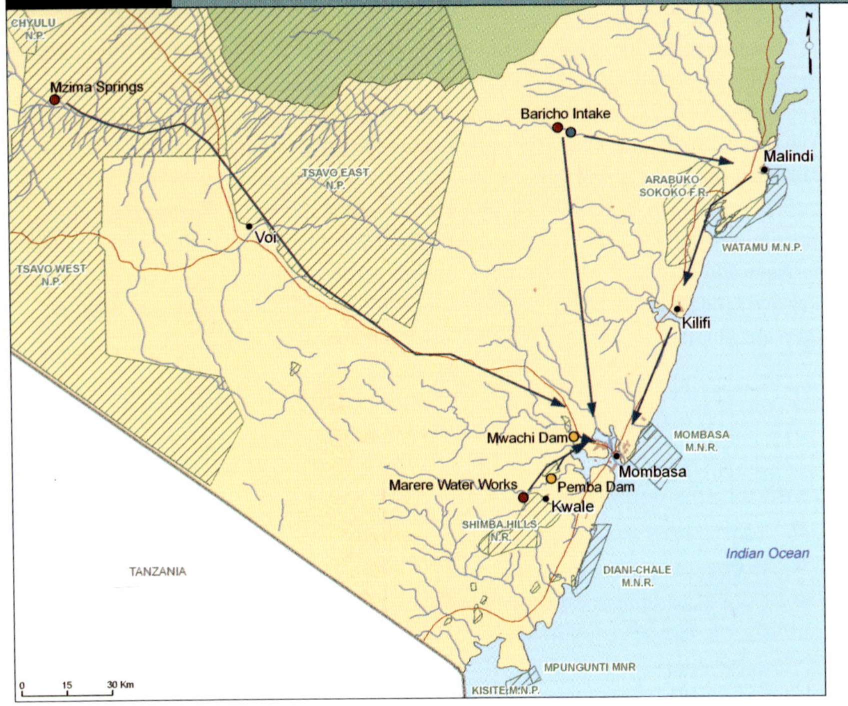

This map shows the water supplies serving Mombasa. A pipeline from Mzima Springs in West Tsavo National Park (about 220 kilometers from Mombasa) transports water to the coast. Marere Dam and Baricho Intake are the other two main sources feeding the coastal water supply system close to Mombasa. Baricho Intake serves the cities of Malindi and Kilifi in addition to Mombasa. Two proposed dams that will bring water to Mombasa (Mwachi Dam and Pemba Dam) are also shown.

WATER PIPELINES
- Existing pipelines

WATER WITHDRAWAL SITES
- Existing supply sites for Mombasa
- Proposed supply sites for Mombasa
- Other important water withdrawal sites

IMPORTANT LAND COVER FEATURES
- Urban areas

DRAINAGE AREAS
- Tana River drainage area
- Athi River drainage area

OTHER FEATURES
- Major roads
- National parks and reserves

WATER BODIES AND RIVERS
- Permanent rivers
- Water bodies

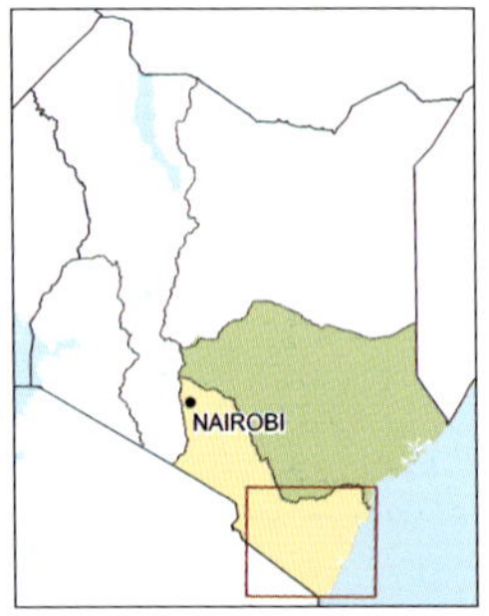

Sources: Cities (SoK and ILRI 2000), water bodies and urban areas (FAO 2000a), parks and reserves (IUCN and UNEP/WCMC 2006), major roads (SoK and ILRI 1997), permanent rivers and Mzima Springs pipeline (NIMA 1997), Athi River and Tana River major drainage areas (MoWD and JICA 1992a), and location of dams, withdrawal points, and pipelines (approximately placed by WRI based on SoK 1971, and on information from MoWD and JICA 1992c, 1992f, 1992g).

a pipeline constructed in 1966, which also serves communities along the corridor) and water works at Baricho and Marere (more recent investments). Mombasa District's demand for water, however, cannot be satisfied entirely by surface water. About 35 percent of the District's demand is met by tapping groundwater sources, and in some areas a majority of households are primarily dependent on groundwater (Munga et al. 2004).

Unfortunately, groundwater supplies in Mombasa District (Map 3.10) are vulnerable to salinity intrusion and pollution from pit latrines and septic tanks as the region currently lacks sufficient sewage treatment to manage the human waste generated in the region. Groundwater from these areas must be treated to be safe for human consumption.

Smaller industrial towns also have trouble providing enough water for industrial activities. According to a report by the Kenya Association of Manufacturers, limited water supply can hamper industrial growth. For example, the report states that Nakuru, home to major industries, is losing business to neighboring towns with more adequate water supplies (Cited in Njuguna-Githinji 1991).

Electricity Generation

Hydropower is the largest source of electricity in Kenya, providing approximately 680 MW or 55 percent of the total installed grid capacity (UNDP et al. 2005). Much of the hydropower comes from large-scale stations and dams on the upper Tana River and the Turkwel River. About 570 MW or 84 percent of Kenya's existing hydropower capacity comes from a succession of dams called the Seven Forks power stations along the upper Tana River (KenGen 2006). Map 3.11 shows the locations of these hydropower dams.

The proposed dams at Mutonga and Grand Falls, just downstream from the existing upper Tana River dams, will likely be the next dams built under Kenya's least cost development plans (UNDP et al. 2005). The Sondu-Miriu hydropower project is currently being constructed to the east of Lake Victoria. Small hydropower systems (generating less than 10 MW each) often provide electricity for off-grid or isolated rural areas. The most important small hydropower sites are in the upper Tana River and a few sites in western Kenya.

Hydropower dams, although contributing significantly to economic development and human well-being, can have negative impacts on populations and ecosystems as well. Dams can affect downstream water supply, displace people, ruin aesthetic and sometimes spiritual landmarks such as waterfalls, and increase threats to fish and other species that depend on rivers for their habitat. Before construction of the Seven Forks dams, the banks of the Tana River flooded naturally during the wet seasons twice a year, helping to sustain the surrounding grasslands, lakes, seasonal streams, and riverine forest and mangrove ecosystems. However, flooding has decreased in volume and frequency since the construction of the five dams (IUCN 2003). An estimated one million farmers, livestock keepers, nomadic and seminomadic pastoralists, and fisherfolk who live along the river and in the Tana Delta depend on the river's remaining seasonal flooding patterns for their livelihoods (IUCN 2003). Investing in appropriate dam design and hydrological management (e.g., timed water releases) could maintain some of these downstream ecosystem benefits but still boost electricity supplies to support Kenya's economic recovery. This could help to achieve a number of development objectives and safeguard the livelihoods of downstream users at the same time (UNEP 2006).

Map 3.11 Water Used for Electricity Generation

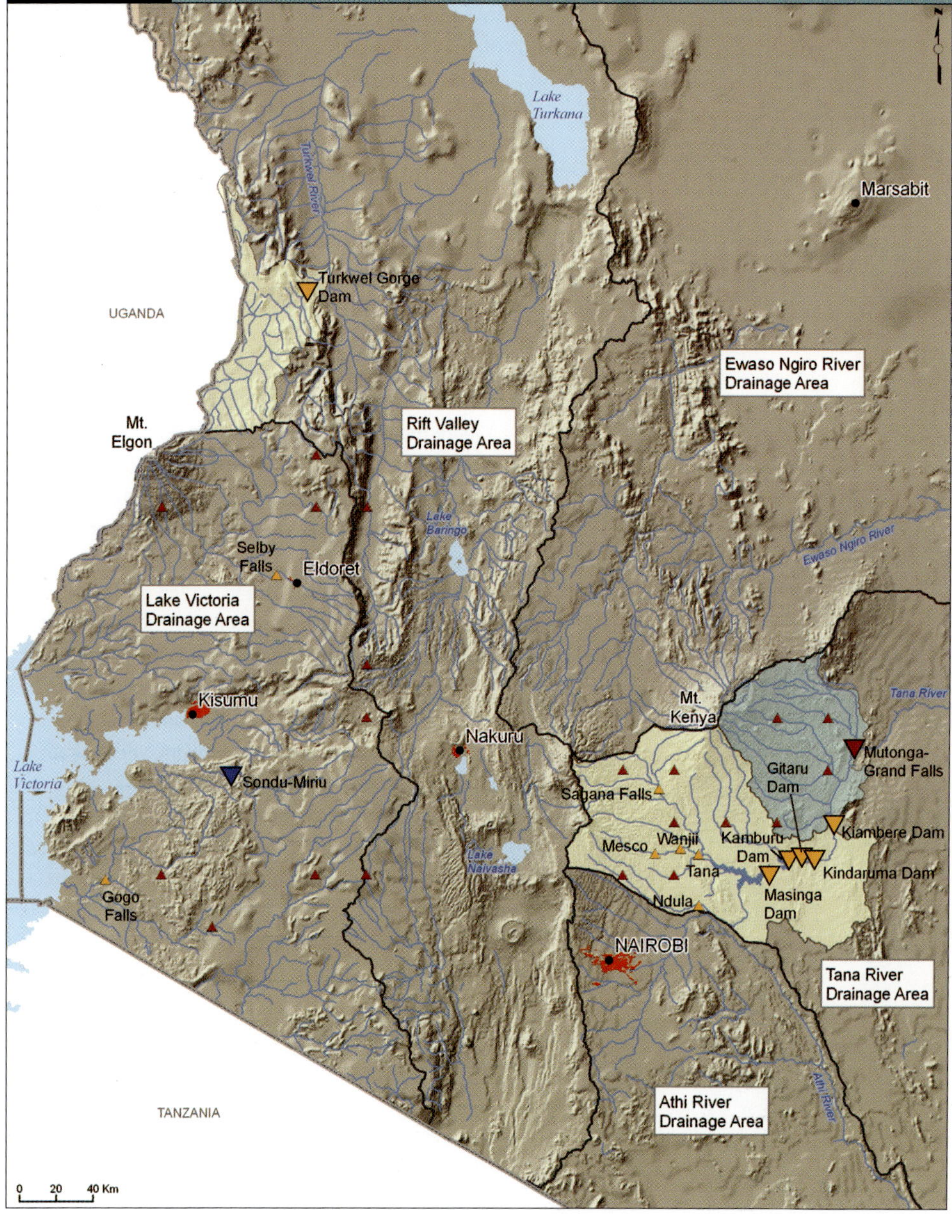

Sources: Cities (SoK and ILRI 2000), water bodies and urban areas (FAO 2000a), permanent rivers (NIMA 1997), 250-meter Digital Elevation Model (SoK, JICA, and ILRI 1996), major drainage areas (MoWD and JICA 1992a), hydropower sites (approximately located by WRI based on SoK 1971, and information from KenGen 2006, and from MoWD and JICA 1992c, 1992h, 1992i), existing small hydropower sites (approximately located by WRI based on SoK 1971, and information from KenGen 2006, and Balla 2006), and proposed small hydropower sites (approximately located by WRI based on SoK 1971, and information from ITDG and ESDA 2005).

HYDROPOWER INFRASTRUCTURE
- Existing hydropower sites
- Hydropower sites under construction
- Proposed hydropower sites
- Existing small hydropower sites
- Proposed small hydropower sites

IMPORTANT DRAINAGE AREAS
- Drainage area important for current dams
- Drainage area important for proposed dams

IMPORTANT LAND COVER FEATURES
- Urban areas

OTHER FEATURES
- Major drainage area boundaries

WATER BODIES AND RIVERS
- Permanent rivers
- Water bodies

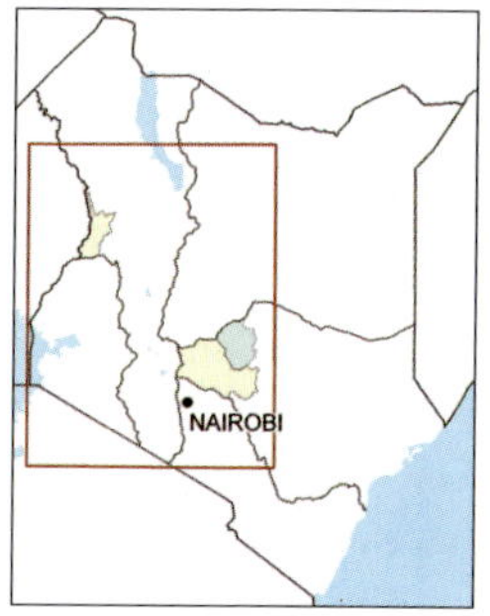

This map shows the spatial distribution of hydropower sites in Kenya. Areas shaded in beige depict the water catchment areas that feed the five existing power stations (indicated on the map by large orange triangles) and reservoirs on the Tana River, as well as one dam on the Turkwel River. Land use practices in these beige-colored areas (i.e., upstream from these dams) can influence the amount of water and sediment flowing into the reservoirs, affecting water quality and the productive lifespan of the hydropower infrastructure. The Sondu-Miriu hydropower project (under construction) is marked by a large blue triangle.

The catchments that feed the site of the proposed dams for the Mutonga-Grand Falls scheme—also intended to help satisfy Kenya's electricity needs—are indicated by the areas shaded in light green. A large red triangle marks the proposed site. The dams would effectively capture the remaining permanent rivers feeding the Tana River from Mount Kenya and significantly impact ecosystems downstream. These include the seasonally flooded grasslands (important for livestock grazing and wildlife), gallery forests along the river's shores (key primate and bird habitats), and coastal ecosystems (valuable for fisheries) in the Tana estuary.

Small orange and red triangles mark the locations of existing and proposed small hydropower sites. A number of the proposed small hydropower sites are considered economically viable and the impact to freshwater systems and associated species and habitats would be limited.

Note: Existing small hydropower sites are operating schemes that were built between 1919 and 1955. There are a number of additional small hydro schemes associated with tea companies, community groups, and a private hospital (Balla 2006).

Crop Production

Since 98 percent of Kenya's cropping is rainfed, most farmers are exposed to the high variability of rainfall within and between years. Only 15 percent of Kenya receives more than 762 millimeters of rain per year, in four out of five years. This is the minimum amount required to grow maize and other non-drought-resistant crops. Another 13 percent of Kenya has more marginal rainfall (508–762 millimeters) requiring special dry farming or irrigation practices to cultivate crops (SoK 2003). But even in high-rainfall areas, sufficient water for a successful harvest is not guaranteed every year—both "long" and "short " rains can be ill timed or not fall at all (e.g., FAO 2000b; KFSSG 2006). Investment in water storage and irrigation infrastructure can reduce the risk of insufficient rainfall for farmers.

Irrigation in Kenya is carried out on both a small-scale, local level and in large-scale irrigation schemes (Map 3.12). Smallholders account for 46 percent of Kenya's irrigation, using it for fruit and vegetable production. Larger commercial firms account for another 42 percent. About 12 percent are public schemes under the National Irrigation Board (FAO 2005).

According to FAO (2005), only 19 percent of Kenya's potential area is equipped for irrigation. The proportion of cropped area which is irrigated is well below the average, at 2 percent compared to 3.7 percent in sub-Saharan Africa as a whole (FAO 2005). This low level of irrigation is due to limited water availability, rising costs of supplying water and building irrigation systems, and poor economic performance of existing irrigation schemes (Onjala 2001).

To satisfy Kenya's future water needs and demands from increased agricultural production, the *Study on the National Water Master Plan* has stressed the importance of investing in water resources development. For example, it has proposed 18 major irrigation schemes and 140 small-scale irrigation schemes for 2010.

Map 3.12 Water Used for Crop Irrigation

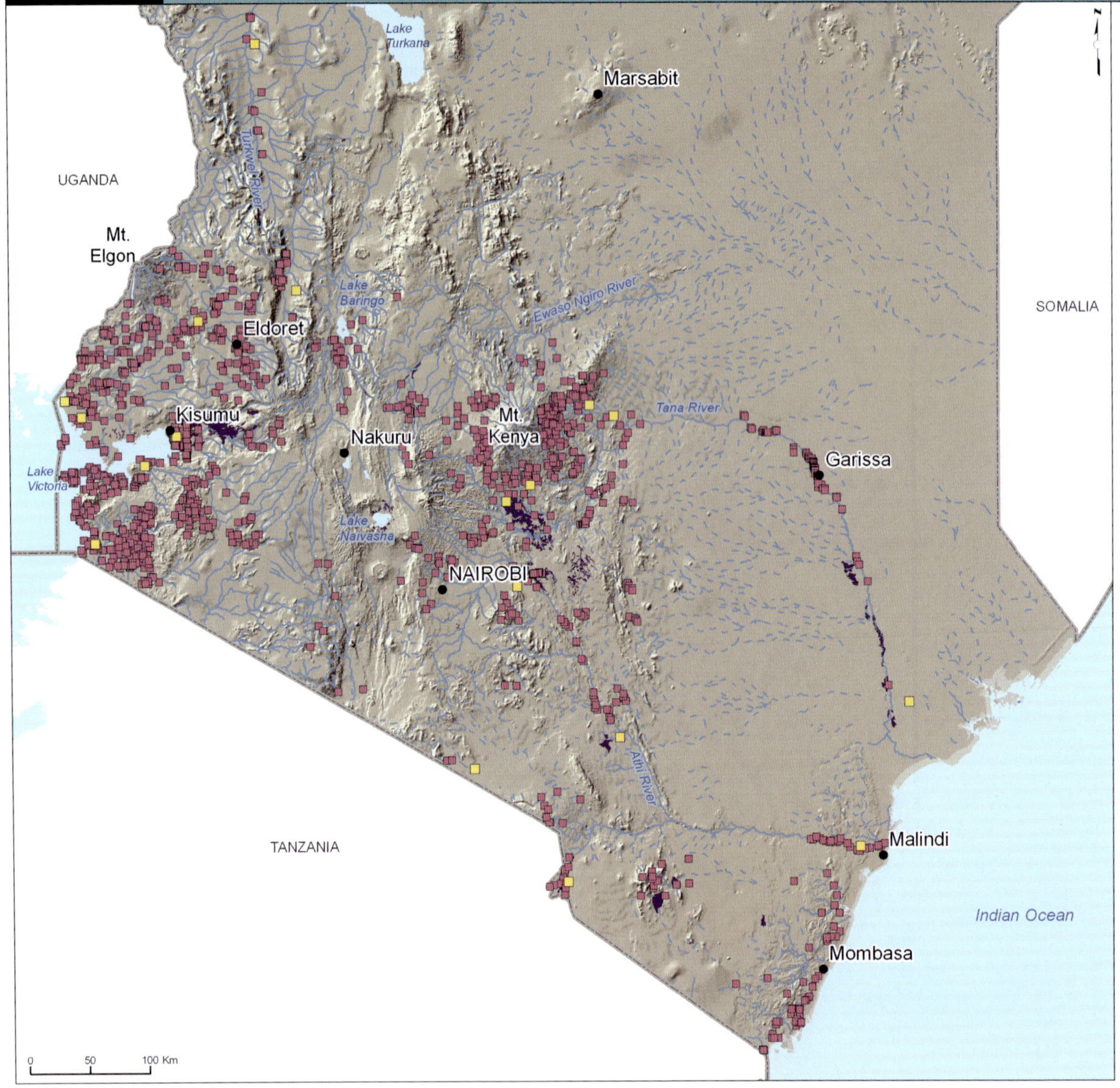

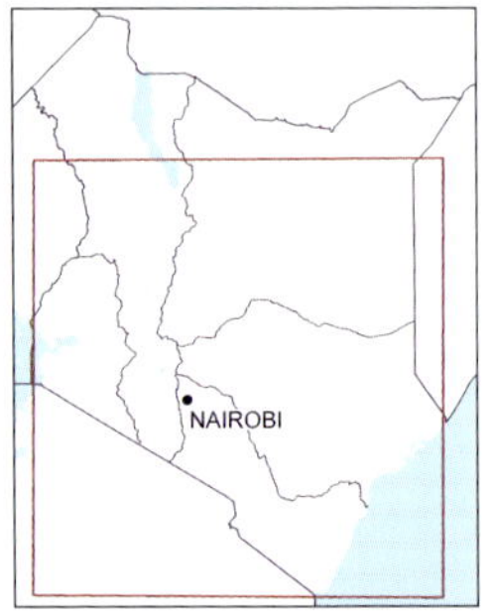

Sources: Cities (SoK and ILRI 2000), water bodies and large-scale irrigation areas (FAO 2000a), permanent and intermittent rivers (NIMA 1997), 250-meter Digital Elevation Model (SoK, JICA, and ILRI 1996), small-scale irrigation and drainage points (IWMI compilation based on MoALD 1995), and proposed large-scale irrigation schemes (MoWD and JICA 1992j and 1992k).

This map shows small-scale irrigation points as well as certain large-scale irrigation schemes in central and southern Kenya. Dark purple shading represents large-scale irrigation systems, with the largest located at the foothills of Mount Kenya. This includes Kenya's largest irrigation investment, the Mwea-Tebere rice irrigation scheme. Covering more than 6,100 hectares, this area produces most of Kenya's rice. Other irrigated areas are located close to Kisumu (where sugar cane is produced) and along the lower Tana River (which produces citrus and rice).

Clusters of small-scale irrigation points, marked by pink squares, are especially prevalent around the shores of Lake Victoria and the base of Mount Kenya. The irrigated areas around the base of Mount Kenya depend mostly on water from the upper Tana and Ewaso Ngiro Rivers, which drain from the top of the mountain.

The map also shows 18 proposed irrigation schemes marked with yellow squares, as outlined by the *Study on the National Water Master Plan*.

IRRIGATION INFRASTRUCTURE

- Existing small-scale irrigation and drainage points
- Proposed large-scale irrigation schemes
- Large-scale irrigation schemes

WATER BODIES AND RIVERS

- Permanent rivers
- Intermittent rivers
- Water bodies

Livestock and Wildlife

Rivers and lakes must maintain a minimum flow to sustain the aquatic and riparian species that depend on them. Fish—an important part of diets and livelihoods for Kenyans who live close to Lake Victoria—depend on an adequate quantity and quality of water to live and reproduce. Kenya's wild animal species also require water; wildlife viewing is central to the country's tourism industry and in some areas, illegal hunting of wildlife provides meat to rural households (see Chapter 4).

Livestock production is also very dependent on adequate water sources. Herding in the arid and semi-arid areas, where over half of Kenya's livestock are produced, relies heavily upon groundwater sources (SoK 2003). It can be difficult to find enough sources of water for livestock due to competing water demands. A typical cow weighing approximately 250 kilograms drinks 20-50 liters of water a day, depending on whether or not the animal is lactating (Peden et al. 2003). Herders with large quantities of livestock often have to travel to distant sources such as small dams, rivers, water pans, and boreholes.

Problems arise when water is scarce, as livestock may wander in search of additional water sources. Cows can pollute river water and spread helminthes (a type of worm carried by snails) when river levels are low and they are forced to walk into the river for water (Peden 2004). During times of drought, there are occasionally clashes between cattle ranchers and pastoralist herders over land rights. Herders often end up moving their livestock into private ranches in order to avoid areas of significant drought, especially in Narok and Kajiado Districts. There are also conflicts over water use between livestock herders and wildlife in these drier areas (Zecchini 2000).

Map 3.13 shows water consumption of major animal species for Kenya's rangeland Districts. It takes into account the distribution of livestock species and wild grazing animals within each subdrainage area and multiplies each animal's weight by its estimated water consumption. Water consumption,

Map 3.13 Average Water Consumption of Livestock and Wildlife by Subdrainage Area, 1994-96

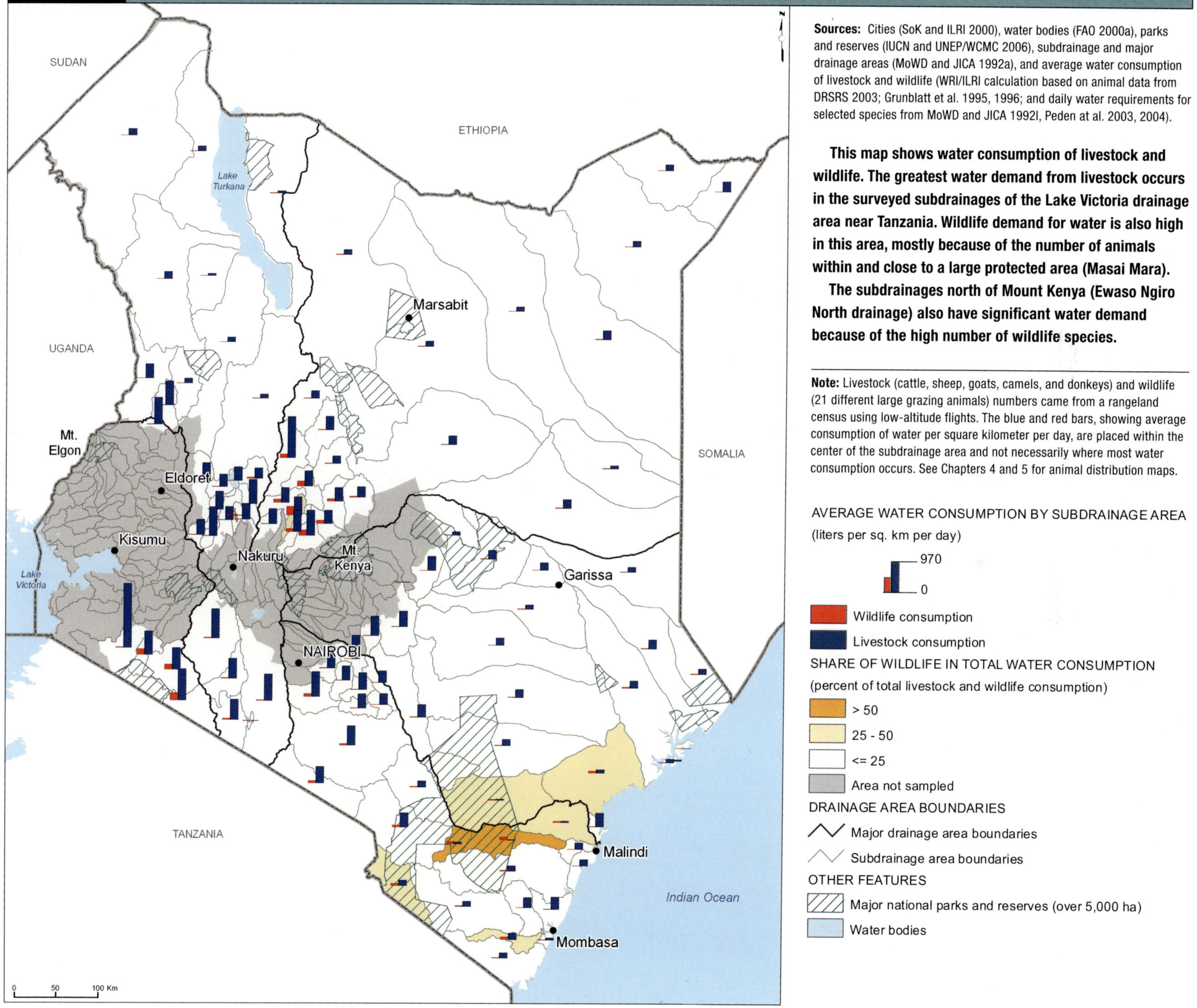

Sources: Cities (SoK and ILRI 2000), water bodies (FAO 2000a), parks and reserves (IUCN and UNEP/WCMC 2006), subdrainage and major drainage areas (MoWD and JICA 1992a), and average water consumption of livestock and wildlife (WRI/ILRI calculation based on animal data from DRSRS 2003; Grunblatt et al. 1995, 1996; and daily water requirements for selected species from MoWD and JICA 1992l, Peden at al. 2003, 2004).

This map shows water consumption of livestock and wildlife. The greatest water demand from livestock occurs in the surveyed subdrainages of the Lake Victoria drainage area near Tanzania. Wildlife demand for water is also high in this area, mostly because of the number of animals within and close to a large protected area (Masai Mara).

The subdrainages north of Mount Kenya (Ewaso Ngiro North drainage) also have significant water demand because of the high number of wildlife species.

Note: Livestock (cattle, sheep, goats, camels, and donkeys) and wildlife (21 different large grazing animals) numbers came from a rangeland census using low-altitude flights. The blue and red bars, showing average consumption of water per square kilometer per day, are placed within the center of the subdrainage area and not necessarily where most water consumption occurs. See Chapters 4 and 5 for animal distribution maps.

which varies by species, is directly proportional to each animal's body weight (MoWD and JICA 1992l; Peden et al. 2003). Some animals, such as eland and impala, can live without drinking water for long periods; other animals, such as elephants, need more regular access to water.

In almost all of the subdrainages in Kenya's rangeland Districts, water demand for livestock is significantly greater than for wildlife. There are only a few subdrainage areas where wildlife consume a larger share of water than livestock. They are within or close to protected areas, which do not permit livestock grazing.

It should be noted that the analysis in Map 3.13 includes only water requirements for drinking water. The amount of water necessary for the production of fodder—either on natural pasture or grown as crops—is about one hundred times greater than the amount necessary for direct consumption by animals (Peden at al. 2003). Incorporating these numbers into the calculation would increase the total amount of water utilized, but would not change the relative relationship between domesticated and wild animals significantly.

Subdrainage areas with both high wildlife and livestock numbers such as the Ewaso Ngiro subdrainage will require special attention to ensure sufficient water supply. It will be especially important for water managers in this area to monitor activities and water withdrawals taking place upstream from these wildlife-rich areas so as to protect the water supply for these animals. In addition, catchments upstream from livestock (for example, areas important for groundwater recharge) need to be managed so that pastoralists further downstream have adequate amounts of water as well.

Over the long term, integrating the water needs of livestock into future development plans will become more important as Kenya's water supply becomes scarcer and demand for livestock products increase. The projections published in the *Study on the National Water Master Plan* estimated that livestock production will be responsible for 15 percent of national water demand in 2010 (MoWD and JICA 1992m).

Map 3.14 Areas Flooded and Prone to Flooding, 2002–06

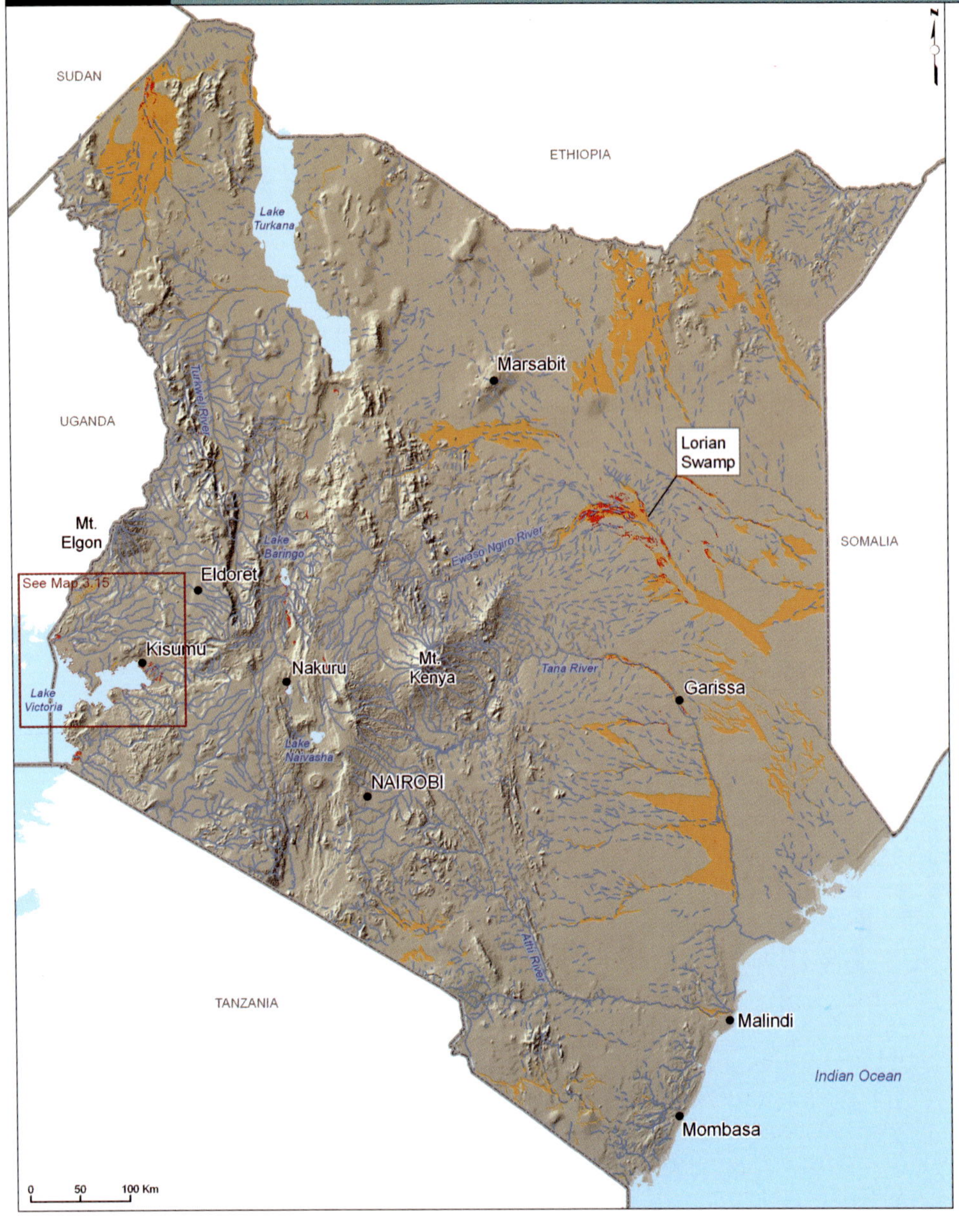

Sources: Cities (SoK and ILRI 2000), water bodies, floodplains, and valley bottoms (FAO 2000a), permanent and intermittent rivers (NIMA 1997), 250-meter Digital Elevation Model (SoK, JICA, and ILRI 1996), and 2002–2006 flooded areas (Brakenridge et al. 2006).

This map shows the areas flooded between 2002–06 (in red), as well as floodplains and low-lying areas prone to flooding (in orange). Floodplains consist of land adjacent to a river channel that is seasonally covered by river water. Readers should note that the flooded areas shown in red are most likely an underestimate of actual flooding. Areas that experienced the most flooding are the shores of Lake Victoria in western Kenya, the banks of the Tana River in eastern Kenya, and the Lorian Swamp in central eastern Kenya, all highlighted on the map. Although the flooding near Lake Victoria does not appear to be extensive from this national map, it is important to understand that population density in that area is high and thus flooding is very destructive.

AREAS FLOODED OR PRONE TO FLOODING
- Flooded areas, 2002 - 2006
- Flood plains and valley bottoms

WATER BODIES AND RIVERS
- Permanent rivers
- Intermittent rivers
- Water bodies

WATER AS A HAZARD: FLOODING

Flooding occurs erratically in Kenya, usually around the season of the "long" rains during the months of March through May. Many floods have affected the western parts of the country in the densely settled Kano Plains, Yala swamp, and other low-lying areas around Lake Victoria. Homes, schools, livestock, and farmlands in other parts of the country have also been destroyed. During the El Niño rains in 1997, for example, flooding affected the city of Nairobi and lower parts of the Tana River, but also the western parts of Kenya, mostly Busia and Nyando Districts (SoK 2003).

In May 2005, devastating floods displaced ten thousand people, especially along the shores of Lake Victoria, as well as in Tana River and Garissa Districts further east. Residents of affected areas reported the flooding to be the heaviest since 1963. Heavy rains also caused flooding in Isiolo District and in the Dadaab refugee camp in north-eastern Kenya, leaving more than 25,000 Somali refugees homeless. Impassable, waterlogged roads seriously hampered efforts to help the victims (ReliefWeb 2005).

Flooding can reduce access to clean water by destroying or polluting drinking water supplies, increasing the chances of contracting waterborne diseases. Stagnant water that remains after flooding can also increase exposure to mosquito-borne diseases such as malaria by providing a medium for mosquitoes to breed. Washed-away bridges and impassable roads can isolate communities for extended periods, leading to food and other shortages.

On the other hand, flooding can sometimes be helpful to both ecosystems and people. About one million people (IUCN 2003) depend on the Tana River's flooding regime for their livelihoods, including nomadic and seminomadic pastoralists, who rely on floodplain grasslands for dry season pasture. Some seasonal fisherfolk and fish traders also depend on the Tana's flooding pattern, as do some farmers, who count on seasonal floods to irrigate their riverbank farms. In addition, birds and wildlife are dependent on the annual flood cycle of the Tana for habitat and forage. Wetlands are often replenished by the flooding as well.

Studying the hydrological response to different types of land cover and land uses in flood-prone areas, implementing better land use planning, and establishing early flood warning systems are possible interventions that could mitigate some of the worst flood impacts.

Map 3.15 **Floods in Western Kenya, 2002–06**

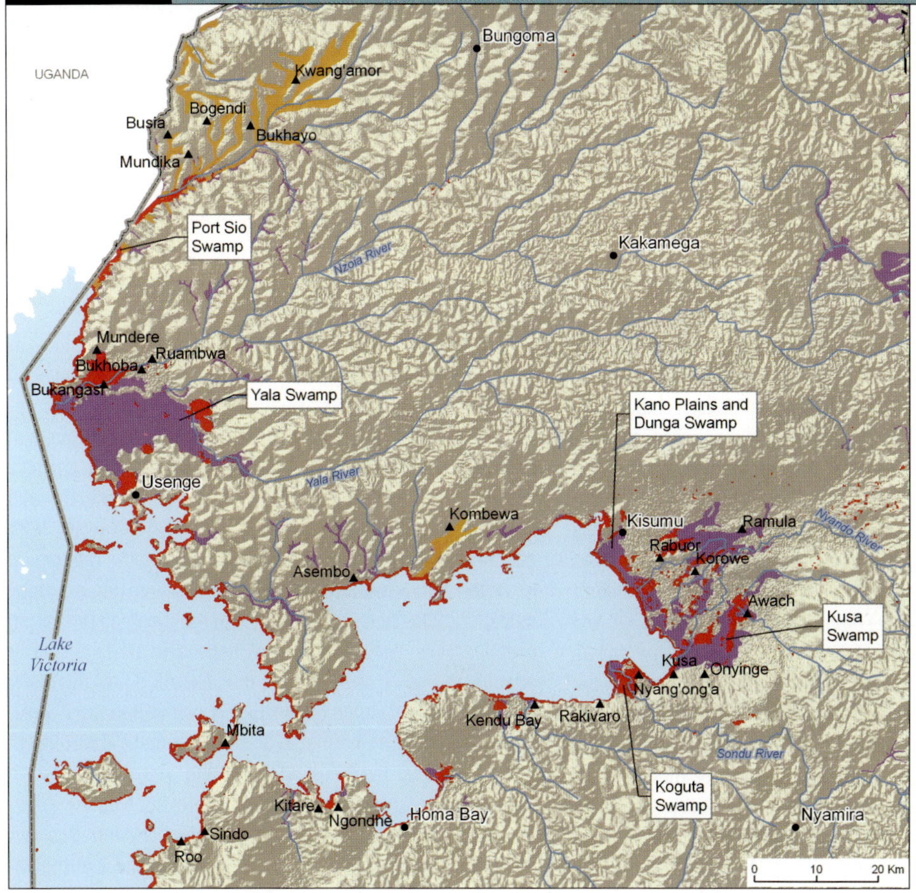

Sources: Cities and market centers (SoK and ILRI 2000), water bodies, wetlands, floodplains, and valley bottoms (FAO 2000a), permanent rivers (NIMA 1997), 90-meter Digital Elevation Model (USGS 2004), and 2002–2006 flooded areas (Brakenridge et al. 2006).

This map focuses on western Kenya for a closer view of flooding that occurred in 2002–06 around the shores of Lake Victoria. Cities, towns, and market centers near floodplains and flooded areas are marked to illustrate flood impacts in these high-density zones. Major swamps are labeled, as well as major rivers such as the Nzoia River, which often floods on its lower reaches.

▲ Market centers

AREAS FLOODED OR PRONE TO FLOODING

Flooded areas, 2002 - 2006

Flood plains and valley bottoms

IMPORTANT LAND COVER FEATURES

Wetlands

WATER BODIES AND RIVERS

Permanent rivers

Water bodies

Box 3.1 Mapping Water-Related Ecosystem Services: Links to National Decision-Making

Kenya has been characterized as a water-scarce country. Decision-makers need to find innovative ways to supply enough water to accommodate the multitude of demands for agriculture, hydropower, tourism, industry, and drinking water, while still supporting plant and animal life. It will also be increasingly important to address the links between poverty and lack of access to improved water supply and sanitation services.

There is a strong relationship between economic status and access to improved water supply and sanitation in Kenya. About 37 percent of rural households rely on open surface water (streams, rivers, ponds, and lakes) for their drinking water (CBS et al. 2004). Public investment in the rehabilitation and expansion of water supply infrastructure has generally benefited urban populations and more affluent communities. But many of the poor who live in informal settlements in urban areas also have no easy and affordable access to potable drinking water.

Kenya's *Economic Recovery Strategy for Wealth and Employment Creation 2003–07* (GoK 2003) proposes many goals related to water and the achievement of economic growth. These include reducing the role of the Kenyan government in the provisioning of water supply and sanitation in favor of more efficient private companies; improving the physical infrastructure of new and existing water schemes; and narrowing the inequality between rich and poor communities in terms of access to treated water and adequate sanitation. The geospatial information presented in this atlas could help decision-makers meet such goals. Geographic indicators of water supply combined with other maps and indicators on human population density, poverty, and physical infrastructure can inform sound water management approaches that also benefit the poor. Below are examples of how maps can assist in the discussion and planning of certain interventions proposed in the Economic Recovery Strategy. Each item begins with a specific goal (in italics) drawn from the Strategy.

- *Improve the physical infrastructure of current water schemes:* Maps of water lines and their status can be used to identify specific locations that need rehabilitation. Combined with census data, planners can estimate how many people are not receiving proper water services due to damaged water lines or dams in need of repair. Delineating flood-prone areas and combining this information with the location of water lines can pinpoint water lines at risk of flood damage. Water and sanitation agencies can publicly release the location of new water infrastructure investments, thus providing communities an opportunity to hold these agencies accountable for their performance and priorities.
- *Increase the poor's access to treated water and sanitation services:* Using census information on sources of drinking water (as shown in Map 3.8) and combining that with poverty maps and additional household data can help prioritize communities with the greatest or most urgent needs. In addition, by overlaying maps of water infrastructure with detailed poverty maps, the water and sanitation sector can select appropriate technologies for poorer areas that require less capital and human resource investment. Constituencies and communities can use regular reports showing where access has improved to examine distributional equity issues and lobby for changes in resource allocation formulas.
- *Rehabilitate existing community water pans, dams, and boreholes in rangeland, in collaboration with the private sector, NGOs, and other development partners, for livestock development and prevention of poverty in arid and semi-arid lands:* Maps highlighting water supplies (as shown in Map 3.8), information on the location of boreholes, dams, and wells (as shown for northern Kenya in Map 5.12, Chapter 5), and maps of livestock density or livestock water demand (see Map 3.13), can all be combined to understand the relationships between water services and livestock development. With additional information on the water needs for tourism, for wildlife, and for other important ecosystem services, planners can identify areas where future water investments may create synergistic benefits or where multiple demands may require careful examination of tradeoffs.
- *Develop new irrigation schemes to promote year-round agriculture and food security, especially in arid and semi-arid lands:* Maps can be used to examine rainfall and farming patterns outside of the highlands to determine which parts of arid and semi-arid lands might be most suitable for development of new irrigation schemes. Maps such as Map 3.12, which shows locations of small and large-scale irrigation, will be useful to create a comprehensive picture of where irrigation efforts are already taking place.
- *Mitigate flooding by constructing dams across rivers, rehabilitating deforested water catchments, constructing dykes, and preparing an early warning system:* Locations most prone to flooding can be mapped (as shown in Map 3.15). With the help of more detailed elevation information; accurate road, housing and population data; and monitoring of weather patterns, rainfall, and flood levels in rivers, an early warning system could alert communities of approaching storms and rising floodwaters.

Another key issue not specifically mentioned in the *Economic Recovery Strategy* but relevant to its goals is the need to examine the competing demands for water resources between upstream and downstream users. Maps can pinpoint rapid land-use changes, cultivation methods, heavy applications of fertilizer and pesticides, discharge of sewage and industrial effluent, and sources of water withdrawals. With additional models (for example, incorporating the magnitude of water withdrawals) or economic valuation (for example, measuring the costs, benefits, or externalities), planners can examine how upstream interventions are affecting water quantity or quality downstream, thus ensuring that the many investments envisioned under the *Economic Recovery Strategy* are not too detrimental to a specific area or community.

As evidenced by the information and maps presented in this chapter, regular data collection efforts such as the *Population and Housing Census, Demographic Health Surveys,* and meteorological monitoring, all provide useful information on water supply, water use, and water-related health impacts. Moreover, significant information has been compiled for the *Study on the National Water Master Plan* (MoWD and JICA 1992m), albeit not in a format that is easily accessible to all stakeholders involved in water and sanitation issues. To strengthen national and local planning, much better integration of these water-specific data with other sector information is needed.

Box 3.2 Creating a Poverty and Demographic Profile for Kenya's 'Water Towers'

Map 3.16 Five 'Water Towers' and Selected Upper Watersheds

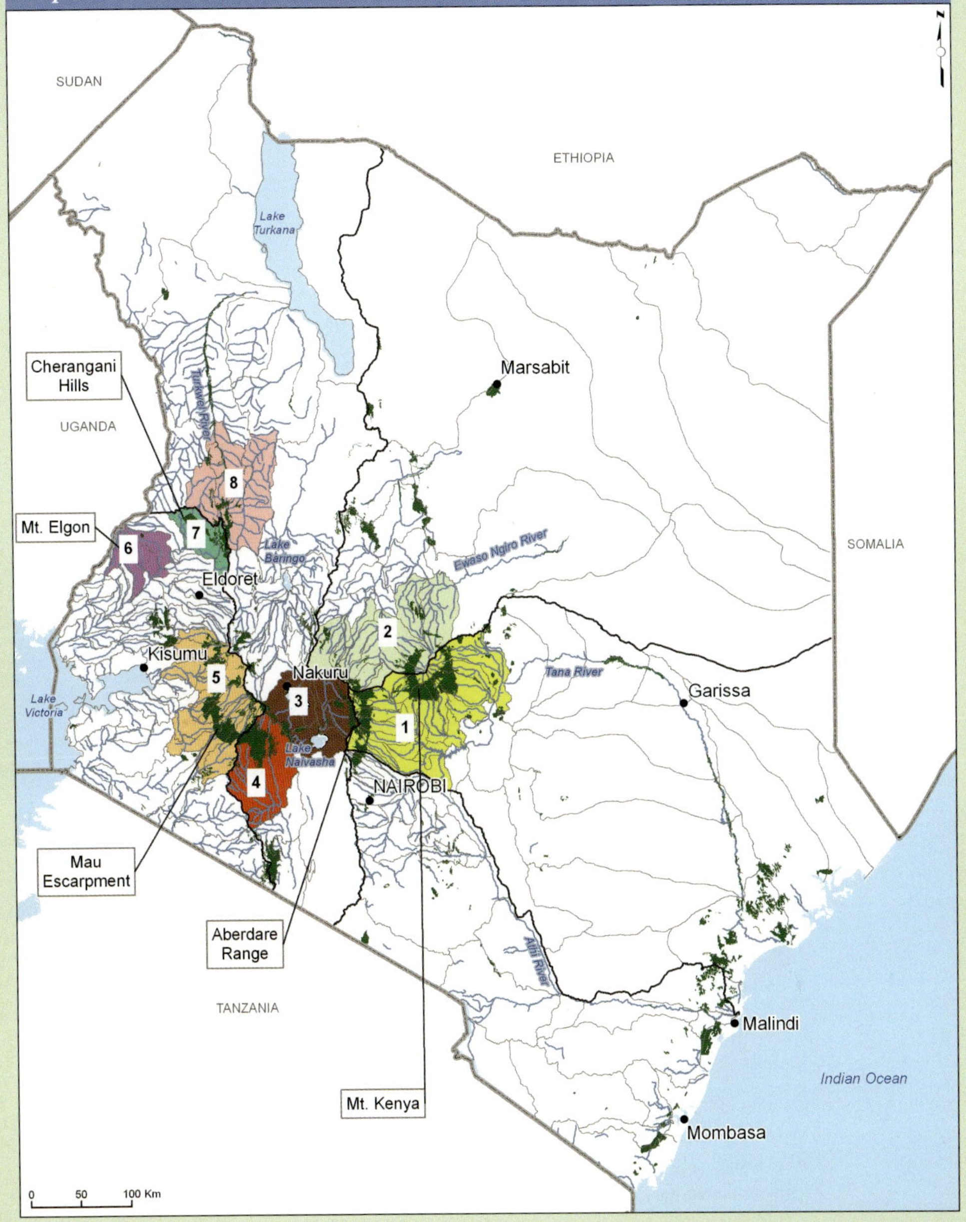

Sources: Cities (SoK and ILRI 2000), water bodies (FAO 2000a), permanent rivers (NIMA 1997), subdrainage and major drainage areas (MoWD and JICA 1992a), upper watersheds for five 'water towers' (WRI delineation based on MoWD and JICA 1992a).

UPPER WATERSHEDS
- Upper Tana River (1)
- Upper Ewaso Ngiro (North) (2)
- Lake Nakuru, Lake Elementaita, and Lake Naivasha tributaries (3)
- Upper Ewaso Ngiro (South) (4)
- Upper western watersheds of the Mau Escarpment (5)
- Upper eastern watersheds of Mount Elgon (6)
- Upper southwestern watersheds of the Cherangani Hills (7)
- Upper northern watersheds of the Cherangani Hills (8)

OTHER FEATURES
- Closed forests

DRAINAGE BOUNDARIES
- Major drainage area boundaries
- Subdrainage area boundaries

WATER BODIES AND RIVERS
- Permanent rivers
- Water bodies

The high rainfall areas in Kenya's mountains are the source of its largest rivers, many of them running year-round. The rivers which drain into the arid and semi-arid lands are an indispensable source of water to grow crops, raise livestock, and support wildlife.

The slopes of these mountains provide a complex bundle of ecosystem services. In general, they are densely settled, particularly the hills below the steeper slopes. Soils are fertile, and the dominant land use is agriculture. The higher elevations include most of Kenya's densest and multilayered tree cover. As of 1995, only 1.7 percent of Kenya's land area had sufficient tree and canopy cover to be classified as closed forest (UNEP 2001). Mount Kenya, the Aberdare Range, the Mau Escarpment, Mount Elgon, and the Cherangani Hills are home to most of these forests, together covering about 1 million hectares (Akotsi and Gachanja 2004). They are sometimes referred to as Kenya's five 'water towers.'

Maps of subdrainages (as shown in Map 3.7) can be used to delineate the upper watersheds (each consisting of various subdrainages) of the major rivers originating from these five mountain ranges. Map 3.16 outlines eight selected upper watersheds. Since all the maps in this volume are available in GIS format, the poverty and population maps in Chapter 2 can be combined with the eight outlined areas to create demographic and poverty profiles. Table 3.1 provides estimates of total population and population density, as well as estimates of the number of poor and the average poverty rate for each of the eight areas. It also shows the distribution of poverty rates among the administrative areas (Locations) falling within each upper watershed. Using this table, the demographic and poverty characteristics for these upper watersheds can be contrasted to understand poverty patterns and target poverty and ecosystem services interventions. For example, downstream users who want to benefit from improved watershed functions need to have sufficient resources to pay for specific land use practices in the uplands, in case planners want to establish a payment-for-ecosystem-services scheme.

What Do the Map and Poverty Profile Show?

- About 7.5 million people live in these eight upper watersheds, which together cover an area of almost 59,000 square kilometers (about 10.1 percent of Kenya's land area). Average population densities range from 19 to 308 persons per square kilometer, with the upper eastern watersheds of Mount Elgon (number 6) and the upper Tana (number 1) being the most densely settled. Limited resource endowments (for example, too little and unreliable rainfall or poor soils, making it difficult to grow crops) and the presence of protected areas (for conserving watersheds or wildlife) are the major reasons for lower population densities in some areas.
- While the eight upper watersheds represent 27.6 percent of Kenya's population, about 23.7 percent of the country's poor live here. The upper eastern watersheds of Mount Elgon, the upper western watersheds of the Mau Escarpment, and the upper northern watersheds of the Cherangani Hills (numbered 6, 5, 8 in the map) have the highest average poverty rates of 55, 51, and 50 percent, respectively.
- The tributaries feeding Lake Nakuru, Lake Elementaita, and Lake Naivasha (number 3) and the upper Ewaso Ngiro (number 2) have low average poverty rates of 36 and 38 percent respectively. These two upper watersheds, plus the upper Tana (number 1) all have clusters of administrative areas with some of Kenya's lowest poverty rates.

Continued

Box 3.2 Creating a Poverty and Demographic Profile for Kenya's 'Water Towers' – *continued*

- Of the eight outlined areas, the greatest number of people live in the upper Tana (3.1 million). Poverty rates for the 222 Locations within this area range from very low to very high covering all four classes in the table. The upper Tana includes a large cluster of the least poor communities but also some very poor administrative areas, most of them in the drier plains below the hills downstream of the Aberdare Range and Mount Kenya.
- This brief comparison shows that poverty and demographic patterns in Kenya's 'water towers' differ. About one quarter of all Kenyans live in the eight selected areas—very close to the total number of people in all of the arid and semi-arid lowlands. The average level of well-being in Kenya's 'water towers,' however, is significantly higher than in the communities further downstream.

Similar profiles could be constructed comparing other water-related maps from this chapter with indicators of human well-being presented in Chapter 2. For example, comparing poverty maps with maps showing high dependence of communities on surface water could help identify areas where poor communities are particularly vulnerable to interruptions in water flows and to water contamination.

Table 3.1 People, Poverty, and Kenya's 'Water Towers'

NAME OF UPPER WATERSHEDS AND MAJOR RIVERS [NUMBER IN MAP]	MOUNTAIN RANGE(S)	AREA	PEOPLE	POVERTY	POVERTY RATE: NUMBER OF LOCATIONS
Upper Tana River [1] Tana River and its tributaries draining Mount Kenya and the Aberdare Range	Mount Kenya, Aberdares	12,474 sq. km 2.1% of Kenya	3.1 million 11.4% of Kenya 250 persons per sq. km	1.3 million 9.2% of Kenya's poor 43% average poverty rate 107 poor individuals per sq. km	< 35% Range: 56 35 - 50% Range: 79 50 - 65% Range: 56 > 65% Range: 31
Upper Ewaso Ngiro (North) [2] Ewaso Ngiro (North) and its tributaries draining the Aberdare Range and Mount Kenya	Mount Kenya, Aberdares	10,541 sq. km 1.8% of Kenya	0.5 million 1.7% of Kenya 44 persons per sq. km	0.2 million 1.2% of Kenya's poor 36% average poverty rate 16 poor individuals per sq. km	< 35% Range: 22 35 - 50% Range: 26 50 - 65% Range: 2 > 65% Range: 1
Lake Nakuru, Lake Elementaita, and Lake Naivasha Tributaries [3] Rivers feeding Lake Nakuru, Lake Elementaita, and Lake Naivasha	Mau Escarpment, Aberdares	5,508 sq. km 0.9% of Kenya	0.8 million 3.1% of Kenya 152 persons per sq. km	0.3 million 2.2% of Kenya's poor 38% average poverty rate 58 poor individuals per sq. km	< 35% Range: 15 35 - 50% Range: 28 50 - 65% Range: 3 > 65% Range: –
Upper Ewaso Ngiro (South) [4] Ewaso Ngiro (South) and its tributaries draining the Mau Forest Complex into the Rift Valley	Mau Escarpment	5,881 sq. km 1.0% of Kenya	0.1 million 0.4% of Kenya 19 persons per sq. km	0.1 million 0.4% of Kenya's poor 49% average poverty rate 10 poor individuals per sq. km	< 35% Range: 1 35 - 50% Range: 7 50 - 65% Range: 12 > 65% Range: –
Upper Western Watersheds of the Mau Escarpment [5] Mara, Sondu Miriu, Nyando, and other rivers draining the Mau Forest Complex	Mau Escarpment	9,826 sq. km 1.7% of Kenya	1.6 million 5.7% of Kenya 160 persons per sq. km	0.8 million 5.5% of Kenya's poor 51% average poverty rate 81 poor individuals per sq. km	< 35% Range: 3 35 - 50% Range: 85 50 - 65% Range: 91 > 65% Range: 7
Upper Eastern Watersheds of Mount Elgon [6] Malakis River and tributaries feeding the Sio and Nzoia Rivers from Mount Elgon	Mount Elgon	2,846 sq. km 0.5% of Kenya	0.9 million 3.2% of Kenya 308 persons per sq. km	0.5 million 3.3% of Kenya's poor[2] 55% average poverty rate 168 poor individuals per sq. km	< 35% Range: 2 35 - 50% Range: 12 50 - 65% Range: 38 > 65% Range: 2
Upper Southwestern Watersheds of the Cherangani Hills [7] Upper tributaries of the Nzoia River flowing from the Cherangani Hills	Cherangani Hills	2,811 sq. km 0.5% of Kenya	0.4 million 1.3% of Kenya 126 persons per sq. km	0.2 million 1.1% of Kenya's poor 46% average poverty rate 57 poor individuals per sq. km	< 35% Range: 1 35 - 50% Range: 35 50 - 65% Range: 7 > 65% Range: –
Upper Northern Watersheds of the Cherangani Hills [8] Tributaries of the Turkwel, Marun, and Kerio Rivers from the Cherangani Hills	Cherangani Hills	8,692 sq. km 1.5% of Kenya	0.2 million 0.8% of Kenya 24 persons per sq. km	0.1 million 0.7% of Kenya's poor 50% average poverty rate 12 poor individuals per sq. km	< 35% Range: – 35 - 50% Range: 24 50 - 65% Range: 21 > 65% Range: 1
Total for Eight Upper Watersheds		**58,579 sq. km** **10.1% of Kenya**	**7.5 million** **27.6% of Kenya** **129 persons per sq. km**	**3.4 million** **23.7% of Kenya's poor** **45% average poverty rate** **58 poor individuals per sq. km**	

Sources: Poverty and demographic estimates (1999) are based on CBS 2002, 2003. Areas for the eight upper watersheds are WRI calculation based on Map 3.16.

Note: All estimates of area, people, and poverty are for the administrative areas (Locations) falling within the upper watersheds outlined on Map 3.16. Data are for 1999 and assume total population of 27.4 million and total number of poor individuals of 14.4 million as estimated by CBS (2003). Kenya's area is 582,650 square kilometers.

SUMMING UP

- From an ecosystem standpoint, water is unique, in that it is linked to all four categories of ecosystem services. *Provisioning services* include: the storage and retention of water in lakes, rivers, and as groundwater; water as an input to grow food, timber, fiber, and fuel; and freshwater for direct consumption. *Regulating services* of freshwater systems and important freshwater habitats (e.g. wetlands) include modifying water flows, recharging and discharging groundwater resources, and diluting or removing pollutants. *Supporting services* include nutrient cycling, soil formation, soil loss, and promoting biodiversity. *Cultural services* include recreational benefits, as well as the spiritual and inspirational roles of water bodies and aquatic habitats.
- Average annual rainfall amounts are distributed very unevenly: about 15 percent of the country receives sufficient rain to grow maize and other non-drought-resistant crops; another 13 percent has more marginal rainfall sufficient only to grow selected drought-resistant crops; and the remaining 72 percent has no agronomically useful growing season.
- Rainfall amounts show distinct seasonal patterns. Areas east of the Rift Valley have two rainy seasons per year. This high variability in seasonal water amounts has contributed to a great diversity of wild plant and animal species. It creates a special challenge for growing crops, however, because none of the two rainy seasons is quite long enough to allow very high yields.
- Rainfall amounts vary greatly from year to year as well. Major droughts and floods have occurred regularly in each decade over the past 30 years.
- Kenya's network of perennial rivers is most dense in the central and western parts of the country, leading to uneven supplies of surface water. The Lake Victoria drainage area supplies the highest share (65 percent) of Kenya's internal renewable surface water per year. The Athi River drainage area provides the lowest share (7 percent).
- The total renewable water resource available per year is 936 cubic meters per person (2004). This designates the country as *water scarce*. Policymakers must pay particular attention to the management of water resources to avoid hindering food production or impeding economic development. Population growth alone will continue to reduce per capita water availability.
- Subdrainages with densely settled urban populations such as Nairobi and Mombasa need to maintain their long distance water transfers to meet growing demand in the future. The same is true for all subdrainage areas in the central part of the Rift Valley north and south of Nakuru. Even in areas with perennial surface water flows, high local demand can outstrip local supply. The *Study on the National Water Master Plan* projects local water deficits for selected subdrainage areas in the upper Ewaso Ngiro, Tana River, and in western Kenya.
- Open surface water is the major source of drinking water for 29 percent of Kenyan households, almost all of them in rural areas. About 32 percent of Kenyan households rely on groundwater for their drinking water. The same proportion of Kenyan households uses piped water (71 percent of urban households and 19 percent of rural households). Families using untreated surface water are relying completely on the regulating services of ecosystems to provide uncontaminated water at sufficient quantities.
- Hydropower is the largest source of electricity providing 55 percent of the total installed grid capacity. About 84 percent of Kenya's existing hydropower capacity is located on the upper Tana River.
- Ninety-eight percent of Kenya's cropping is rainfed. Thus, the high variability of rainfall within and between years poses a significant risk for most farmers. Irrigation, covering the remaining 2 percent of cropland, is carried out by smallholders (46 percent), larger commercial firms (42 percent), and by public schemes (12 percent). Only 19 percent of Kenya's potential area is equipped for irrigation.
- In almost all of the subdrainage areas in Kenya's rangeland Districts, water demand for livestock is significantly greater than for wildlife. Only in a few subdrainage areas within or close to protected areas do wildlife consume a larger share.
- What constitutes an ecosystem service for one group may be a disaster for another. For example, floods can have both negative and positive impacts depending on the context. Floods regularly destroy homes, schools, and crops, and kill people and animals. This is especially true in western Kenya in the densely settled low-lying areas around Lake Victoria. On the other hand, flooding can sometimes be helpful to both ecosystems and people. About one million people in the lower Tana River depend on the river's flooding regime for their livelihoods. In addition, birds and wildlife depend on the annual flood cycle of the Tana for habitat and forage.

In Chapter 4

CONTENTS

LIST OF MAPS

AUTHORS AND CONTRIBUTORS

Patti Kristjanson (ILRI)
Paul Okwi (ILRI)
Nancy Mutunga (FEWS NET)
James Oduor (ALRMP)
James Nyoro (Tegemeo)
Diana Lee-Smith (ILRI)
Allan Kute (WFP)
Geoffrey Kimathi (WFP)
Polly Akwanyi (WFP)
An Notenbaert (ILRI)
Michael Arunga (ILRI)
Isabelle Baltenweck (ILRI)
Amos Situma (DRSRS)
Janet Nackoney (WRI)
Florence Landsberg (WRI)
Greg Mock (consultant)
Emily Cooper (consultant)
Norbert Henninger (WRI)
Dan Tunstall (WRI)
Hyacinth Billings (WRI)
Carolina de Rosas (WRI)
Jamie Worms (WRI)

WHAT THIS CHAPTER SHOWS

This chapter examines the principal domestic sources of food in Kenya, including crop production, livestock, fishing, and hunting-gathering. A detailed livelihood map gives an overview of how Kenyan households use natural resources, wage labor, and other urban employment to make a living. Maps of cropping intensities show that Kenya's rainfed agriculture reflects the country's rainfall patterns, with a significant proportion of farmers being exposed to the risks of unreliable rainfall or prolonged drought. A detailed view of central and western Kenya, where more than 90 percent of croplands are located shows that farmers dedicate large shares of their cropland to food crops in selected high-potential Districts such as Trans Nzoia, Uasin Gishu, Lugari, upper Nandi, and Nakuru (maize and other cereals), Narok (wheat), and lower Kirinyaga (rice). Food crop shares are also high in the more marginal cropping areas—but here agriculture is dominated by lower-yielding maize—for example, along Lake Victoria and large parts of Laikipia, Machakos, Mwingi, Kitui, Makueni, Taita Taveta, Kwale, Kilifi, and Malindi Districts. Livestock production in Kenya also displays distinct spatial patterns: high dairy output and surpluses primarily in central Kenya; milk deficits in large parts of Nyanza and Western Provinces; and pastoral and agropastoral livestock rearing in the arid and semi-arid lands. The chapter concludes with a set of maps on fishing and hunting-gathering of wild animals and plants.

Food

Obtaining food, the most basic human need, is an activity that is always closely linked to natural resources. This chapter covers four dominant sources of food and livelihoods in rural Kenya: crop production, livestock, fishing, and hunting-gathering. Using indicators such as the presence and level of an activity (cropping, livestock rearing, fishing, etc.), and its contribution to cash income, this chapter explores the distribution of different livelihood strategies throughout Kenya, and how these patterns are influenced by ecosystems and the resources they provide. In some cases, changes in the resources available—for instance declining fish catches and crop yields—have begun to influence livelihoods, or may in the near future. Changes in land-use patterns—the creation of permanent water points in Turkana, for example, and increasing reliance on wage labor—also have repercussions on the environment and the people within it.

FOOD AND LIVELIHOODS

Sources of Food

Kenyans obtain most of their food from a few prominent sources. Agriculture provides an important source of subsistence as well as cash income for food for rural households. Maize, the staple food for most Kenyans, is the most widely grown cereal crop. Other major food crops include beans and cassava, and cereal crops such as wheat, millet, and sorghum. Kale, tomatoes, onions, potatoes, carrots, and cabbage are important minor crops.

Kenyan livestock consist of chickens, cattle, camels, pigs, sheep, and goats. These animals provide meat, milk, and eggs, and are an important source of protein and micronutrients, especially for children. Livestock play a particularly vital role as a food source in the semi-arid and arid lands that cover more than 80 percent (SoK 2003) of Kenya, where it is difficult or impossible to grow most crops.

Fishing provides food and a way to earn cash income for many Kenyans living near major bodies of water, particularly Lake Victoria. However, fishing plays a fairly small role in much of the country.

Finally, hunting wildlife and gathering nuts, fruits, and tubers in Kenya's forests and savannas remain important to many, as has been true for thousands of years. These wild resources become particularly critical in times of drought, stress, and hunger—whenever other resources become unreliable.

Predominant Strategies for Food and Livelihoods

The spatial distribution of different ecosystem types greatly influences the choice of livelihood strategies that Kenyan families pursue. Livelihood strategies can range from focusing predominantly on livestock products such as meat and milk in rangeland ecosystems, to a mix of livestock, food, and cash crops in areas with adequate rainfall and soils. In some areas of the country, fishing, hunting, and gathering are all important sources of food and livelihoods—typically in forest, rangeland, and freshwater ecosystems. In urban ecosystems, a large percentage of households rely on wages and other income sources to purchase food, but agriculture still plays an important role in the daily activities of many urban families (see Box 4.1).

While subsistence food production is still widespread in Kenya, most households attempt to diversify their food and income sources. A recent survey covering each Sublocation across Kenya asked experts to describe the predominant strategies for obtaining food, clothes, and shelter for the majority of families in that Sublocation (ALRMP et al. 2006). Map 4.1 organizes these data into major livelihood zones, which are grouped into six broad classes reflecting various levels of ecosystem modification and net returns to land and labor:

- Forests or mixed fishing;
- Pastoral or agropastoral;
- Marginal mixed farming;
- High-potential mixed farming;
- Cash cropping or irrigated cropping;
- Wage labor or urban livelihoods.

Map 4.1 Predominant Livelihood Strategies Pursued by Households, 2003-05

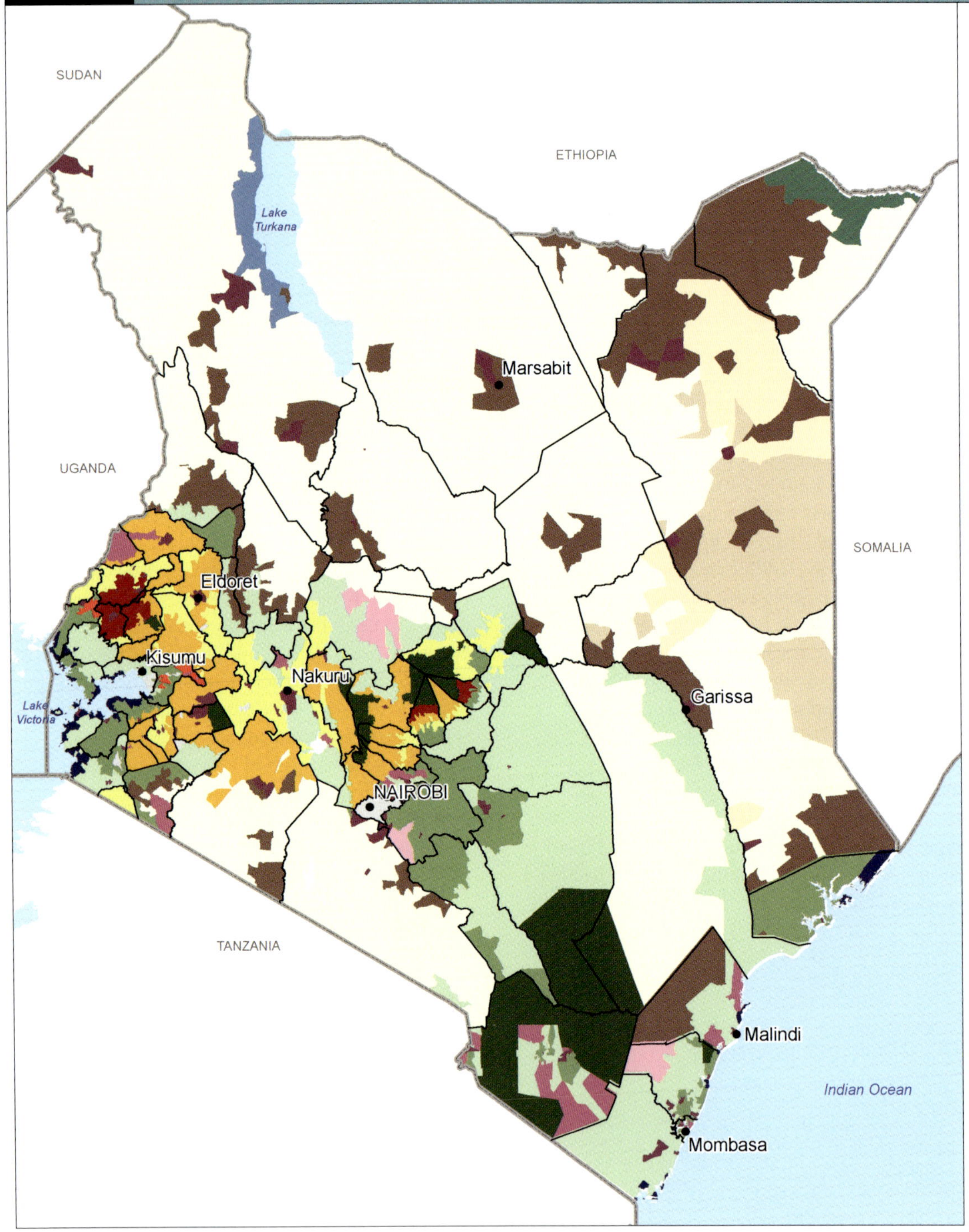

Sources: Administrative boundaries (CBS 2003), cities (SoK and ILRI 2000), water bodies (FAO 2000), and livelihood zones (ALRMP et al. 2006).

In most of Kenya's arid and semi-arid areas, pastoral livelihood strategies dominate. This involves moving livestock periodically to follow the seasonal supply of water and feed. Depending on the availability of water, feed, and capital, families may chose certain mixes of species, as in areas close to Somalia where mixed herds of goats and sheep are common. Cropping combined with pastoral livestock raising (agropastoral strategies) are clustered along the margins where rainfed agriculture is possible and around more permanent water sources such as the mountains close to Marsabit and along the Tana River near Garissa. They are often close to trading and market centers (shaded in dark purple), which provide some employment and wage opportunities.

In the majority of central and western Kenya, high-potential agricultural lands are dominated by a mix of dairy cattle, food, and cash crops (shaded yellow and orange). Mixed farming along the shores of Lake Victoria, in the croplands east and southeast of Nairobi, and in the coastal hinterlands is more marginal (shown in two shades of green). In many of these areas, rainfall is more erratic or soils are less fertile. Here, yields and incomes coming from a mix of livestock and food crops are generally lower.

Fishing, sometimes combined with pastoral livestock raising or food crop cultivation (shown in different shades of blue), is much more localized. It is the dominant livelihood strategy for communities along the shores of Lake Victoria, Lake Turkana, and the Indian Ocean.

In some areas, the link from ecosystems to livelihoods is more indirect. Families in the rangelands northwest of Mount Kenya (Laikipia District), for example, depend more on casual wage labor on large ranches (shown in pink); in parts of the coastal hinterlands, plantation labor, mining, and other wage labor are important (shown in dark pink).

Note: Data on livelihood zones are based on questionnaires sent to key food security experts in all 71 Districts (generally about 6-10 persons). In some cases where further clarification was necessary, questionnaires were sent to experts below District level (Division). This group of experts classified each of Kenya's 6,632 Sublocations by their predominant livelihood strategy and other livelihood characteristics.

FORESTS OR MIXED FISHING
- Large protected area or forest
- Fishing and pastoral
- Fishing and food crops

PASTORAL OR AGROPASTORAL
- Pastoral (all species)
- Pastoral (sheep and goats)
- Pastoral (cattle, sheep, and goats)
- Agropastoral

MIXED FARMING - MARGINAL
- Food crops and livestock
- Food crops, cash crops, and livestock

MIXED FARMING - HIGH POTENTIAL
- Food crops and livestock
- Food crops, cash crops, and livestock

CASH CROPPING OR IRRIGATED CROPPING
- Cash cropping - low potential
- Cash cropping - high potential
- Irrigated cropping

WAGED LABOR OR URBAN LIVELIHOODS
- Casual waged labor (ranching)
- Casual waged labor (plantation, mining, and others)
- Trade business, and employment centers
- Urban centers

OTHER FEATURES
- District boundaries
- Water bodies

Box 4.1 Urban and Periurban Agriculture

Urban agriculture went unnoticed until the 1980s, when the first pieces of research revealed startling numbers. In Kenya, one third of urban dwellers were growing subsistence crops and raising livestock, and two thirds were farming in either urban or rural areas, or both. The numbers in the rest of East Africa were the same or higher (Urban Harvest 2004).

Agriculture is a major livelihood strategy of the urban poor in their struggle against hunger and poverty. Studies in Kenya have since confirmed that urban dwellers (particularly women) who grow crops or livestock, feed their children better than those who do not. Some of the major issues surrounding urban and periurban agriculture include the following:

Public health. Municipalities are very worried about the potential public health hazards of urban agriculture. One significant risk is associated with pathogens, toxic chemicals, or heavy metals that are often present in waste water or solid waste used in urban farming. Additional hazards include overuse of agrochemicals in densely populated areas, creation of vector breeding sites, and air pollution. For example, polluted air, largely from petrol fumes, deposits lead in the soil and on the leaves of plants. In addition to heavy metals such as lead, cadmium, and zinc, there are complex organic compounds produced by numerous indoor and outdoor sources in urban areas, including vehicles, industrial emissions, appliances, and woodfuel burning. The urban poor frequently burn plastic and other materials to get rid of waste or even to use as fuel.

Zoonotic diseases. Zoonotic diseases are those that are transmitted between animals and people. Keeping animals allows poor residents to feed their children milk and eggs, but people are not always aware of the health risks of living so close to animals. Households may also know the risks but feel they can't do anything about them. In poor and crowded urban areas—where sanitation systems are often inadequate or missing altogether—livestock diseases can jump to humans. Cysticercosis, an infection caused by the pig tapeworm, is a good example. A person does not have to eat pork or keep pigs to become infected with cysticercosis; poor hygiene, or consumption of contaminated food or water can cause infection if someone accidentally ingests eggs from human tapeworm carriers.

Nutrient cycling. It has been said that the biggest agricultural productivity problem in sub-Saharan Africa is soil infertility, yet urban areas are vast sinks of nutrients which are being wasted. Nairobi produces 635,000 tons of solid waste in a year, 70 percent of it organic, containing thousands of tons of nitrogen, phosphorus, and potassium. Almost all of this material is wasted, lying in landfills—or worse—blocking up drains. Research shows that less than one half of one percent of urban waste is being composted for agricultural use (Urban Harvest 2004). And by whom? The urban poor.

As for manure, in 2003 Nairobi had 24,000 head of dairy cattle, but virtually none of the manure from these cows was sold. Nairobi does, however, export livestock fodder from urban to rural areas in the dry seasons, when the grass is depleted in the countryside.

Wastewater produced by urban agriculture is similarly nutrient-rich. It is a potentially valuable resource, but can also carry dangerous levels of heavy metals. The goal is to find water-management systems that can help farmers to safely use the nutrients in wastewater while preventing the heavy metals from making their way into food.

Pigs in Kibera, Nairobi. Raising pigs in the urban environment poses a serious risk of spreading the pig tapeworm to nearby human residents.

Cattle in Soweto-Kahawa, Nairobi. Some livestock in Nairobi slums are well looked after and stall-fed on urban-grown napier grass. This photo was taken in Soweto-Kahawa, where manure and compost from domestic waste are also cleaned up and recycled for crop growing.

Importance of the Agriculture and Food Sector

The agricultural sector, which includes livestock products and food and nonfood crops, is the dominant source of food and livelihoods in Kenya. In 2004, it contributed 26 percent of Kenya's gross domestic product (53 percent, if indirect links to other economic sectors are counted), 60 percent of total export earnings, 45 percent of government revenue, and 62 percent of jobs in the formal economy. Accounting for employment in the informal sector, the share of Kenyans depending on agricultural resources for their livelihoods rises to almost 80 percent (RoK 2006; CBS 2004, 2005). These official figures do not include the value of food that is hunted or gathered, nor do they value products such as animal blood that are part of pastoralists' diets.

Recent surveys of smallholder farming households throughout Kenya highlighted the following trends in the food sector (Jayne et al. 2000):

- *Diversification.* Incomes and livelihood strategies of rural farm households are highly diversified. Maize accounts for only 14 percent of total household income, on average, and does not exceed one quarter of total income, even in the highly productive maize areas of the northern Rift Valley. Other crops such as tea, vegetables, fruits, sugarcane, coffee, and root crops generally account for more than 20 percent of household income. Households that have traditionally relied solely on livestock for their livelihoods are also diversifying into cropping and other income-earning activities (Kristjanson et al. 2002).
- *Importance of non-farm and non-land income.* Smallholders currently derive between 25 and 70 percent of their income from non-farm sources, such as wage labor. Small rural farms in Kenya no longer rely mostly on cereal crops for their livelihoods. Similar trends are being seen in more remote areas that were traditionally pastoral and would now be considered agropastoral.
- *Small farm sizes.* Farm sizes have been declining with increased population pressure, from 0.53 hectares per farmer in 1960, to 0.20 hectares—less than half as much—in 2000 (FAO 2006, as cited in Jayne et al. 2000). This has made it much less viable to earn a living from crops with a low value per hectare.
- *Importance of cash crops.* Crops with the highest net returns to land and labor vary widely across Kenya, but generally tend to be those grown solely for cash income—horticultural crops, sugar, tea, and coffee. The exceptions are a few high-potential maize areas that include Trans Nzoia and Uasin Gishu in the North Rift Valley, where maize—not typically grown for export—is an important cash crop.
- *Most households must rely on the market for food security.* Most rural smallholders outside the "grain basket" of Rift Valley and Western Provinces, even in the high-potential agricultural zones, are net buyers of maize throughout the year. While almost all rural households grow maize to help feed their families, it is typically insufficient to meet their requirements, and households must use income earned from livestock, cash crops, or off-farm sources to purchase most of the maize they consume.

Croplands

Croplands are the primary source of food and livelihoods for the majority of Kenyans. Kenyan landscapes where cropping is present can be roughly divided into agropastoral areas, cropland-dominated areas, and urban and periurban areas. Agricultural activities are carried out in parts of all of these areas, but the intensity, type, and location of crops varies within and between them. Since most Kenyan farming relies exclusively on rainfall, the spatial extent of croplands is closely linked to the country's annual and seasonal rainfall patterns (see Maps 3.1 to 3.5 in Chapter 3). Kenya's croplands are concentrated in the higher and more reliable rainfall zones (the highlands, Lake Victoria basin, and a narrow coastal strip) and in areas adjacent to year-round freshwater sources such as the lower Tana River. Farming intensity, or the percentage of land under cultivation, varies significantly across Kenya, with the areas of highest rainfall and soil quality being able to support the highest percentages of cropland (Map 4.2).

At a national level, the total area under cultivation continues to increase, although at slower rates (FAO 2006). At a more local scale, this expansion includes new cropping in degazetted forest lands (Matiru 1999), conversion of "wetter" rangelands in Narok and Trans Mara Districts (Serneels and Lambin 2001; Lamprey and Reid 2004; Norton-Griffiths et al. in press), and fast growth of horticultural crops, such as fruits and vegetables. In the last 20 years, the greatest sustained growth in farm area expansion has been in crops with relatively high value per unit of land. This includes horticultural crops, sugarcane, and until recently, tea (Jayne et al. 2000).

Anecdotal evidence suggests that some of the crop expansion is into marginal lands where there is a high risk of crop failure because of low and variable rainfall levels. A significant proportion of Kenya's cropland is already planted each year in areas with a high likelihood of insufficient rains. Map 4.3 approximates these areas by delineating croplands that receive less than 800 millimeters of rainfall a year. In most low-rainfall areas, households rely upon a combination of mixed crops, livestock rearing, and other activities for their livelihoods. However, they still remain highly reliant upon the weather. Farmers in Kitui, Makueni, and Mwingi Districts, for example, are greatly dependent upon the second rainy season (the "short rains") to ensure they harvest at least one crop per year.

Map 4.2 **Intensity of Cultivation, 2000**

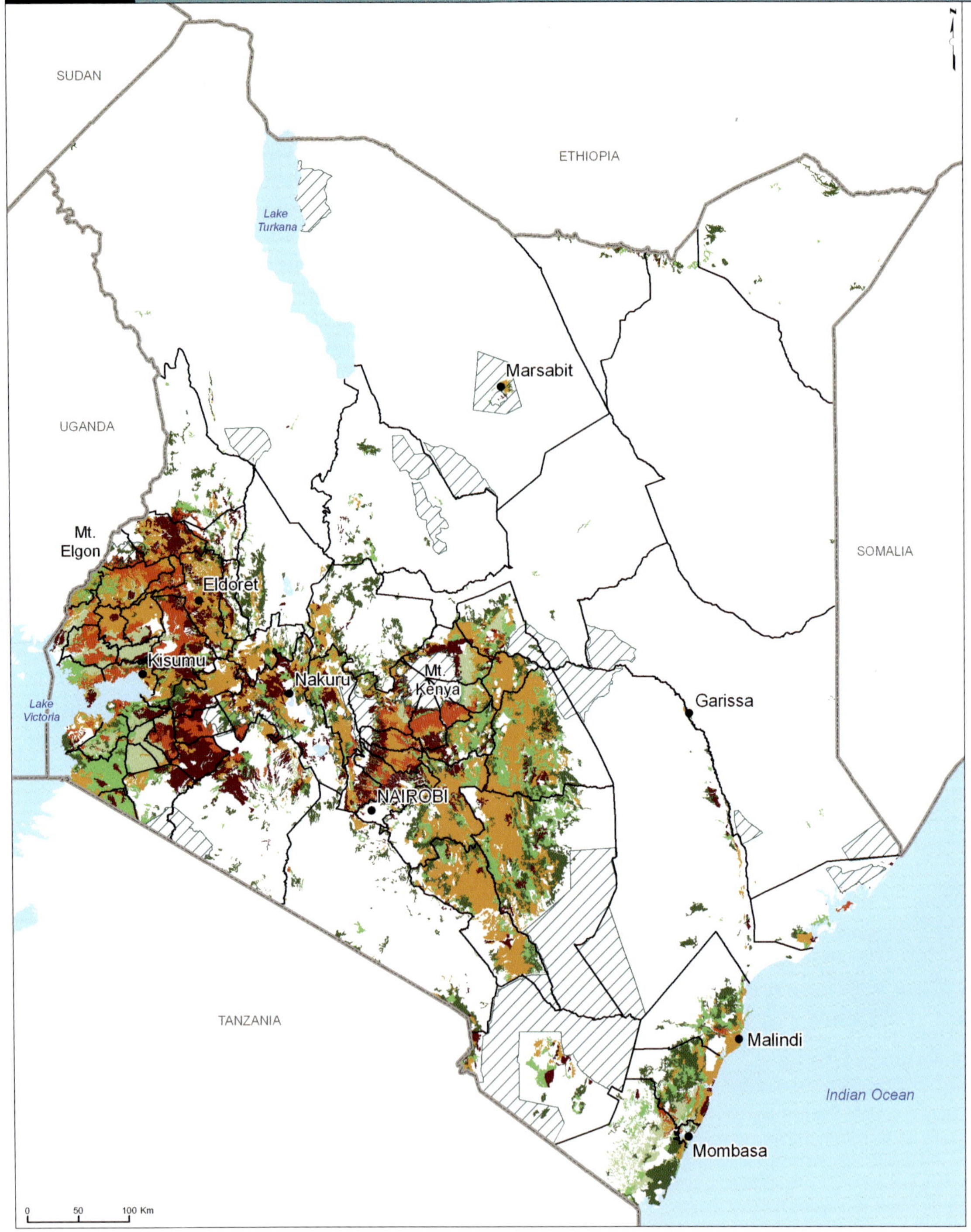

Sources: Administrative boundaries (CBS 2003), cities (SoK and ILRI 2000), water bodies (FAO 2000), parks and reserves (IUCN and UNEP/WCMC 2006), croplands (FAO 2000), and cropland intensity (WRI calculation based on *Africover* legend for croplands in FAO 2000).

Intensively farmed land—areas of more than 80 percent cropland—represent only a small proportion of Kenya's agroecosystems. These densely cropped areas (shown in dark brown) are found predominantly in the highlands of central and western Kenya and in small patches of lowlands. They include intensively produced crops such as wheat, tea, sugarcane, irrigated rice, and high-yielding maize (see Map 4.4 for a more detailed view).

The majority of Kenya's agroecosystems consist of landscapes with 50 or 60 percent active cropland (shown in lighter green and orange), mixed with less intensively managed land. The latter can include, for example, forests or woodlands that can support mixed activities such as wood extraction and livestock grazing.

Note: The standardized Land Cover Classification System of *Africover* (FAO 2000) can be used to show to what degree the spatial units (polygons) within the *Africover* map are "natural and semi-natural areas" or "managed" (cultivated) areas. The *Africover* classification system and the associated rules used to interpret the satellite imagery allow the creation of six discrete classes of cropland intensity, reflecting a stepwise gradient from the lowest (only 15 percent of the polygon is covered by cropland) to the highest category (more than 80 percent of the polygon is cultivated). The *Africover* map does not provide sufficient information to create a continuous legend ranging from zero to 100 percent. The map cannot show cropping that falls below the 15 percent threshold.

Map 4.3 Intensity of Cultivation in Low-Rainfall Areas, 2000

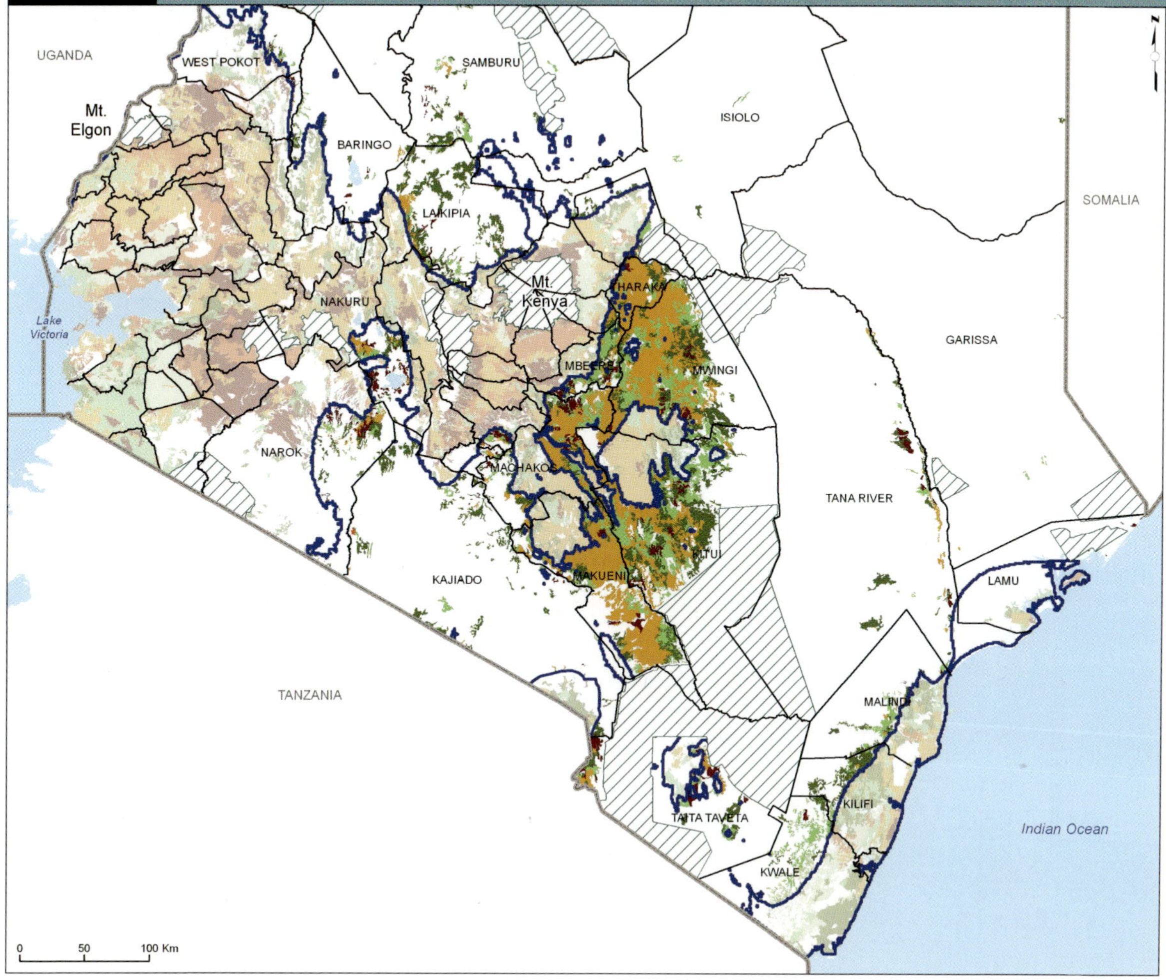

Sources: Administrative boundaries (CBS 2003), cities (SoK and ILRI 2000), water bodies (FAO 2000), parks and reserves (IUCN and UNEP/WCMC 2006), 800 millimeter average annual isohyet (Hijmans et al. 2005), croplands (FAO 2000), and cropland intensity (WRI calculation based on *Africover* legend for croplands in FAO 2000).

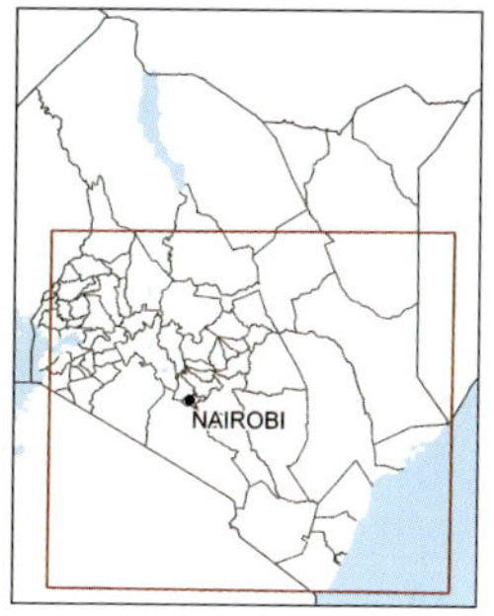

Kenya's croplands can be delineated into areas that receive, on average, less than 800 millimeters of rain per year (shown in brightly colored zones) and those with higher annual rainfall (shown in faded colors). Annual rainfall of 800 millimeters, evenly distributed across the year, is sufficient to grow maize. The risk of crop failure increases, however, when this amount is split over two rainy seasons separated by a longer period with very little rainfall (bimodal rainfall patterns). With the exception of Rift Valley and the western highlands, maize is grown in two distinct seasons—the "short" and the "long" rain seasons. Most of the areas with less than 800 millimeters of rainfall are in Kitui, Makueni, Mwingi, and the lower Machakos Districts. They also include the cropland-rangeland boundaries in Samburu and Laikipia Districts and the coastal hinterlands of Malindi, Kilifi, and Kwale Districts.

Note: See note for Map 4.2.

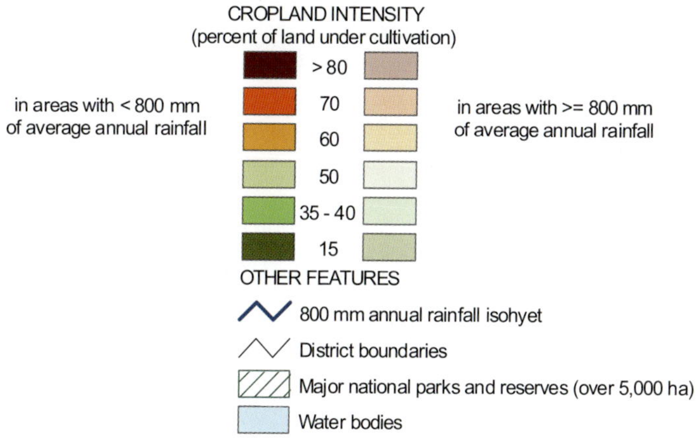

FOOD CROPS: MAJOR SUPPLY AREAS

The following section looks at major patterns and trends in agriculture as a food source in Kenya. While the majority of agricultural land is dedicated to food production—maize, in particular, is grown on a wide scale—food crops do not occupy all cropland in Kenya. In conjunction with some of the earlier maps, planners can look at where food crops are being grown, under what rainfall conditions, and the percentage of cropland they cover. While Kenyans generally grow both food and cash crops on all croplands with sufficient rainfall, there are a few locations where cash crops are dominant, occupying more than 75 percent of the cropland in that area. In the hills below the Aberdare Range and Mount Kenya, for example, tea and coffee dominate. Tea is the predominant crop in agroecosystems in selected areas further west as well, such as Bomet, Buret, Kericho, and Nyamira Districts.

This section also looks at what types of food Kenyans are growing, and what the trends in production have been—in terms of both crop area and yield—for the major food crops over the past 15-20 years. Maize is a staple crop in Kenya, primarily as a food source and to a lesser extent for household income. The graphs present trends in maize production and the locations of high output. Despite increases in crop area and demand that has risen with population growth, yields have been declining in recent years, leading to an increasing reliance on imported maize. Finally, Box 4.1 briefly examines some of the major issues surrounding the important and often underestimated role played by urban and periurban agriculture.

Food Cropping in Central and Western Kenya

Kenyan government agencies collect detailed data on areas regularly planted with maize and other crops; this chapter highlights the degree to which farmers' crops are oriented toward food and nonfood production in central and western Kenya. Nonfood crops, sometimes referred to as cash crops, mainly include tea, coffee, sugarcane, tobacco, and sisal. Map 4.4 depicts the spatial pattern of food and nonfood cropping. The map covers both high-potential production systems (mixed farming systems and cereal-dairy systems in the eastern and western highlands) and more marginal cropping areas, for example, areas in the Districts directly bordering Lake Victoria, and most land in Makueni, Mwingi, Kitui, and southern Machakos Districts.

In the majority of the croplands of central and western Kenya, farmers plant more food than other crops. Areas with little food cropping include the important coffee- and tea-growing areas. Here, special zones were established in 1986 and are now managed by the Kenya Tea Zone and Conservation Corporation. The purpose of these zones is to grow tea, establish intensively managed fire wood plantations (for drying tea), and improve livelihoods, thus creating an area where local communities put less resource pressure on the bordering gazetted forests. Every year the Corporation employs 2,000-10,000 people to harvest tea leaves (SoK 2003).

National Trends of Selected Food Crops

Trends in maize production. In terms of cropped area, maize is Kenya's most important food crop. For a large proportion of the population—both urban and rural—it is also the primary source of calories. Maize consumption is estimated at 98 kilograms per person per year, or around 2.7 to 3.1 million metric tons per year. The crop accounts for roughly 25 percent of gross farm output from the small-scale farming sector (Nyoro et al. 2004).

While most smallholder farms produce some maize for home consumption, maize that reaches the commercial market comes mostly from large-scale farms (the top 10 percent of farms in terms of size). These large commercial farms produce over 80 percent of the domestically marketed maize in Kenya (Jayne et al. 2000). Although the remaining 20 percent comes from smallholder farms, only a small proportion of these farms actually send much maize to the marketplace. Thus, Kenya's marketed maize output comes from a relatively small portion of the total farm population.

Figures 4.1 and 4.2 look at trends over the past two decades in areas planted and production levels of maize and other major crops in Kenya. In terms of quantity, Kenya's maize production peaked during the mid- to late-1980s, and has since stagnated due to declining yields. From 1985-2003, maize output fluctuated between 1.7 and 3.0 million tons per year, with an average of 2.5 million tons over the period. Maize is planted throughout the country, from high-yielding areas to riskier, semi-arid zones. Yields vary dramatically, from around 500 kilograms per hectare in semi-arid areas to greater than 2,500 kilograms per hectare in the high-potential maize zone (De Groote et al. 2005). Low-potential areas include Kajiado, Makueni, and Mwingi Districts; parts of Meru; and parts of Machakos. The high-potential maize zone includes the Districts of Trans-Nzoia, Uasin Gishu, Bomet, Nakuru, Bungoma, Lugari, Nandi, Kericho, and highland areas of Kakamega. Even in high-potential areas, yields are typically very low, with little or no chemical fertilizers applied. Indeed, there may be more appropriate crops to plant that could increase both food security and profitability. However, the decision to grow maize for food reflects national consumption and dietary patterns that have been established over decades. If a shift in the choice of crops does occur, it will likely be a gradual change.

Figure 4.1 Area under Selected Food Crops, 1985 – 2003

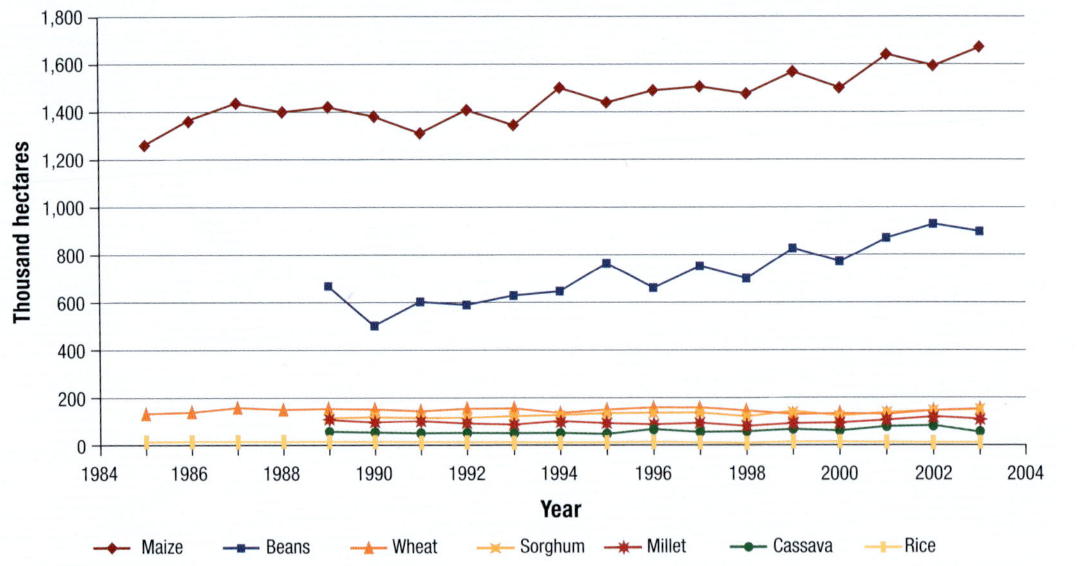

Sources: CBS (1986, 1995, 2001, and 2005) and FAO (2006).

In 2003, maize covered the largest share of Kenya's croplands, with a harvested area of 1.67 million hectares. This was an increase from around 1.25 million hectares in 1985. Beans were second in area with 0.89 million hectares in 2003. Wheat and sorghum covered around 150,000 hectares each, followed by 108,000 hectares of millet. Cassava stood at 50,000 hectares and irrigated rice at 10,000 hectares for the same year. While the area under cassava and sorghum has grown slightly, the area under wheat and millet has remained more or less the same over the period. The total area planted with rice—a fairly minor cereal crop in Kenya—decreased by 25 percent. It has since increased after expansion in coastal Districts and rehabilitation of rice schemes around Lake Victoria (Mutunga 2006).

Map 4.4 Food Crops as Percentage of all Cropland in Central and Western Kenya, 1997

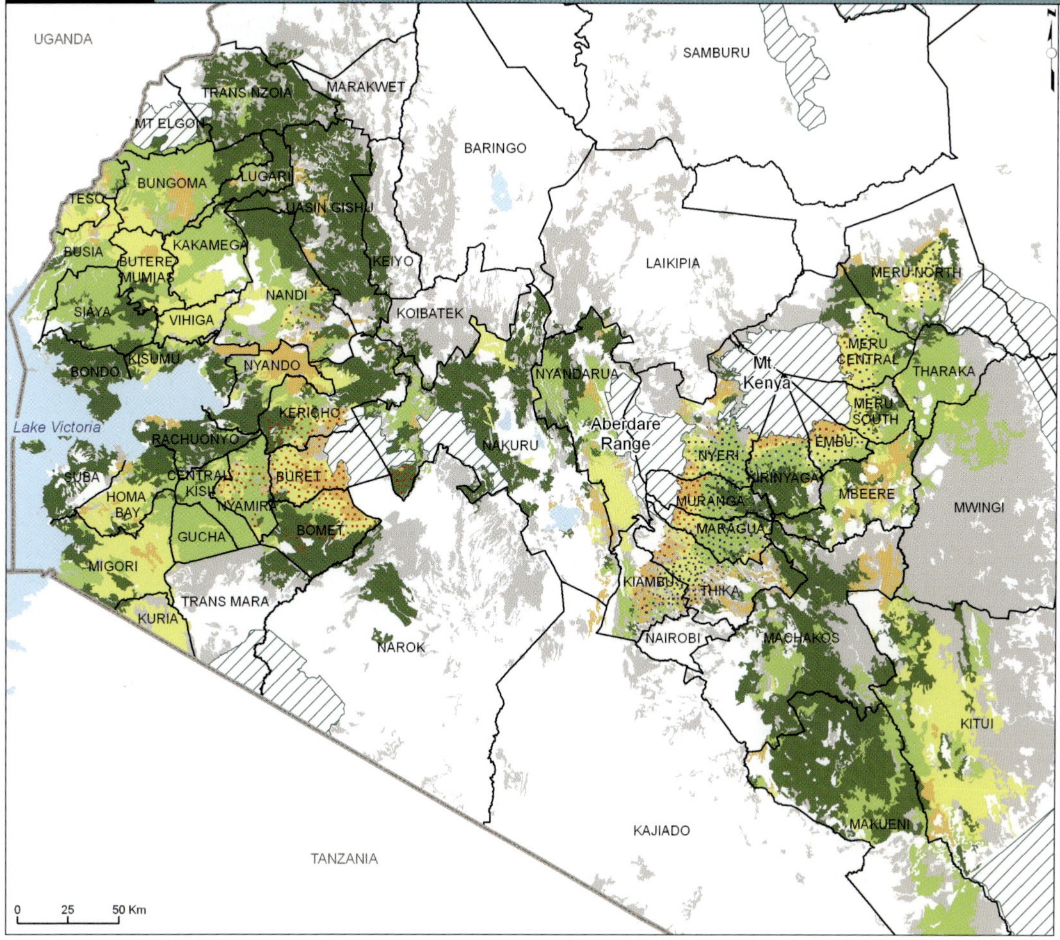

Sources: Administrative boundaries (CBS 2003), water bodies (FAO 2000), parks and reserves (IUCN and UNEP/WCMC 2006), cropland areas (FAO 2000), and percentage food crops in sampled croplands (WRI calculation based on ICRAF and DRSRS 2001).

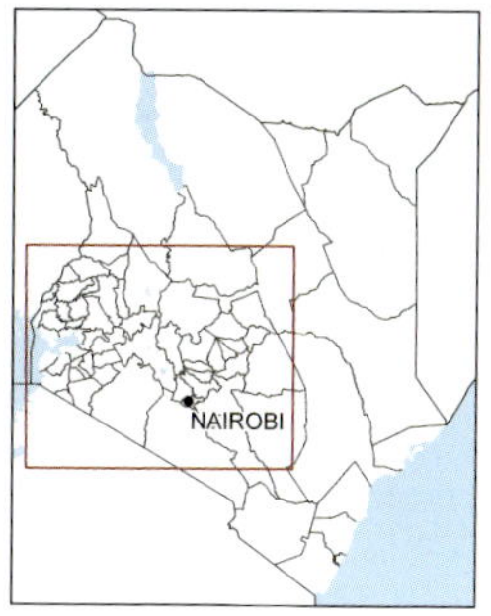

This map depicts the share of cropland that is dedicated to food crops, irrespective of the overall cropland intensity shown in Map 4.2. By using only two categories (food and nonfood) and grouping the data into four broad data ranges, the map is relatively robust to the seasonal changes in specific crop choices caused by differences in rainfall, prices, demand, and labor availability.

Spatial patterns of food cropping do not necessarily mirror those of cropland intensity. Areas where more than 75 percent of farmers' cropland is dedicated to food crops (shown in dark green) are concentrated in high-potential Districts such as Trans Nzoia, Uasin Gishu, Lugari, upper Nandi, and Nakuru (maize and other cereals); Narok (wheat); and lower Kirinyaga (rice). High food-crop shares also occur in more marginal cropping areas such as the Districts bordering Lake Victoria and large parts of Machakos and Makueni Districts (but here low-yielding maize is the major contributor).

The lowest shares of food crops (25 percent, shaded in orange) cover the tea-growing areas (depicted by clusters of red points) along the Aberdare Range; Mount Kenya; and parts of eastern Bomet, Buret, Kericho, and Nyamira Districts. Areas with a food share of 25-50 percent (shown in yellow) include the coffee-growing zones of the Aberdare Range and Mount Kenya in Central Province (shown with clusters of dark blue points). In the west, for example, in Siaya, Kakamega, and Migori Districts, low shares of food crops are typically paired with sugarcane or tobacco crops. Areas with low shares of food crops (shown in yellow and orange) in Kitui District may be temporary, reflecting large shares of fallow cropland during the 1997 season of the aerial surveys.

FOOD SHARE
(percent of sampled cropland)

- > 75
- 50 - 75
- 25 - 50
- <= 25
- Cropland not sampled or food share unknown

NON-FOOD CROPS

- Tea (large share in sampled cropland)
- Coffee (large share in sampled cropland)

OTHER FEATURES

- District boundaries
- Major national parks and reserves (over 5,000 ha)
- Water bodies

Note: The map combines detailed crop information from 5,747 aerial photos for a growing season in 1997, each providing a sample point of detailed crop information. These samples are averaged to spatial units (polygons) of croplands from Kenya's most recent land-cover map (FAO 2000).

With a growing gap between production and consumption, maize imports have become increasingly important. Kenya went from being a net exporter of maize from 1986-87 through 1990-91, to a significant net importer from 1997-98 to present. Imports have ranged from 75,000 to 1.1 million metric tons per year, the latter number reflecting high demand because of drought-related crop failure (Nyoro et al. 2004). These official figures, however, do not include the considerable percentage of the maize trade that happens informally. Unrecorded dealings with Kenya's neighboring countries—especially imports from Uganda, and to some extent from Tanzania—are estimated to have been as high as 150,000 tons per year in the early 1990s.

Trends in horticulture. Fruits and vegetables are important for both consumption and income in many rural households across Kenya. These crops have a relatively high value per unit of land, and have witnessed a great expansion in farm area over the past decade. In 2003, horticultural production of fruits (primarily mangoes, papayas, bananas, passion fruits, pineapples, oranges, coconuts, and macadamia nuts) and vegetables (kale, cabbage, carrots, tomatoes, avocadoes, French beans, and indigenous vegetables) together covered an area of about 250,000 hectares (Figure 4.3).

Between 1989 and 2003, the area under vegetables and fruit crops grew by about a third. Growth in output and value were also significant over this period. This is also reflected in the export statistics, which have shown tremendous growth in the last decade. However, over 90 percent of all fruit and vegetables produced during this period were consumed domestically. While most smallholders across Kenya (with the exception of arid regions) produce horticultural products, fewer than 2 percent of them produce for the export market, and Kenya exports little produce to regional markets (Muendo et al. 2004).

Figure 4.2 National Production of Maize, Wheat, and Rice, 1985 – 2004

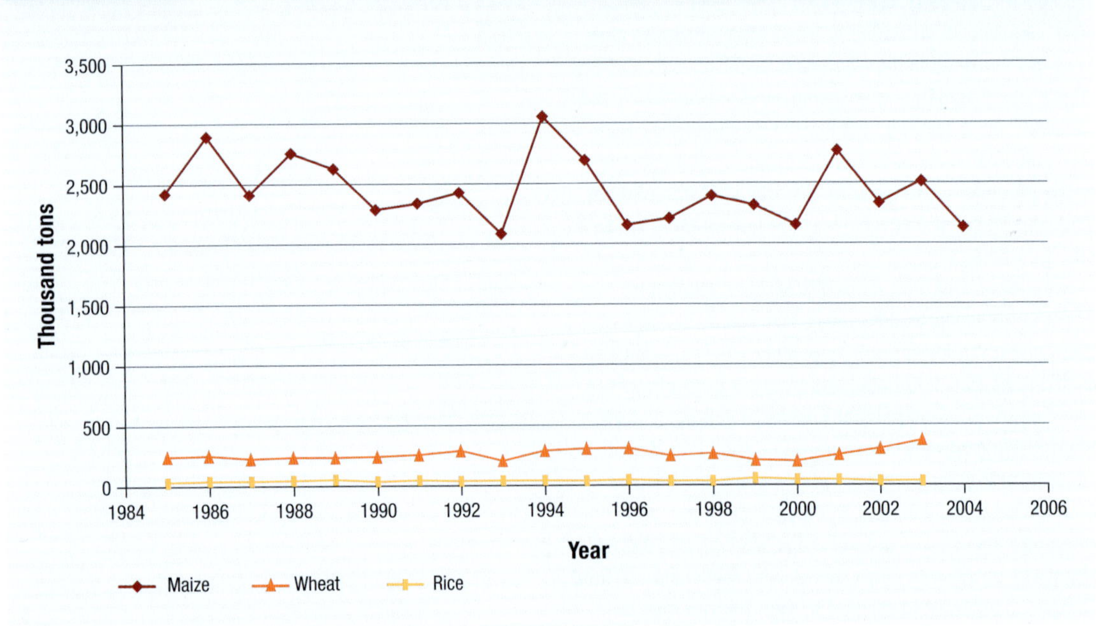

Sources: CBS (1986, 1995, 2001, and 2005) and FAO (2006).

While the total area under maize cultivation in Kenya has slowly but steadily increased, total maize production has stagnated. Yields have declined from 1.84 tons per hectare in the 1985-1990 period, to 1.71 tons in the 1990-1995 period, to 1.58 tons per hectare in the 1996-2004 period (Nyoro et al. 2004). Wheat and rice production data show no major changes in yields over the past decade. The trend line of total output parallels that of the area estimates.

Figure 4.3 National Production of and Area under Fruit and Vegetables, 1989 – 2003

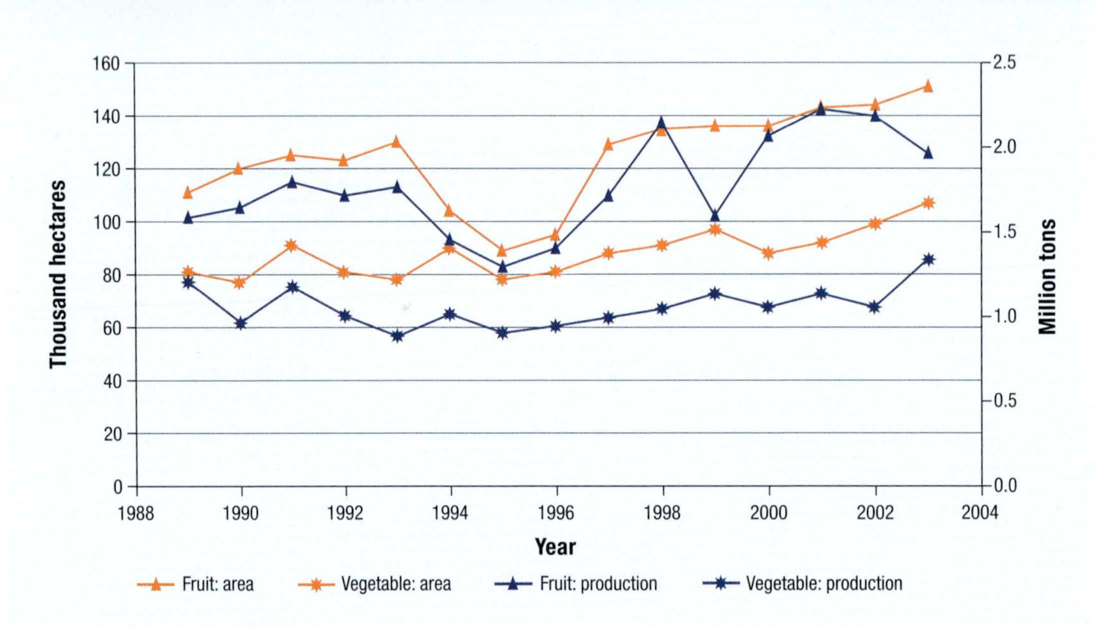

Sources: CBS (1986, 1995, 2001, and 2005)

Between 1989 and 2003, the area of cropland under fruit production increased from 110,000 hectares to around 150,000 hectares, with a slight dip between 1993 and 1997. Vegetable production followed a similar trend, increasing from around 80,000 hectares in 1989 to over 100,000 hectares in 2003. Fruit and vegetable production data in million tons generally echo these area trends.

LIVESTOCK PRODUCTS

Livestock play a part in the livelihoods of Kenyans in almost every corner of the country. In the large sections of the country too dry to support much agriculture, pastoral households rely extensively on livestock for their living. In the more central areas, where dairy cattle can be kept alongside more intensive cropping, milk production is one of the most important livestock-related activities. Milk is also a critically important commodity from a health and nutrition standpoint. Maps 4.5 and 4.6 examine the levels of milk production in areas of central and western Kenya, showing where those levels exceed or are insufficient to meet the needs of current population levels.

In the rangelands, livestock products contribute to most areas of household life. They provide multiple sources of food, are the major source of cash income in many areas, and serve as the primary source of savings for most pastoral households. Maps can be used to identify places where livestock production plays an especially important role. Maps 4.7 and 4.8 portray livestock densities and the share of cash income that livestock contributes.

Dairy in Central and Western Kenya

In a country where starches form the bulk of people's diets, milk is an especially important food. It is a source of high quality protein and micronutrients generally lacking in cereal-based diets, and is particularly important for children and child-bearing women. Kenyans love milk; they consume more of it than almost anyone else in the developing world. On average, each Kenyan drinks about 100 liters of milk a year, four times the average for sub-Saharan Africa (Staal 2004a).

Map 4.5 **Milk Production per Square Kilometer in Central and Western Kenya, 2005**

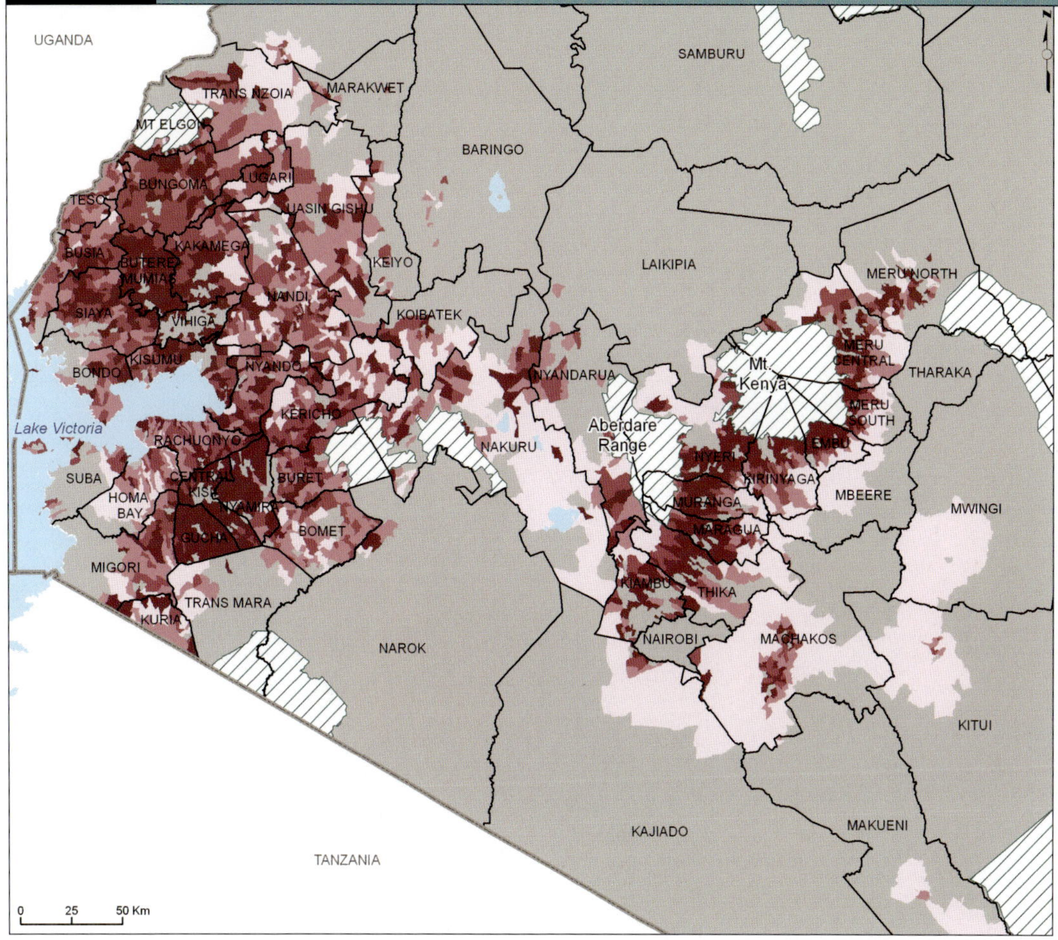

Sources: Administrative boundaries (CBS 2003), water bodies (FAO 2000), parks and reserves (IUCN and UNEP/WCMC 2006), and milk production per square kilometer (Baltenweck et al. 2005).

The output of milk, measured by the density of milk production (liters per square kilometer), varies across the highland and western regions. The highest densities (more than 100,000 liters per square kilometer per year) can be seen in the densely settled and farmed foothills east of the Aberdare Range and south and southeast of Mount Kenya. Similar high production densities are found in Gucha, Central Kisii, and Nyamira Districts, as well as in Butere-Mumias District. The drier lowland areas of Mbeere, Mwingi, Machakos, and Makueni Districts have lower outputs per square kilometer.

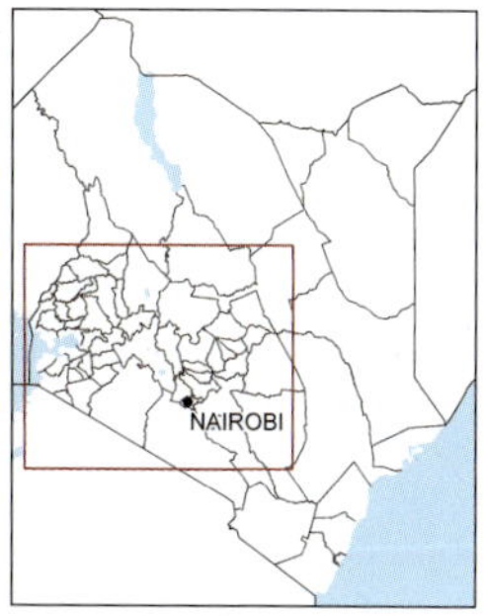

MILK PRODUCTION
(liters per sq. km per year)
- > 100,000
- 75,000 - 100,000
- 50,000 - 75,000
- <= 50,000
- No data

OTHER FEATURES
- District boundaries
- Major national parks and reserves (over 5,000 ha)
- Water bodies

One million people are supported by the dairy sector in Kenya. There are over 600,000 dairy-producing households, the majority of which are smallholder farmers. These households generate an estimated 365,000 wage-paying jobs in addition to the family labor involved (Staal 2004b). An additional 40,000 people make a living transporting, selling, and processing milk, and providing farmers with fodder and other inputs.

Small-scale milk production has been found to be highly efficient: dairy smallholdings in Brazil, Kenya, and India sometimes earn higher profits per liter than larger farms (Delgado et al. 2003; Steve et al. 2006). In addition, dairy cattle enhance smallholder crop farming systems throughout Kenya: their manure adds nutrients to the soil, maintaining fertility and boosting crop yields.

This section looks at milk production in the central and western regions of Kenya, and at the existence of milk "surpluses" and "deficits" in these regions. Calculations of milk production (Map 4.5) were done by assessing the number of dairy cattle in an administrative area, and extrapolating out liters of milk per area. Demand for milk was calculated simply by estimating the milk needs per person, and applying that number to the population density of each area. Areas with more milk produced than needed by the population are considered "surplus" areas, while those with more demand than can be met by current production are considered to be in "deficit" (Map 4.6). Many areas of the country produce more milk than they need locally. Hence, establishing good transportation and marketing systems for dairy could go a long way toward increasing the availability of milk in deficit areas.

Map 4.6 Milk Surplus and Deficit Areas in Central and Western Kenya, 1997

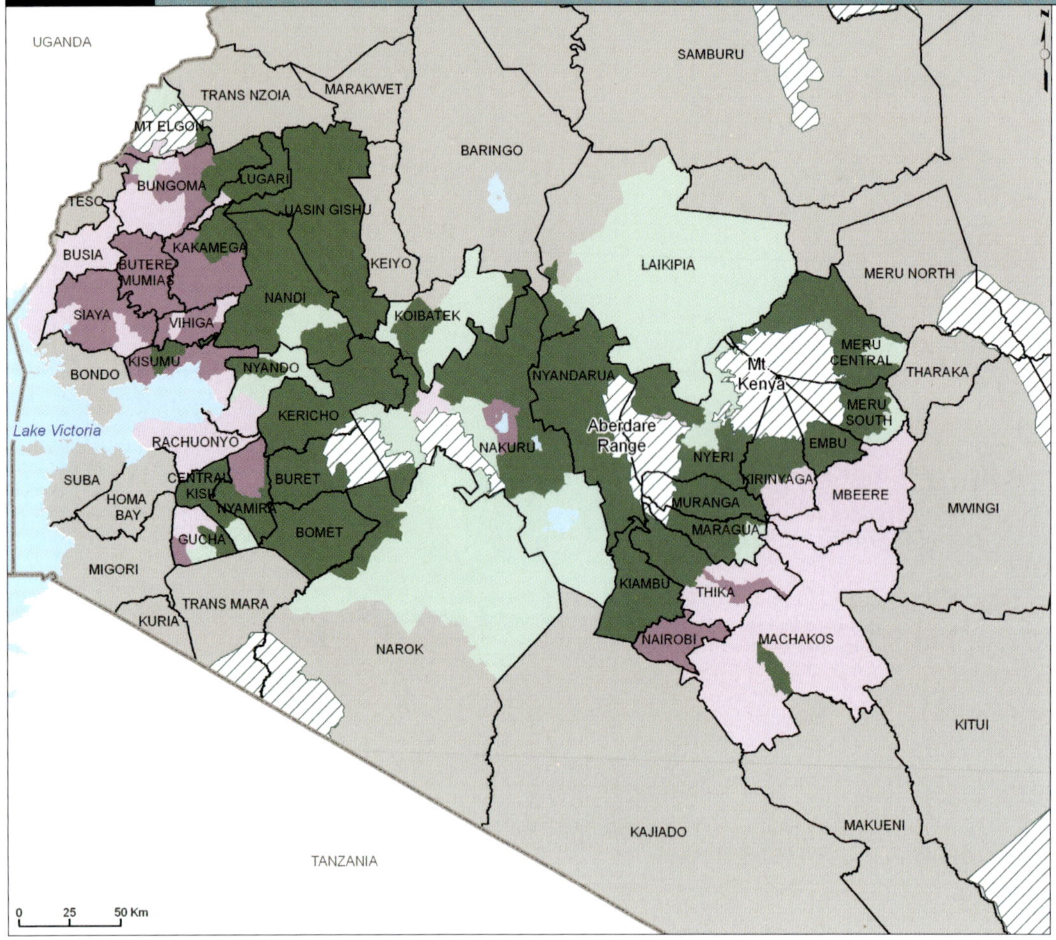

Sources: Administrative boundaries (CBS 2003), water bodies (FAO 2000), parks and reserves (IUCN and UNEP/WCMC 2006), and milk surplus and deficit areas (Baltenweck et al. 2005).

There are many areas of Kenya—particularly in the central highlands and Rift Valley—that produce more milk than they need locally. Milk surpluses are more closely related to population than production levels. Nakuru District has large sections with milk surpluses (shown in light and darker green), despite having generally lower levels of production per unit area. Areas north of Lake Victoria, in Nyanza and Western Provinces, which have high levels of milk output per square kilometer (Map 4.5), do not produce enough milk to meet the needs of the local population. Note that this pattern does not necessarily reflect a high milk output per cow. In many of these areas, milk-producing households are settled densely in a small area, resulting in a high aggregated milk output per square kilometer. For example, many farmers in Siaya and Kisumu Districts rely on low-yielding indigenous breeds. Milk deficits are also found in drier areas such as Machakos and Mbeere Districts, where milk production per unit area is low.

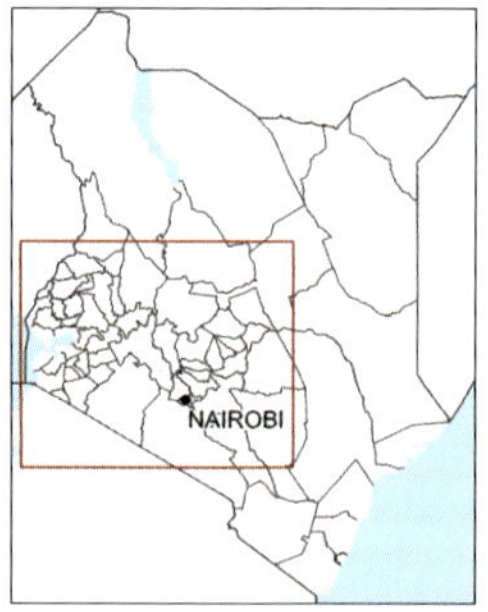

MILK PRODUCTION SURPLUS AND DEFICIT

- High surplus (> 10,000 liters per sq. km per year surplus)
- Slight surplus (<= 10,000 liters per sq. km per year surplus)
- Slight deficit (<= 10,000 liters per sq. km per year deficit)
- High deficit (> 10,000 liters per sq. km per year deficit)
- No data

OTHER FEATURES

- District boundaries
- Major national parks and reserves (over 5,000 ha)
- Water bodies

Livestock in the Rangelands

Kenya's semi-arid and arid rangelands cover more than 80 percent of its land area (SoK 2003), corresponding closely to Kenya's lowest human population densities and higher livestock densities. Whereas dairy cattle predominate in the more central areas, the rangelands are primarily pastoral. Camels are an important livestock species in the northern areas, while cattle, sheep, and goats are found throughout the rangelands. In large parts of the drier rangelands, livestock are shifted to follow the availability of fodder and rain. In some parts of the country, these patterns have begun to change with the introduction of fixed water points (see Map 5.8 in Chapter 5 which shows water sources and livestock densities for the northern rangelands).

The maps in this section use two indicators to tell the story of livestock management in Kenya. This section looks first at where in the rangelands livestock are being raised (Map 4.7). Distinct patterns emerge: in parts of the Rift Valley and Districts such as Machakos, Mwingi, and Kitui, low-density livestock rearing occurs, this is spread consistently across the area, overlapping with marginal cropping activity, or—in Turkana, for example—reflecting the presence of fixed water points. In the more arid areas, livestock distribution becomes patchier as pastoral systems take over; people and animals move around more and there is little or no cropping. The other indicator examined is the contribution of livestock to household incomes (Map 4.8). The data on cash income come from "expert opinion" for small administrative areas within each District, giving a sense of the relative importance of different activities and products for livelihoods and subsistence.

Map 4.7 **Livestock Density in the Rangelands, 1994-96**

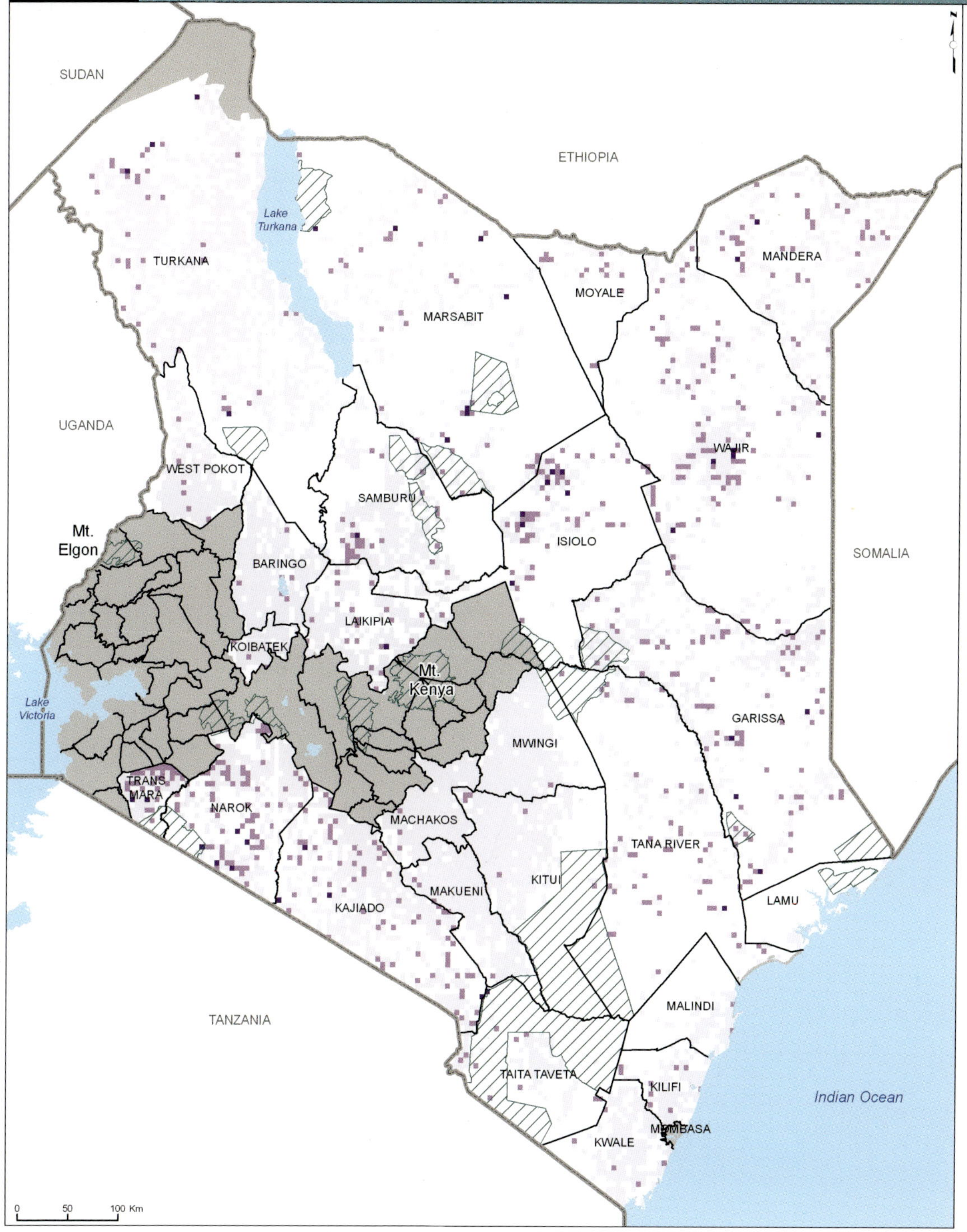

Sources: Administrative boundaries (CBS 2003), water bodies (FAO 2000), parks and reserves (IUCN and UNEP/WCMC 2006), and livestock density (ILRI calculation based on DRSRS 2003, Grunblatt et al. 1995, Grunblatt et al. 1996).

The map shows some of the highest livestock densities in Trans Mara and parts of Narok Districts—in rangelands that receive more rainfall and are close to the cropping boundaries for most crops. Livestock can be found throughout most rangelands of West Pokot, Baringo, Machakos, Makueni, Kitui, and Mwingi Districts, and the coastal area in Kwale and Kilifi Districts. Densities are much lower here and represent either marginal croplands or agropastoral areas (Map 4.1).

Note: Species numbers are aggregated (using a *tropical livestock unit* which is equivalent to an animal weight of 250 kilograms) to squares of 5 kilometers by 5 kilometers and then averaged by square kilometer. Livestock includes cattle, sheep, goats, camels, and donkeys that were observed during low-altitude flights.

LIVESTOCK DENSITY
(tropical livestock unit per sq. km)
- > 175
- 50 - 175
- 0 - 50
- No observation of animals
- Non-rangeland areas or no data

OTHER FEATURES
- District boundaries
- Major national parks and reserves (over 5,000 ha)
- Water bodies

Livestock provide a range of important products to rural households: meat, milk, and blood are consumed or sold, as are hides and skins. Manure is applied to crops in areas of the rangelands where sufficient water is available for limited cropping; closer to cities, it is also sold. Finally, livestock serve both as a bank account and an insurance strategy in the pastoral rangelands. Herd loss is a major risk factor in these areas. With limited alternative insurance or investment options available, herd accumulation is an important means of managing risk in pastoral households (Gebru and McPeak, 2004).

Map 4.8 Share of Cash Income from Livestock, 2003-05

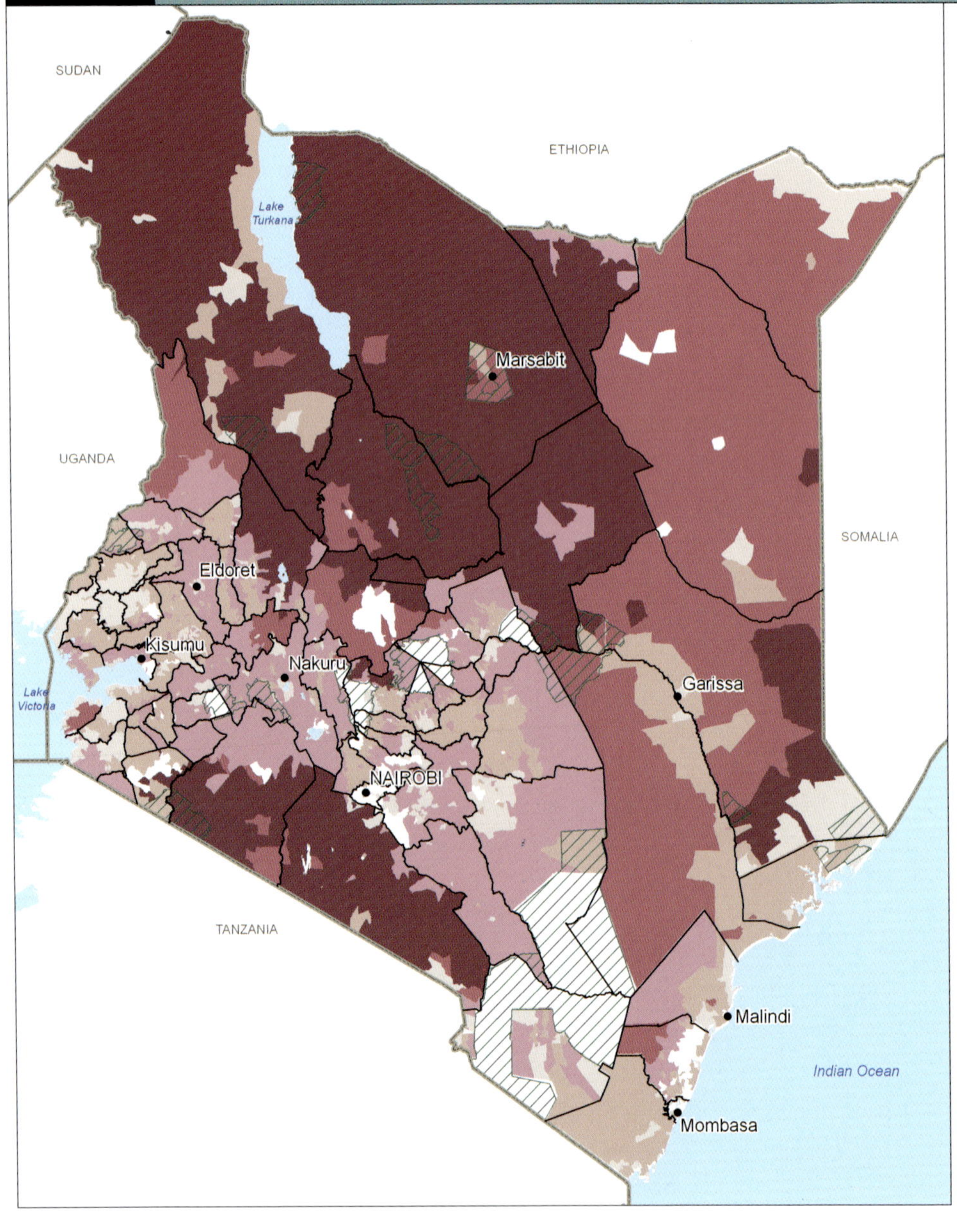

Sources: Administrative boundaries (CBS 2003), cities (SoK and ILRI 2000), water bodies (FAO 2000), and share of cash income from livestock (ALRMP et al. 2006).

The map indicates that livestock activities play a role almost everywhere in Kenya, but they are a particularly important source of cash income in the drier parts of the country, where there are few other sources of income. Areas where more than 50 percent of cash is from livestock include the semi-arid and arid lands of southern, eastern, northern, and northeastern Kenya. It is also notable that the percentage of total cash income coming from livestock is substantial throughout much of central Kenya, ranging from 25 to 50 percent. Along the coast and around Lake Victoria (areas with very high poverty incidence), livestock generally contribute less than 25 percent of total cash income.

Note: Data on livelihood zones are based on questionnaires sent to key food security experts (generally about 6-10 persons) in each District. In some cases where further clarification was necessary, questionnaires were sent to experts below District level (Division). This group of experts classified each of Kenya's 6,632 Sublocations by their predominant livelihood strategy and other livelihood characteristics including different sources of cash income.

CASH INCOME FROM LIVESTOCK
(percent of total income)

- > 75
- 50 - 75
- 25 - 50
- 10 - 25
- 0 - 10
- No cash income from livestock

OTHER FEATURES

- District boundaries
- Major national parks and reserves (over 5,000 ha)
- Water bodies

FOOD FROM THE WILD

Wild animals and plants are an important source of food and livelihoods in most parts of Kenya. Gathering nuts, fruits, and tubers; collecting honey; and hunting wildlife—including rodents, guinea fowl, and other birds, as well as larger animals such as antelope—remain important to many. These wild resources become particularly critical in times of drought, hunger, or whenever other resources become unreliable.

Case studies and general observations suggest that the use of wild animals and plants is common in the daily lives of many Kenyans, but the magnitude of the harvesting efforts, their importance during different seasons, and their significance to particular groups (very poor households, women, etc.) are less well understood at the national level. Kenya's official statistical system collects information on the quantity and value of its fisheries. Hunting and gathering activities, however, are greatly underreported because most products are either directly consumed in households or sold through local informal markets. The fact that hunting of most wildlife is illegal accentuates this information deficit.

Understanding the relationship between harvesting rates for fish, wildlife, or plants and the rate at which these ecosystem products are replenished is essential to sustaining their use over the long term. When harvest rates exceed replacement rates, it undermines the capacity of ecosystems to continue to provide these products, thus jeopardizing livelihoods and food security.

The following sections present maps that show where fishing (both from freshwater bodies and the Indian Ocean), and hunting and gathering activities are important. The gathering activities presented in the maps do not solely include food, but also other products such as building materials and traditional medicines, which either provide direct benefits to families or boost cash income.

The maps rely on recent surveys in which experts characterized small administrative areas by the dominant activities contributing to livelihoods and food security of the majority of families in that area (ALRMP et al. 2006). It is a first approximation of where fishing, and hunting and gathering are important for livelihoods. However, reliance on wild animals and plants may be even more significant for certain groups or at certain times, a fact easily masked by the administrative averages shown in the maps.

In almost all areas throughout Kenya, hunting and gathering plays some role in people's livelihoods (see Map 4.11). While other activities such as cropping, raising livestock, fishing, or wage labor may account for a greater share of the day's activities, hunting and gathering can still contribute significantly to a family's cash income, especially in communities located in the arid and semi-arid regions of the country. Even in more crop-dominated Districts, hunting and gathering can contribute as much as 20 to 40 percent of a family's cash income in selected communities.

Fishing is concentrated in communities close to Kenya's major lakes, permanent rivers, and along the Indian Ocean and plays no role in the rest of the country. In some of these communities, it can contribute more than 50 percent of families' cash incomes.

Fishing and Fish Farming

Fish provide an important source of food for Kenyans, particularly along lakes, rivers, and the coast. Map 4.9 highlights areas where fishing makes an important contribution to livelihoods. As seen in Table 4.1, the vast majority (96 percent) of fishing activity in Kenya is freshwater fishing, and most of that occurs on Lake Victoria.

Declining fish stocks are a serious problem for the nearly 40,000 people who fish for a living. Total fish production in the country decreased from 214,712 metric tons in 1999 to 164,261 metric tons in 2001 (MoLFD 2001).

The development of aquaculture, or fish farming, has been fairly stagnant in recent years, but has started to attract renewed interest recently. The Department of Fisheries is actively promoting the transfer of pond management technology from its research farm to interested fish farmers through participatory on-farm trials (MoLFD 2001).

The Kenyan coastal region covers seven Districts, and its territorial waters cover 12 nautical miles. In this area more than 5,400 artisanal fishers operate, largely during the September to March period when the waters are generally calm (MoLFD 2001). The majority of the fishing vessels used are dugout canoes powered by wind, sails, and paddles.

Table 4.1 Fish Landings, Fisherfolk, and Fishing Vessels By Major Supply Area, 2001

	FISH LANDINGS					
	QUANTITY		VALUE			
SOURCE	(TONS)	(PERCENT)	(KSH MILLION)	(PERCENT)	FISHERFOLK (NUMBER)	FISHING VESSELS (NUMBER)
Lake Victoria	151,804	92	7,253.1	92	33,037	10,014
Lake Turkana	3,787	2	49.6	1	234	78
Lake Baringo	117	–	2.9	–	75	25
Lake Jipe	65	–	2.5	–	65	35
Lake Naivasha	5	–	0.3	–	57	19
Tana River Dams	232	–	8.1	–	372	124
Fish Farming	998	1	98.8	1		
Other Areas	802	–	36.6	–		
Total Freshwater	**157,810**	**96**	**7,452.1**	**94**	**33,840**	**10,295**
Marine Fishing	6,451	4	461.9	6	5,463	1,881
TOTAL	**164,261**	**100**	**7,913.8**	**100**	**39,303**	**12,176**

Source: MoLFD (2001).

Note: Total value of all fish landings (Ksh 7.9 billion) equals US$ 113.1 million (at US$ 1 = Ksh 70).

Map 4.9 Share of Cash Income from Fishing, 2003-05

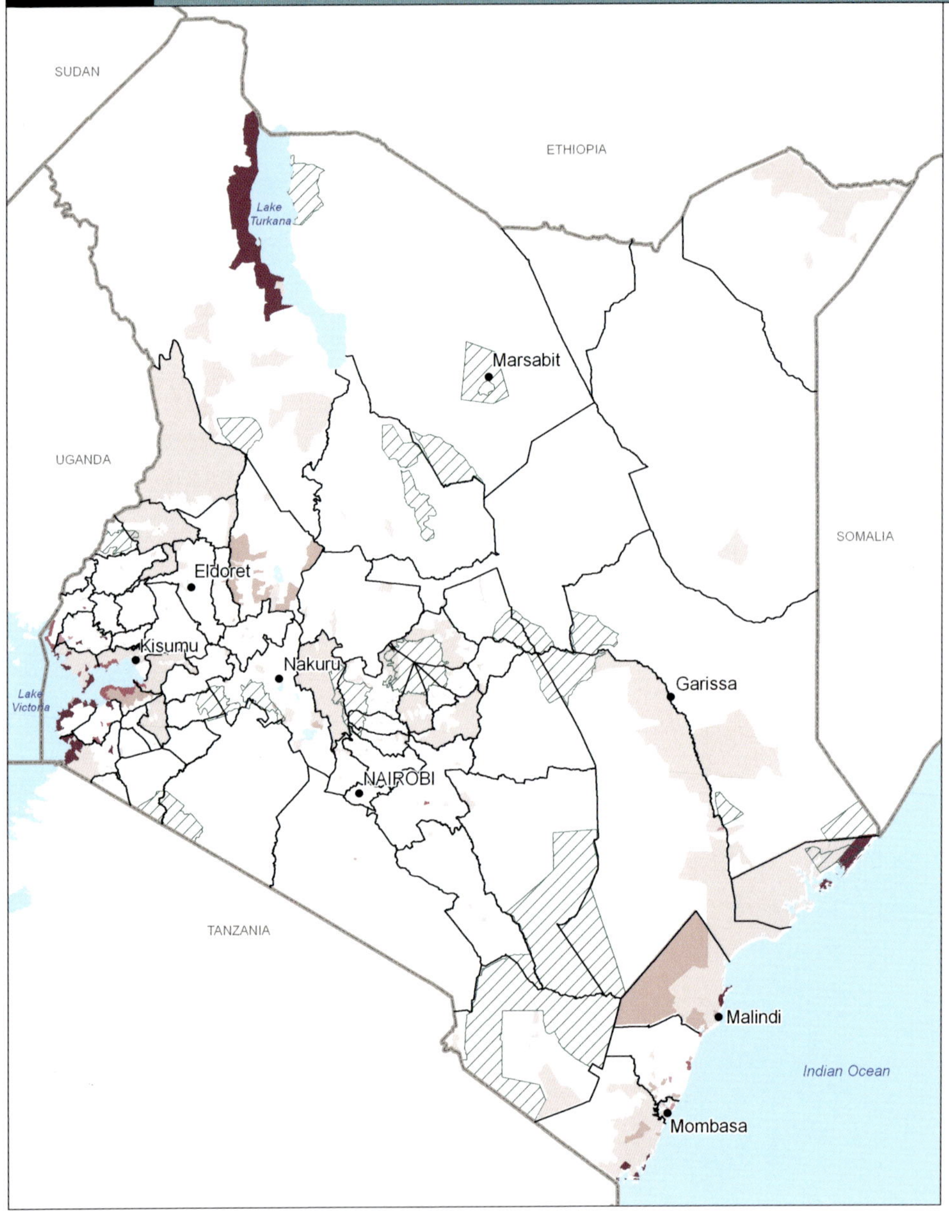

Sources: Administrative boundaries (CBS 2003), cities (SoK and ILRI 2000), water bodies (FAO 2000), and share of cash income from livestock (ALRMP et al. 2006).

There are very few areas in Kenya where a substantial amount of income comes from fishing. The areas where it predominates (shown in dark purple) are found along the shores of Lake Victoria south of Kisumu, along the western shore of Lake Turkana, and at marine fishing sites in Malindi and towns further north on the coast. Elsewhere, fishing typically provides less than 10 percent of total cash income. It contributes less than 5 percent of cash income along most of the southeastern coast, with slightly higher levels further inland of Malindi and south of Mombasa.

Note: Data on livelihood zones are based on questionnaires sent to key food security experts (generally about 6-10 persons) in each District. In some cases where further clarification was necessary, questionnaires were sent to experts below District level (Division). This group of experts classified each of Kenya's 6,632 Sublocations by their predominant livelihood strategy and other livelihood characteristics including different sources of cash income.

CASH INCOME FROM FISHING
(percent of total income)
- > 50
- 25 - 50
- 10 - 25
- 5 - 10
- 0 - 5
- No cash income from fishing

OTHER FEATURES
- District boundaries
- Major national parks and reserves (over 5,000 ha)
- Water bodies

Hunting and Gathering

While a minority of Kenya's 72 tribes have always hunted wild animals for food (mostly buffalo, impala, gazelle, giraffe, and monkeys), it appears that illegal hunting of wild animals may be on the increase, and that relatively inexpensive "bushmeat" (selling for around Ksh 80 ($US 1.14) per kilogram) is now widely available to poor consumers for purchase or barter. The antipoaching staff at the Kenya Wildlife Service reports that as many as 1 million animals are now dying in illegal snares each year, and in the past five years, 48,900 snares were recovered throughout Kenya's protected areas (Pflanz 2005).

A recent survey compared the use and trade of illegal bushmeat in Kitui and Samburu Districts. The survey found that 80 percent of Kitui households consume an average of 14 kg of bushmeat each month, representing the bulk of all meat consumed (TRAFFIC 2000). The value of this meat is equivalent to about one third of a typical household's monthly income. Bushmeat is less than half as expensive as domestically raised meat in Kenya. Affordability was the main reason rural Kenyan households indicated bushmeat as their most important protein source. The study also found that the poorer the household, the greater its reliance on bushmeat (TRAFFIC 2000).

In Samburu District, monthly household consumption was significantly lower (1.1 to 1.4 kg) than in Kitui District. Hunting was exclusively for home consumption (unlike in Kitui, where 25 percent was traded) and families ate larger antelope species which are relatively abundant in local hunting areas. Kitui households had to rely more on a local supply of small mammals and birds, supplemented by traded bushmeat for larger, higher-priced species.

One reason for these differences in bushmeat consumption patterns is the abundance of particular animal species. Another reason stems from differences of culture and history. For example, Samburu and Maasai pastoralists, who in the past relied only to a limited extent on bushmeat, have begun to utilize the resource more in recent years as human population pressure has increased and the standards of living based on livestock production has declined (Nkedianye 2003; TRAFFIC 2000).

Map 4.10 shows the density of wildlife (specifically large animals that graze in the open) in the rangeland areas of Kenya. It is difficult to tell precisely where bushmeat is being taken, but by using this map of large mammals as a proxy for bushmeat hunting, it is possible to tell where hunting is likely to play a large role in livelihoods. Map 4.11 shows the percentage of total household income that comes from hunting and gathering activities. Because it includes gathering activities in addition to hunting, its spatial patterns differ somewhat from the map of wildlife density.

Map 4.10 **Wildlife Density in the Rangelands, 1994-96**

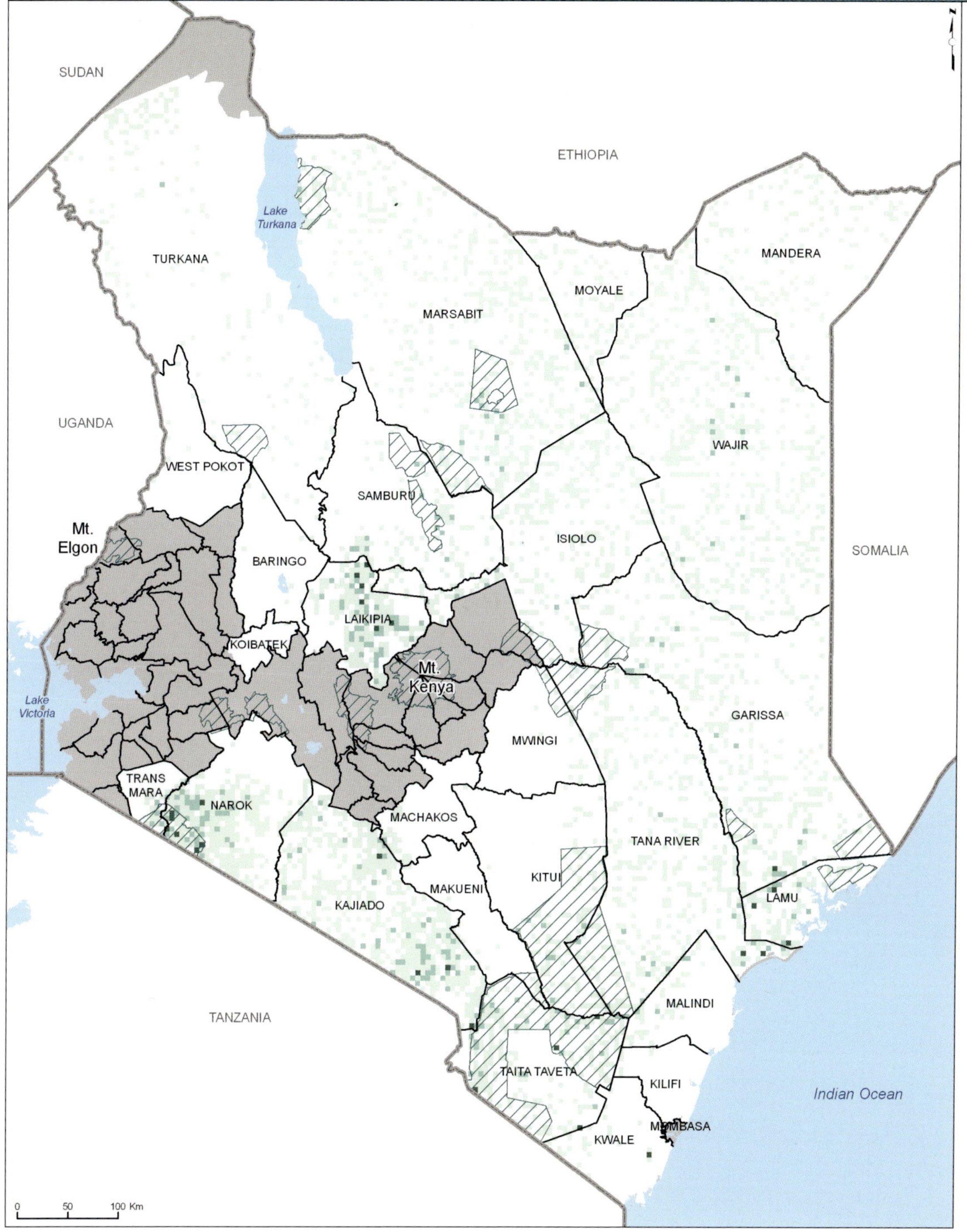

Sources: Administrative boundaries (CBS 2003), water bodies (FAO 2000), parks and reserves (IUCN and UNEP/WCMC 2006), and 1994-96 wildlife density (ILRI calculation based on DRSRS 2003, Grunblatt et al. 1995, Grunblatt et al. 1996).

Areas of high wildlife density are particularly concentrated in Narok, Kajiado, and Laikipia Districts, close to some of the best-known national parks and reserves or large privately held ranches (see Chapter 5 for a full discussion). As mentioned in the text, there are also higher concentrations of large mammals in Samburu than in Kitui District.

Note: Species numbers are aggregated (using a *tropical livestock unit* which is equivalent to an animal weight of 250 kilograms) to squares of 5 kilometers by 5 kilometers and then averaged by square kilometer. The wildlife counts include 21 different large grazing animals that can be observed during low-altitude flights.

WILDLIFE DENSITY
(tropical livestock unit per sq. km)
> 100
20 - 100
0 - 20
No observation of animals
Non-rangeland areas or no data
OTHER FEATURES
District boundaries
Major national parks and reserves (over 5,000 ha)
Water bodies

Map 4.11 **Share of Cash Income From Hunting and Gathering, 2003-05**

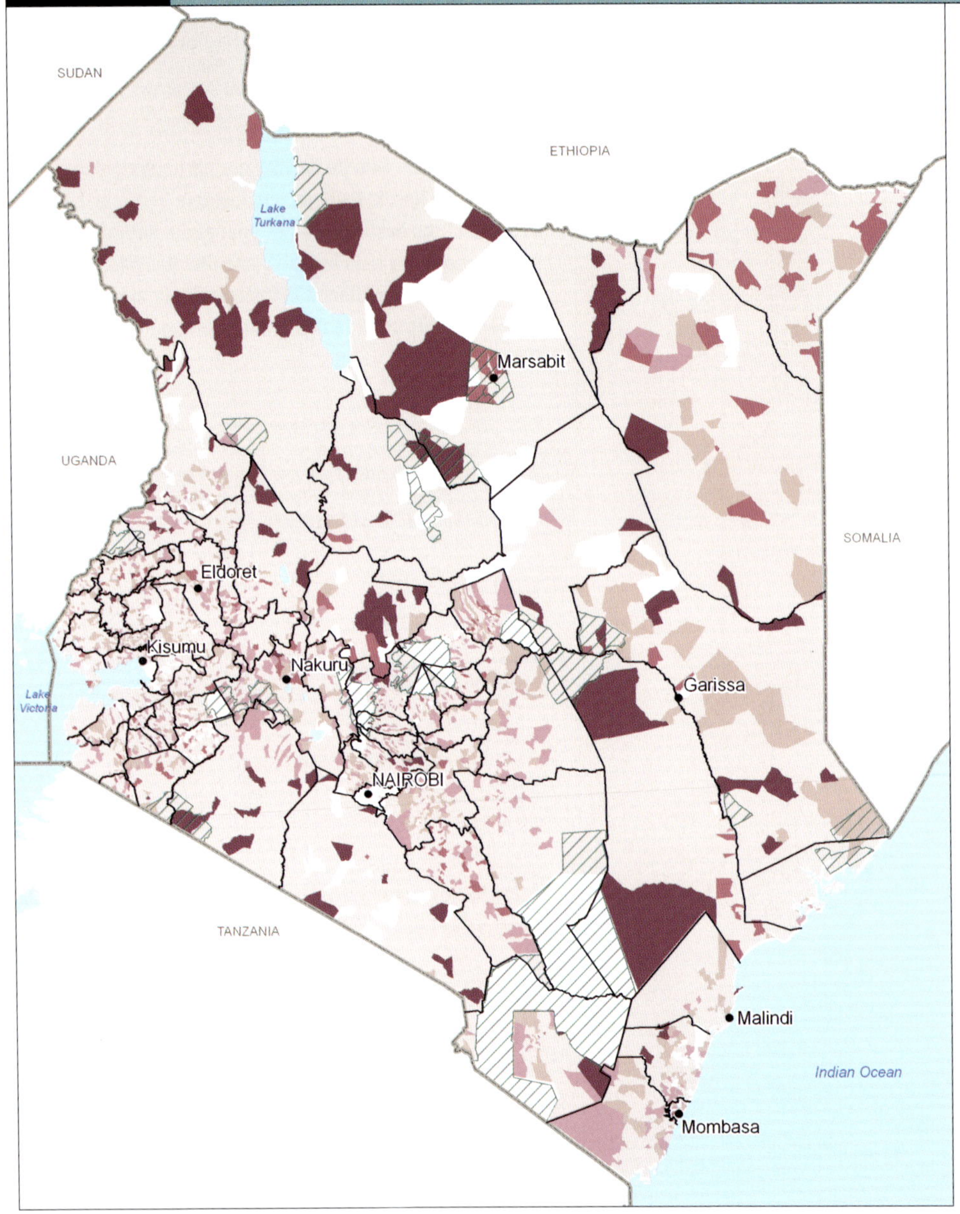

Sources: Administrative boundaries (CBS 2003), cities (SoK and ILRI 2000), water bodies (FAO 2000), and share of cash income from livestock (ALRMP et al. 2006).

The map demonstrates two important things: first, income from hunting and gathering plays a role almost everywhere in Kenya (there are very few white areas, where such income contributes nothing to family income). Second, the prevalence of these activities is mixed; areas where hunting and gathering provide a very large percentage of cash income (>60 percent) are scattered across different regions of the country. Several of these areas border the edges of major national parks, but others, such as sections in the northern rangelands close to Lake Turkana, are further afield. Throughout the highlands, and in Nyanza and Western Provinces, there is a diverse mix of reliance on hunting and gathering activities. (Note that the map does not include income from collection and sale of woodfuel.)

Note: Data on livelihood zones are based on questionnaires sent to key food security experts (generally about 6-10 persons) in each District. In some cases where further clarification was necessary, questionnaires were sent to experts below District level (Division). This group of experts classified each of Kenya's 6,632 Sublocations by their predominant livelihood strategy and other livelihood characteristics including different sources of cash income.

CASH INCOME FROM HUNTING AND GATHERING
(percent of total income)

- > 60
- 40 - 60
- 20 - 40
- 10 - 20
- 0 - 10
- No cash income from hunting and gathering

OTHER FEATURES

- District boundaries
- Major national parks and reserves (over 5,000 ha)
- Water bodies

Box 4.2 Linking Food and Livelihood Maps to National Decision-Making

Kenya's *Economic Recovery Strategy for Wealth and Employment Creation, 2003–2007* states that economic recovery must build on investments and improvements in the agriculture sector (GoK 2003). With this in mind, many of the nation's agricultural policies are currently under revision. Spatial indicators of food production patterns such as those presented in this chapter, when combined with other maps and indicators on credit, road infrastructure, and poverty, can help to inform this policy revision and contribute to implementing the *Economic Recovery Strategy*. Below are some suggestions for how the information in this atlas can contribute to a few of the specific interventions (highlighted in italics) proposed in the *Economic Recovery Strategy*:

- ***Diversify enterprises and crop uses:*** Maps can show where production and use of nontraditional crops coincide with high poverty levels and good road access. This can be compared to the crops and food sources farmers in these areas currently rely on to help devise diversification strategies.
- ***Promote dairy, goats, and other small stock:*** Areas with high poverty densities, high poverty rates, and a production shortfall of milk for local markets (as shown in Maps 4.5 and 4.6) could be the most promising areas to boost dairy outputs and at the same time improve nutritional and income levels of households in poorer communities.
- ***Support development of facilities for milk processing:*** By combining maps of milk production, milk demand, road infrastructure, and poverty levels, planners could locate milk-processing facilities to boost market integration in areas with high poverty levels.
- ***Establish new irrigation infrastructure:*** Information on existing irrigation efforts (as in Map 3.12, Chapter 3, showing large and small-scale irrigation points), combined with maps of irrigation potential and other water uses, can identify areas that have fewer trade-offs (and potential conflicts) with other water users. In combination with information on levels of food security and poverty, new irrigation infrastructure could target less food-secure and poorer communities.

In addition to helping to implement the *Economic Recovery Strategy*, geospatial information on food production can contribute to achieving the country's multiple targets under the Millennium Development Goals (MDGs), such as reducing poverty and hunger. The recent assessment on the needs and costs to achieve the MDGs in Kenya (MoPND et al. 2005) established that, in order to meet the hunger eradication goal as set in the MDGs, Kenya must invest some Ksh 154 billion (US$ 2.2 billion at US$ 1 = Ksh 70) in areas such as improving soil fertility, water harvesting and utilisation, extension services, rural roads and energy, schools and preschool feeding programs, agricultural research, and capacity building over the next ten years. Making such investments will benefit from a solid information base to ensure that resources target the right households and areas. For example, decision-makers could use food mapping to shed light on the following questions:

- Which areas are degraded and could most benefit from increased soil fertility? How do these areas coincide with different livelihood strategies?
- Where could better water harvesting techniques reduce vulnerability to crop failures?
- On what sources of food do people currently rely? What kind of crops are farmers growing now and what could they grow in the future?
- Where would rural access roads help poor communities to become more competitive with locally produced goods?

Box 4.3 Creating a Poverty and Demographic Profile for Croplands With Different Magnitudes of Food Cropping

Map 4.12 Food Cropping in Five Provinces, 1997

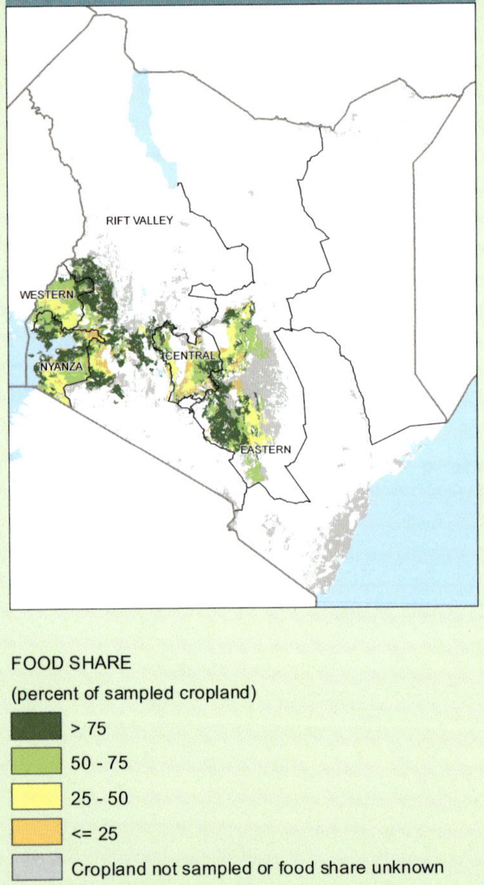

The decision to grow cash crops or food crops greatly affects food security, income levels, and the well-being of farmers. Understanding these crop choices at an aggregated country level could improve policies to revitalize Kenya's agriculture. Or it might better target agricultural programs, thereby increasing access to agricultural inputs for communities with high poverty rates or those at greater risk of food insecurity.

Household surveys show that most Kenyan farming families choose a highly diversified mixture of crops. Almost all families grow maize, but maize does not contribute more than one quarter of total income for most households. It is generally combined with other food crops and cash crops that provide higher returns to land and labor, such as horticultural crops, sugar, tea, and coffee. The decision to keep a high share of cropland in food crops—especially when it includes the staple crop maize and very few other crops—could indicate subsistence farming, which is generally associated with higher poverty rates.

Combining maps that show the degree to which farmers have dedicated their cropland to food versus cash crops (Map 4.4) with poverty maps could provide insights into possible relationships between the intensity of food cropping and poverty rates. Since the underlying data for these maps are in GIS format, they can also be used to create a poverty profile for different food cropping zones. Map 4.12 shows the level of food cropping for five Provinces. Table 4.2 classifies the land area of each Province into six classes that show whether areas are cropped and to what degree croplands are covered with food crops. For each of the classes, the table provides estimates of total population and population density, and estimates of the number of poor and the average poverty rate.

What Do the Map and Poverty Profile Show?

- At this aggregation, there does not appear to be a simple, straightforward association of high poverty with the choice of farmers to maintain a high share of food crops versus cash crops. The differences in the poverty rates between the five Provinces—ranging from 32 percent in Central Province to 64 percent in Nyanza Province—are much greater than the differences in poverty rates between the different food cropping zones within each Province.
- Nonetheless, in all but one Province, poverty rates tend to be slightly higher for the areas where farmers grow more than 75 percent food crops. The exception is the Rift Valley. Rift Valley Province includes Kenya's more productive cereal growing areas and cereals grown for cash income.
- This suggests that additional information on the number and types of specific crops grown (for example whether food crops are high-value vegetables or dryland cereal crops) is required to illuminate the spatial patterns of food cropping and poverty.

Similar profiles can be constructed overlaying other food-related or livelihood maps from this chapter with indicators of human well-being presented in Chapter 2. For example, comparing poverty maps with maps showing selected livelihood strategies, such as hunting and gathering, or fishing, could help to identify areas where poor communities are particularly vulnerable to ecosystem degradation and loss of environmental income.

Table 4.3 People, Poverty, and Food Cropping

PROVINCE	AREAS WITHOUT CROPLAND AND FOOD SHARE IN SAMPLED CROPLAND AREAS	AREA (SQ. KM)	NUMBER OF PEOPLE (000)	AVERAGE POPULATION DENSITY (NUMBER OF PEOPLE PER SQ. KM)	NUMBER OF POOR (000)	AVERAGE POVERTY RATE (PERCENT)	KSH NEEDED PER MONTH TO REACH POVERTY LINE[1] (MILLIONS)
EASTERN							
	No Cropland	118,134	670	6	359	54	93.2
	Cropland not sampled	15,141	716	47	425	59	120.8
	Food Crops 0 - 25%	2,411	152	63	89	58	24.9
	Food Crops 25 - 50%	5,485	852	155	483	57	133.2
	Food Crops 50 -75%	7,596	699	92	424	61	122.0
	Food Crops > 75%	8,729	1,077	123	667	62	199.0
	TOTAL 9 Districts	**157,495**	**4,166**	**26**	**2,445**	**59**	**693.1**
CENTRAL							
	No Cropland	3,675	351	96	110	31	16.3
	Cropland not sampled	2,001	435	217	128	29	19.7
	Food Crops 0 - 25%	1,383	414	299	123	30	15.1
	Food Crops 25 - 50%	1,624	587	361	187	32	23.4
	Food Crops 50 -75%	2,745	1,062	387	338	32	44.0
	Food Crops > 75%	1,796	382	213	138	36	21.8
	TOTAL 6 Districts	**13,224**	**3,231**	**244**	**1,023**	**32**	**140.3**
RIFT VALLEY							
	No Cropland	145,696	1,969	14	968	49	245.6
	Cropland not sampled	16,961	1,122	66	505	45	111.3
	Food Crops 0 - 25%	3,156	242	77	123	51	29.8
	Food Crops 25 - 50%	3,320	438	132	221	50	52.4
	Food Crops 50 -75%	3,514	400	114	195	49	42.8
	Food Crops > 75%	11,978	1,852	155	865	47	218.5
	TOTAL 6 Districts	**184,625**	**6,022**	**33**	**2,877**	**48**	**700.3**

Continued

Table 4.3 People, Poverty, and Food Cropping — continued

PROVINCE	AREAS WITHOUT CROPLAND AND FOOD SHARE IN SAMPLED CROPLAND AREAS	AREA (SQ. KM)	NUMBER OF PEOPLE (000)	AVERAGE POPULATION DENSITY (NUMBER OF PEOPLE PER SQ. KM)	NUMBER OF POOR (000)	AVERAGE POVERTY RATE (PERCENT)	KSH NEEDED PER MONTH TO REACH POVERTY LINE[1] (MILLIONS)
NYANZA							
	No Cropland	806	208	258	134	65	65.9
	Cropland not sampled	804	202	252	132	65	47.6
	Food Crops 0 - 25%	682	125	183	73	58	20.5
	Food Crops 25 - 50%	2,519	662	263	411	62	129.6
	Food Crops 50 -75%	3,627	1,604	442	1,004	63	303.3
	Food Crops > 75%	4,107	1,064	259	712	67	251.4
	TOTAL 12 Districts	**12,544**	**3,866**	**308**	**2,466**	**64**	**818.3**
WESTERN							
	No Cropland	1,061	126	119	78	62	23.0
	Cropland not sampled	416	106	254	61	58	16.3
	Food Crops 0 - 25%	435	138	318	82	60	23.0
	Food Crops 25 - 50%	2,224	1,077	484	646	60	190.4
	Food Crops 50 -75%	3,079	1,148	373	668	58	181.9
	Food Crops > 75%	1,242	410	330	245	60	68.7
	TOTAL 6 Districts	**8,457**	**3,006**	**355**	**1,781**	**59**	**503.3**
	TOTAL 39 Districts	**376,346**	**20,290**	**54**	**10,593**	**52**	**2,855.4[2]**

Sources: Poverty and demographic estimates (1999) are WRI/ILRI calculation based on CBS 2002 and CBS 2003. Area without cropland, cropland not sampled, and food crop area percentages are WRI calculation based on data for Map 4.4 (ICRAF and DRSRS 2001; FAO 2000).

Note: [1] The poverty gap measures the average expenditure shortfall (gap) for the poor in a given administrative area relative to the poverty line. It is a crude estimate of the minimum amount of resources needed to eradicate poverty (see Chapter 2).
[2] The total amount needed to close the poverty gap for one month in the 39 Districts (Ksh 2.9 billion) equals about US$ 40.8 million (at US$ 1 = Ksh 70).

SUMMING UP

- Crop production, livestock, fishing, and hunting-gathering are important sources of food and livelihoods in rural Kenya.
- In terms of total area and numbers, smallholders dominate Kenya's rainfed agriculture. Most rural households grow maize to help feed their families and rely on the market for food security (between 25 and 70 percent of smallholder income is from non-farm sources). A significant proportion of Kenya's crops are planted in areas with a high likelihood of insufficient rains.
- A mix of dairy cattle, food, and cash crops dominates high-potential agricultural lands in central and western Kenya. Similar mixed farming along Lake Victoria and large parts of Laikipia, Machakos, Mwingi, Kitui, Makueni, Taita Taveta, Kwale, Kilifi, and Malindi Districts is more marginal. Here rainfall is more erratic or soils are less fertile, resulting in lower yields and incomes.
- Croplands with high shares of food crops (more than 75 percent) are concentrated in high-potential Districts such as Trans Nzoia, Uasin Gishu, Lugari, upper Nandi, and Nakuru (maize and other cereals); Narok (wheat); and lower Kirinyaga (rice). High shares of food crops (low-yielding maize) are also prevalent in the more marginal croplands mentioned above.
- In terms of cropped area, maize is Kenya's most important food crop. It is planted throughout the country, from high-yielding areas to riskier, semi-arid zones. Large-scale farms produce over 80 percent of the domestically marketed maize. National average maize yields have declined over the past two decades.
- Fruits and vegetables—high value crops—have greatly expanded in farm area over the past decade and reached 250,000 hectares in 2003.
- One third of urban dwellers in Kenya are growing subsistence crops and raising livestock in urban areas, and two thirds are farming in either urban or rural areas, or both.
- On average, each Kenyan drinks about 100 liters of milk a year, produced by 600,000 households, primarily from central and western Kenya. The central highlands and Rift Valley have a milk surplus, while large parts of Nyanza and Western Provinces do not produce enough milk to meet local demand.
- In the more arid rangelands, livestock are shifted to follow the availability of fodder and rain (pastoral livestock raising). Cropping combined with pastoral livestock raising (agropastoral) tends to occur around more permanent water sources or where intermittent rainfall is sufficient in a good rainfall year to grow some crops.
- Nearly 40,000 people fish for a living—sometimes combined with livestock raising or food cropping—in selected areas along Lake Victoria, Lake Turkana, and the Indian Ocean. About 92 percent of the fish landed in Kenya is from Lake Victoria.
- Gathering nuts, fruits, and tubers; collecting honey; and hunting wildlife—including rodents, guinea fowl, and other birds, as well as larger animals such as antelope—all are important sources of food. While data on hunting for wild animals are incomplete, household survey data for two Districts suggest that bushmeat provides important and affordable protein to families. An estimated 1 million animals are dying in illegal snares every year.

In Chapter 5

CONTENTS

LIST OF MAPS

AUTHORS AND CONTRIBUTORS

Mohammed Said (ILRI)
Robin Reid (ILRI)
Dan Omolo (ILRI)
Sandra van Dijk (ILRI)
Norbert Henninger (WRI)
Janet Nackoney (WRI)
Florence Landsberg (WRI)
Karen Holmes (consultant)
Patrick Wargute (DRSRS)
Wycliffe Mutero (KWS)
Jaspat Agatsiva (DRSRS)
Richard Bagine (KWS)
Ololtisatti Ole Kamuaro (ILRI)
Shem Kifugo (ILRI)
Ville Vuorio (ILRI)
Greg Mock (consultant)
Dan Tunstall (WRI)
Hyacinth Billings (WRI)
Jo Tunstall (consultant)
Carolina de Rosas (WRI)
Jamie Worms (WRI)

WHAT THIS CHAPTER SHOWS

Biodiversity underpins all ecosystem processes and is the foundation of Kenya's rich natural heritage. This chapter presents maps depicting both the breadth of Kenya's biodiversity and current pressures and trends affecting it. An initial set of maps outlines Kenya's endowment of mammals and the areas important for bird conservation. Maps showing agricultural cultivation and development of water sources in the rangelands then depict landscape-level pressures on Kenya's biodiversity. While agriculture is a significant factor in biodiversity decline, not all cropping has to be detrimental. A set of maps shows where farmers in central and western Kenya are making a positive contribution to biodiversity through crop selection and farming practices. The next series of maps shows how Kenya's diversity of rangeland species has been affected by human pressures. Maps showing the distribution of large grazing mammals in the rangelands reveal severe contractions in the size and distribution of wildlife populations since the 1970s, identifying areas that have witnessed declines in wildlife numbers and wildlife density. However, the maps and tables also indicate that not all the news about Kenya's rangeland species is bad. Two examples show a reversal in trends: the recovery of wildlife populations in Laikipia District and the stabilizing of elephant numbers. These demonstrate that national-level policies, community-based conservation efforts, and changes in local land use patterns can lead to outcomes that are more supportive of Kenya's unique biological endowment.

5

Biodiversity

Biodiversity—the full variety of plants, animals, and microorganisms found on Earth—is the source of many benefits crucial to human well-being. It provides the underlying conditions necessary for the delivery of ecosystem services (MA 2003). For instance, biodiversity provides the basis for crop varieties grown for food and fiber, as well as for traditional medicines. Biodiversity is also an important source of livelihoods, especially for poor people in rural areas, and underlies important sectors of the economy, such as nature-based tourism.

Kenya is fortunate to be endowed with a rich natural heritage. The country is home to over 6,500 plant species, more than 260 of which are found nowhere else in the world. With more than 1,000 bird species and over 350 species of mammals, Kenya ranks second highest among African countries in species richness for these animal groups (Biggs et al. 2004).

Perhaps most distinctive about Kenya's extraordinary biological endowment is its diversity of large mammals. Because Kenya straddles the boundary between Africa's northern and southern savanna zones (Bigalke 1978), more species of large mammals are concentrated in its rangelands than in virtually any other African country. For instance, distinct species of both giraffe and zebra inhabit Kenya's northern savannas as well as its southern savannas.

Box 5.1 Biodiversity and Ecosystem Services

Biodiversity, which includes variation at the level of genes, species, and entire ecosystems, is a necessary condition for ecosystems to function. The type and quantity of services people receive from ecosystems depend on the particular assemblages of plants, animals, and microorganisms present in a given ecosystem. Moreover, options for growing improved crop varieties or for obtaining genetic materials to develop new medicines depend strongly on the diversity of life forms supported by the surrounding environment.

While it is relatively straightforward to monitor changes in the provision of individual ecosystem services, ecologists understand far less about how to optimize a bundle of diverse ecosystem services and ensure that ecosystems can continue to provide these services over the long term. Studies indicate a relationship between the level of biodiversity, the types and quantities of specific ecosystem services, and the stability (sometimes referred to as resilience) of the system to provide these services. For example, scientists now know that grasslands containing greater numbers of plant species are often more productive and more capable of surviving periods of drought (Tilman et al. 1996). Studies have also demonstrated that greater biodiversity can act as a barrier to invasion by disruptive alien species (Kennedy et al. 2002).

Such evidence implies that the preservation of biological diversity is essential to maintain stable ecosystems. However, ecologists find it difficult to gauge how much biodiversity a system can afford to lose—for example, to boost the production of a single service, such as crop production or fish catch—without jeopardizing the entire system's stability. It is this unseen connection between biodiversity and the capacity of ecosystems to produce valued services that makes biodiversity a foundation of human health and well-being.

The Costs of Biodiversity Loss: The Example of Lake Victoria

The introduction of alien species into Lake Victoria—including fish species such as Nile tilapia and the Nile perch, as well as the invasive water hyacinth plant—provides an object lesson in how biodiversity underlies healthy ecosystem functioning. Within a few short years, these alien species have drastically changed the Lake Victoria ecosystem, diminishing its capacity to produce the ecosystem services required for local livelihoods, and dramatically shifting the distribution of human benefits derived from the lake.

The fish species introduced into Lake Victoria proved to be efficient predators of the native fish species, thus reducing overall biodiversity even as total fish catch grew considerably (Achieng 1990). Commercial fishers and processors profited from the increased catch of Nile perch, and Kenya's export earnings grew by some US$ 280-400 million per year due to perch exports. However, small-scale fishers, who could not afford the fishing and processing equipment required to fish for perch, were shut out of the new market and sustained serious losses to their livelihoods. In addition, the diets of many local people suffered as the availability of native fish species plummeted (Revenga et al. 2000).

At the same time, the proliferation of the water hyacinth plant began to choke local waterways, restricting transport and the ability of local fishers to access the lake. Despite these damages, the invading plants may have helped to prevent total extinction of local fish species by providing the smaller fish with hiding places to escape their new predators (Ogari 2001). Nonetheless, the drastic changes in the Lake Victoria ecosystem have caused some ecologists to question the long-term stability of the lake (Kaufman 1992). The lake's ecology has become a greatly simplified system of predators and prey with the Nile perch on top of the food chain, massive expansion of the invasive water hyacinth, and more frequent episodes of anoxic (i.e., low oxygen) conditions and algal blooms associated with pollution from land-use changes in the surrounding catchment area.

This chapter presents information on Kenya's biodiversity: its condition and trends, its importance to the economy and people, and potential mechanisms for safeguarding the country's biodiversity assets for generations to come. Special attention is devoted to large mammals—such as elephants, wildebeest, and zebras—because Kenya is such a special place for these animals, and also because of the excellent long-term data sets available on the spatial distribution of large mammals in Kenya. The questions addressed by this chapter include:

- Where are there high concentrations of species diversity for mammals and birds in Kenya?
- How have human activities affected biodiversity in Kenya?
  - What has been the impact of agricultural conversion and forest loss in the Kenyan highlands?
  - How has infrastructure development affected wildlife in Kenya's rangelands?
- What is the spatial distribution of wildlife populations in Kenya's rangelands?
  - How have these spatial patterns changed over time?
  - Where is rangeland wildlife in trouble, where is it recovering, and why?

SELECTED INDICATORS OF BIODIVERSITY

The following pages present mapped indicators of the diversity and distribution of Kenya's mammal and bird species. The focus on these animal groups reflects, in part, the availability of detailed, long-term data sets. Obtaining equivalent data on other categories of Kenyan biodiversity (such as insects, plants, aquatic species, etc.) is far more difficult.

To help orient the reader and provide context, Map 5.1 depicts the major ecosystem types found in Kenya. Each kind of ecosystem supports distinctive assemblages of plant and animal species; for instance, forest-dwelling animals and plants often are quite different from species that inhabit the desert (although there is usually some degree of overlap).

Map 5.1 **Major Ecosystem Types, 2000**

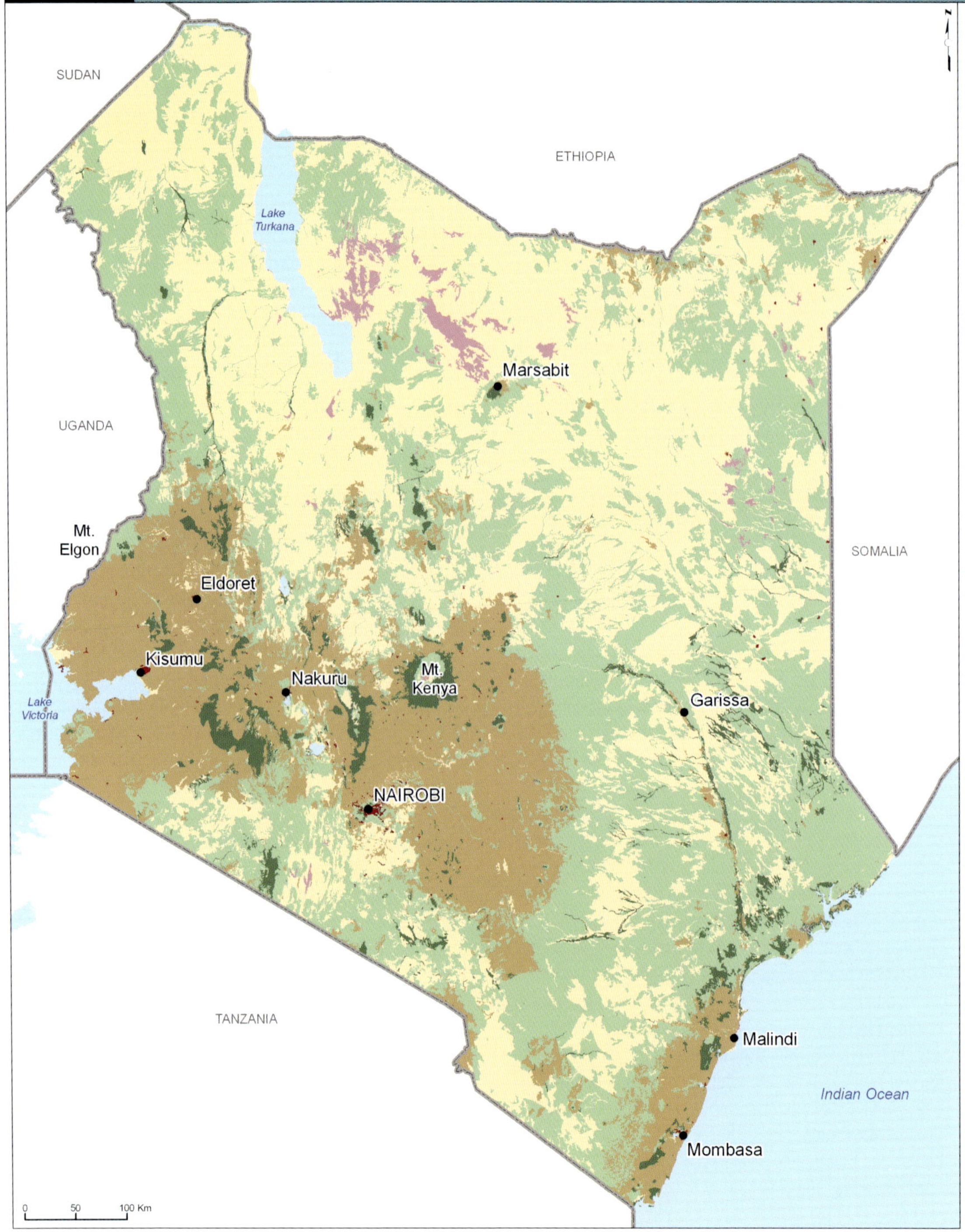

Sources: Cities (SoK and ILRI 2000), water bodies (FAO 2000), and major ecosystem types (FAO 2000).

Kenya's biodiversity resources vary across the major ecosystem types found in the country. The predominant ecosystems (75 percent of Kenya) are bushland and woodland (light green-shaded map areas) and savanna and grassland (pale yellow-shaded map areas), which support a variety of wildlife, including Kenya's distinctive herds of grazing animals and its large carnivorous cats. Much of the land area that historically was covered by montane forest or coastal dry forest (White 1982) has been converted to cropland (shown in light brown), with significant impacts for primates and other forest-dwelling species. Kenya's lakes, rivers, and other aquatic ecosystems are also important sources of biodiversity.

Note: Forest is the aggregate of two categories in the *Africover* legend (closed and multilayered trees). Bushland and woodland combine various *Africover* classes such as open trees, thickets, and shrublands. Savannas are grasslands with shrubs or sparse trees. Bare areas include areas covered by rocks or rock fragments.

More than 80 percent of Kenya's land area is classified as arid or semi-arid land (SoK 2003), ranging from desert landscapes to rangelands and other sparsely vegetated areas that support grasses, shrubs, and a few trees. Densely forested areas now make up only about 1.7 percent of the landscape (UNEP 2001). Agroecosystems have become the dominant land use in the Kenyan highlands and elsewhere, accounting for about 19 percent of the country's land area. Most Kenyans live in these densely populated croplands. Ecosystems characterized by human settlements and the built environment cover only about 0.2 percent of Kenya's land area, but about a third of the population lives in these urban areas (CBS 2001).

Distribution of Mammal Species

Examining the number of species in a given area is one of the simplest ways to measure biodiversity. Map 5.2 depicts the number of mammal species expected in various ecosystems and habitat types across Kenya.

From the map, it is easy to see that wildlife are unevenly distributed across the countryside. Most (but not all) wildlife species prefer to live in places with plentiful rainfall, high-quality soils, and more abundant food sources—that is, in the Kenyan highlands, forests, and wet savannas. Areas where rainfall is lower and soils are less fertile generally support fewer species per unit of land area. However, the vegetation in such areas, though sparser, often provides a perfect habitat for small animal species and can even sustain modest populations of some large animals. Wildlife depend on specific kinds of ecosystems and habitats, creating distinctive spatial patterns of species distribution. Large numbers of species often are concentrated in certain locations that feature the preferred habitat types, abundant sources of food, and sufficient rainfall.

Understanding the spatial distribution of species diversity is important for assessing current and potential effects on wildlife from the ever-expanding reach of human activities and settlements into formerly undisturbed habitat. For instance, the expansion of large-scale mechanized agriculture and human settlements in the dry-season wildlife range can interfere with the annual migration of hundreds of thousands of wildebeest and zebras to Masai Mara National Park from the Serengeti plains of Tanzania (Serneels and Lambin 2001, Lamprey and Reid 2004, Norton-Griffiths et al. in press). (See Chapter 6 for detailed information on migration routes.)

Note that the following map does not indicate the *actual* presence of species on the ground, but rather the *predicted* number and spatial distribution of species based on an extensive database of African mammals (IEA 1998). The data underlying this map reflect the habitat preferences of 281 mammal species in Africa, and were used to estimate potential ranges for these species throughout the continent, adjusting for the effects of nearby human settlements and cropping activities. These data are useful for broad comparisons across significant expanses of land area.

Map 5.2 **Predicted Diversity of Selected Mammal Species**

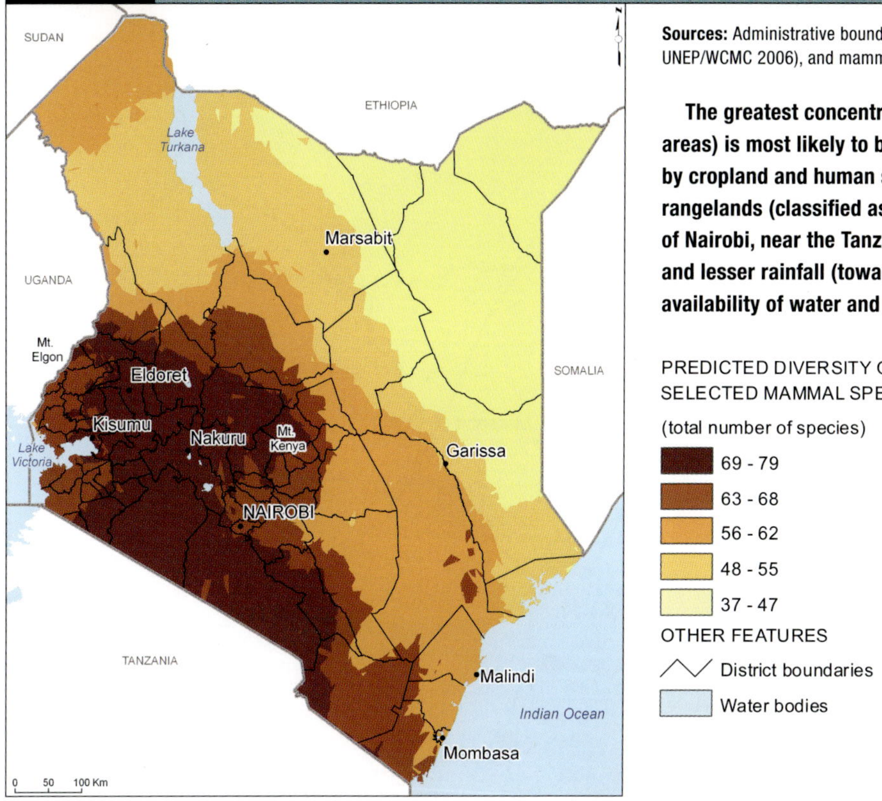

Sources: Administrative boundaries (CBS 2003), cities (SoK and ILRI 2000), water bodies (FAO 2000), parks and reserves (IUCN and UNEP/WCMC 2006), and mammal diversity (IEA 1998).

The greatest concentration of mammal species (more than 69 species, indicated by the dark brown-shaded areas) is most likely to be found in Kenya's central and western highlands—areas that are now dominated by cropland and human settlements. Predicted species diversity are at similarly high levels in the adjoining rangelands (classified as bushland, woodland, savanna, or grassland in Map 5.1), such as those located south of Nairobi, near the Tanzanian border. Numbers of mammal species are smallest in areas of lower elevation and lesser rainfall (toward the border with Ethiopia and Somalia), signaling mammals' dependence on the availability of water and specific climatic and habitat conditions.

Key Sites for Bird Diversity

Kenya is one of the richest countries in Africa in terms of diversity of bird life. About 1,090 different bird species are found here; some are full-time residents, while others are migrants within Africa, or between Africa and Asia or Europe. Some 11 species are endemic to Kenya, that is, they are found nowhere else in the world (African Bird Club 2006).

The birds of Kenya depend on various habitats. According to the African Bird Club, some 230 species rely exclusively on Kenya's forest habitats, and 110 species require habitats undisturbed by human activities to reproduce successfully (African Bird Club 2006). Wetlands are another habitat type critical for maintaining the diversity of Kenya's birds, including ducks, egrets, flamingoes, geese, herons, ibises, pelicans, and storks.

Countrywide data on the spatial distribution of specific bird species and populations were not readily available for use in this report. Thus, the maps in this chapter depict various sites that are generally important for conservation of bird diversity in Kenya, including Important Bird Areas (IBAs), Endemic Bird Areas (EBAs), and wetlands (see Map 5.3).

Important Bird Areas are globally important sites for bird conservation designated by BirdLife International and country partners (such as Nature Kenya, Kenya Wildlife Service, National Museums of Kenya, and Kenya Forest Department). They must meet certain criteria for international significance, such as the presence of key bird species that are vulnerable to global extinction or the presence of exceptionally large numbers of migratory birds or other irreplaceable bird populations. Conservation experts have identified 60 IBAs in Kenya (Bennun and Njoroge 1999), covering some 5.7 million ha (10 percent of the country's land area). These areas play a critical role in ensuring the survival of local and migratory bird species.

Of these 60 sites, only 35 are located inside parks, sanctuaries, reserves, or other protected areas (Bennun and Njoroge 1999). Thus, the survival of local and migratory species relies heavily on coexistence with people in landscapes that have been significantly altered by human activities. A recent assessment of the conservation status of Kenya's IBAs indicated that many are in decline—a finding that bodes ill for Kenya's rich bird diversity (Ng'weno et al. 2004). Indeed, some 27 bird species in Kenya have been listed as "critically endangered, endangered, or vulnerable" in the IUCN *Red List* (IUCN 2006).

A second category of key sites for bird diversity in Kenya consists of the Endemic Bird Areas (EBAs)—sites where two or more species of "restricted range" (less than 50,000 sq km) occur together (BirdLife International 2006). Most of Kenya's EBAs are located outside of parks and other protected areas, and all overlap or border densely settled, intensively farmed landscapes.

Also of great significance for bird diversity are Kenya's wetlands. Most of these are seasonal rather than permanent, and most are not legally protected as parks, wildlife reserves, or sanctuaries, particularly in the dry northern and eastern parts of the country. The largest wetland areas are found in Kenya's semi-arid and arid lands, with fewer and smaller wetlands located in agroecosystems. For instance, remnants of wetlands are located in the farmed landscapes north of Nairobi and southwest of Mount Kenya, but these are hard to distinguish on a national-scale map. (However, they can be seen on finer-scale maps of the upper Tana River in Chapter 8.)

Map 5.3 **Areas Important for Bird Conservation and their Status, 2003-04**

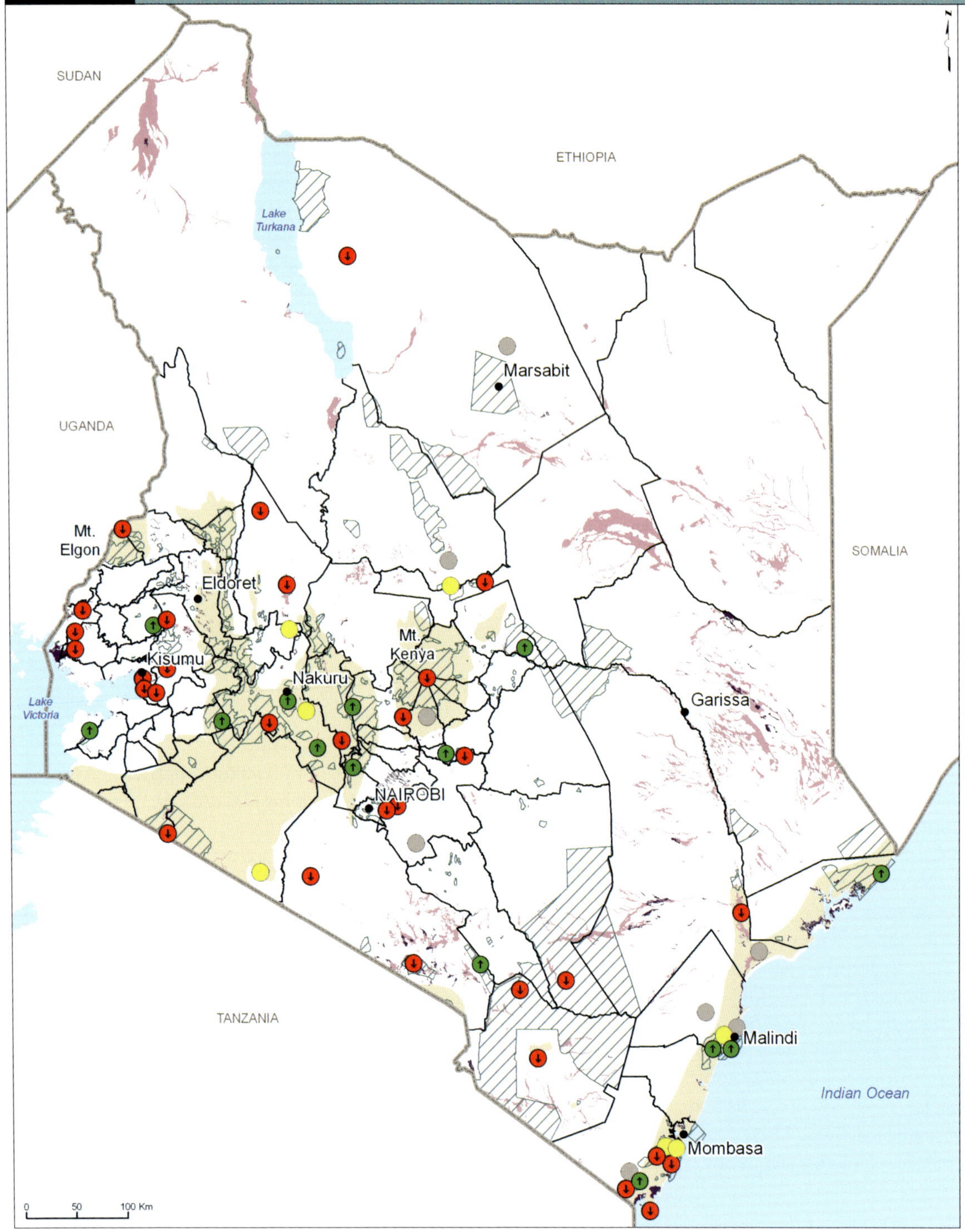

Sources: Administrative boundaries (CBS 2003), cities (SoK and ILRI 2000), water bodies (FAO 2000), parks and reserves (IUCN and UNEP/WCMC 2006), centroid of Important Bird Areas (Fishpool and Evans 2001), status of Important Bird Areas (Ng'weno et al. 2004), Endemic Bird Areas (Stattersfield et al. 1998), and wetlands (FAO 2000).

Kenya's 60 Important Bird Areas (IBAs, represented by points in different colors reflecting their conservation status) encompass most ecosystem types and a broad range of habitat conditions. Other key sites for bird species diversity are Kenya's eight Endemic Bird Areas (EBAs), shown as beige-shaded map areas), which are concentrated in the western and central highlands, in coastal forests and lowlands, and in parts of the southern rangelands. Critical habitat for many species (including egrets, herons, and Kenya's famous flamingoes), is provided by seasonal and permanent wetlands (purple-shaded map areas), most of which are located outside of parks or other protected areas.

A 2003-04 assessment of the status of Kenya's IBAs (Ng'weno et al. 2004) found that half were in decline (mapped as red points), about a quarter were improving (green points), and eight were stable (yellow points), with the status of the remaining sites unknown (grey points).

Note: The map depicts each Important Bird Area by a point in the center of its associated area. Some IBAs are much smaller than the point shown in this national map and others cover a much larger area, such as IBAs associated with the large protected areas of Masai Mara or the two Tsavo National Parks. IBAs range from 1 hectare to more than 1 million hectares in size (Bennun and Njoroge 1999).

STATUS OF SELECTED BIRD AREAS (IBAs)
- Improving
- Stable
- Declining
- Unknown

AREAS IMPORTANT FOR BIRD CONSERVATION
- Endemic bird areas
- Permanent wetlands
- Seasonal wetlands

OTHER FEATURES
- District boundaries
- National parks and reserves
- Water bodies

HOW PEOPLE AFFECT KENYA'S BIODIVERSITY

One of the greatest challenges to Kenya's rich wildlife heritage is that concentrations of high biodiversity often overlap with the places where people prefer to live as well. In East Africa, interactions between people and wildlife have been taking place for thousands of years in landscapes rich in large mammals. However, in recent decades, human activities and settlements have brought unprecedented change to Kenya's ecosystems. The two primary ways in which people are modifying the natural landscapes are by conversion of forests, rangelands, and other natural systems to agricultural cultivation; and development of new water sources in rangelands. The following sections examine landscape-level indicators of human modification of Kenya's ecosystems.

Intensity of Cultivation

Kenya's croplands are concentrated in zones of greater, more reliable rainfall, that is, the central and western highlands, the Lake Victoria basin, and a narrow strip of coastal lowlands. Across the remainder of Kenya's land area, the climate and soils are too dry and risky for rainfed agriculture.

Human conversion of forests and rangelands into managed, farmed landscapes often brings fundamental changes in the plant and animal communities found there. In general, agroecosystems feature fewer species and less biodiversity than the natural systems they replace. Conversion to agriculture also changes hydrological patterns, that is, surface and subsurface water flows. These changes are most drastic when the land is converted to highly mechanized agriculture, which typically features large fields of a single cereal crop, such as maize, rice, or wheat.

It is important to note, however, that within Kenya's croplands, farmers use the land at different levels of intensity (see Map 5.4). Even in the most intensively cultivated landscapes, some remnants of natural vegetation remain. Depending on how these fragments are managed and incorporated into the larger agricultural landscape, small-scale farmers can make a potentially significant contribution to maintaining Kenya's biological heritage.

Map 5.4 Intensity of Cultivation, 2000

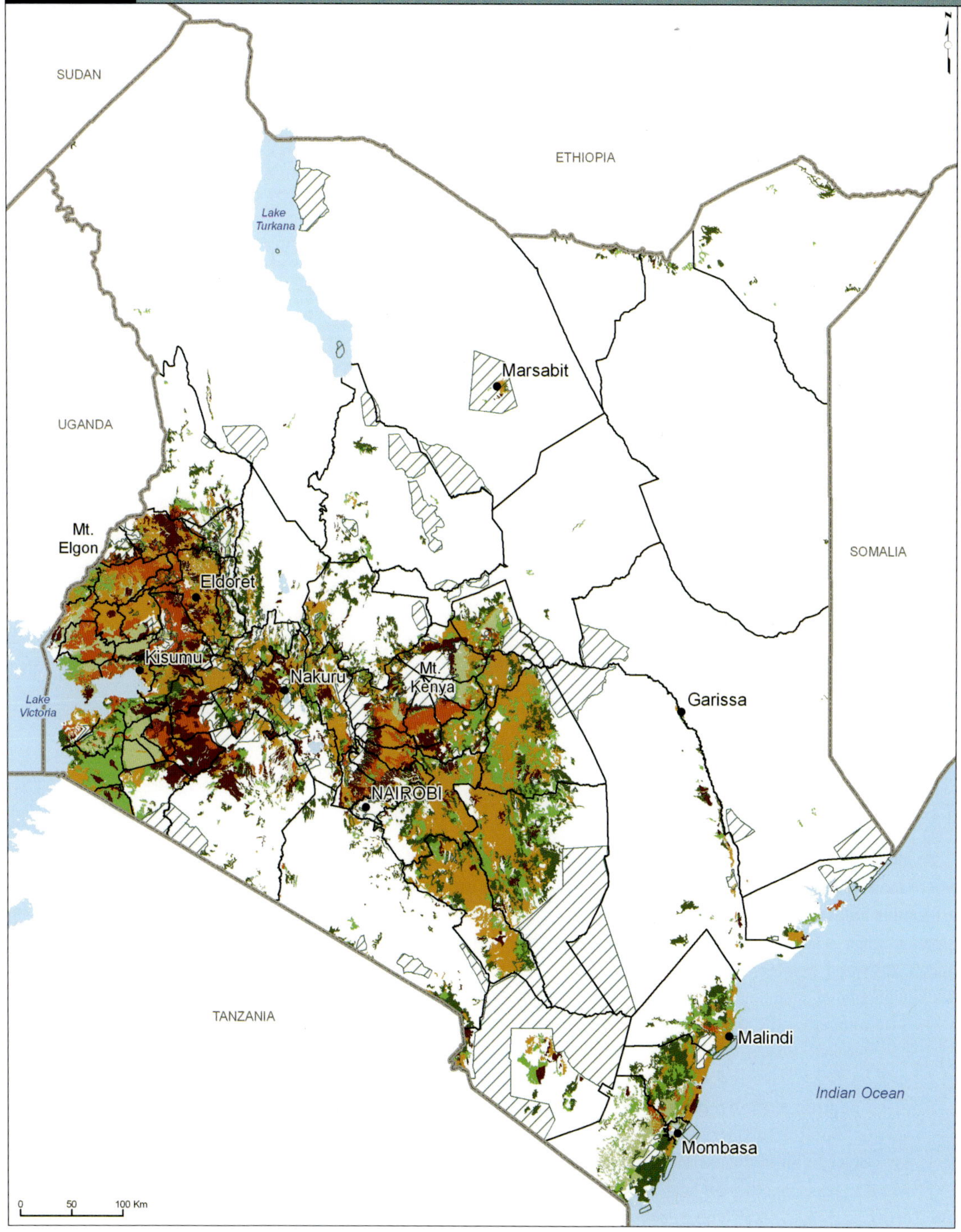

Sources: Administrative boundaries (CBS 2003), cities (SoK and ILRI 2000), water bodies (FAO 2000), parks and reserves (IUCN and UNEP/WCMC 2006), croplands (FAO 2000), and cropland intensity (WRI calculation based on *Africover* legend for croplands in FAO 2000).

The most intensively farmed landscapes in Kenya (brown-shaded map areas, with more than 80 percent of local land area under cultivation) are found mostly in the central and western highlands and in small patches in the coastal lowlands. Most agroecosystems have 50 to 60 percent of their land area under cultivation (light green or orange-shaded map areas), with farmers' fields interspersed with patches of less-managed landscapes, such as forests, woodlands, and other natural habitats.

Note: The standardized Land Cover Classification System of *Africover* (FAO 2000) can be used to show to what degree the spatial units (polygons) within the *Africover* map are "natural and semi-natural areas" or "managed" (cultivated) areas. The *Africover* classification system and the associated rules used to interpret the satellite imagery allow the creation of six discrete classes of cropland intensity, reflecting a stepwise gradient from the lowest (only 15 percent of the polygon is covered by cropland) to the highest category (more than 80 percent of the polygon is cultivated). The *Africover* map does not provide sufficient information to create a continuous legend ranging from zero to 100 percent. The map cannot show cropping that falls below the 15 percent threshold.

CROPLAND INTENSITY
(percent of land under cultivation)

- > 80
- 70
- 60
- 50
- 35 - 40
- 15

OTHER FEATURES

- District boundaries
- National parks and reserves
- Water bodies

Agriculture and Biodiversity Conservation

Agricultural landscapes are often associated with reduced levels of biodiversity. However, this need not be the case. It is possible for farmers to grow their crops and manage their land in ways that contribute to, rather than detract from, the conservation of native plant and animal species.

Maps 5.5-5.7 depict landscape-level indicators that are related to biodiversity-friendly farming and land management methods. Based on analysis of remote sensing data and aerial photography of central and western Kenya (a priority area for both biodiversity conservation and agricultural production), we selected three indicators—average number of crops grown, extent of tree cover in croplands, and average size of farmers' fields. When combined, these indicators portray the quality of habitat provided by agricultural landscapes for native highland plants and animals.

- Farmers contribute to agricultural biodiversity when they grow **multiple crop species**, either simultaneously or sequentially over the course of a single season (Map 5.5). In some parts of the highlands, farmers grow up to eight different crop species at one time. Analysis of remote sensing data and aerial photography reveals such concentrations of "polycropping" throughout the highlands.
- The **extent of tree cover in croplands** (Map 5.6) also can have an important bearing on habitat quality. The presence of more trees in agricultural landscapes—in hedgerows, woodlots, and forest remnants—clearly benefits bird life dependent on these habitats (Reid et al. 1997, Wilson et al. 1997). Trees also provide shade for streams that flow through croplands, which helps maintain lower water temperatures that promote replenishment of fish stocks. In several areas of the highlands, tree cover in farmlands exceeds 30 percent; at this proportion, the trees themselves can make a significant contribution to plant biodiversity as well as providing wildlife habitat. Fruit trees and trees in hedgerows and woodlots also provide an important source of food, fuel, forage, and building materials for farmers.

Map 5.5 Average Number of Crops Grown in Croplands of Central and Western Kenya, 1997

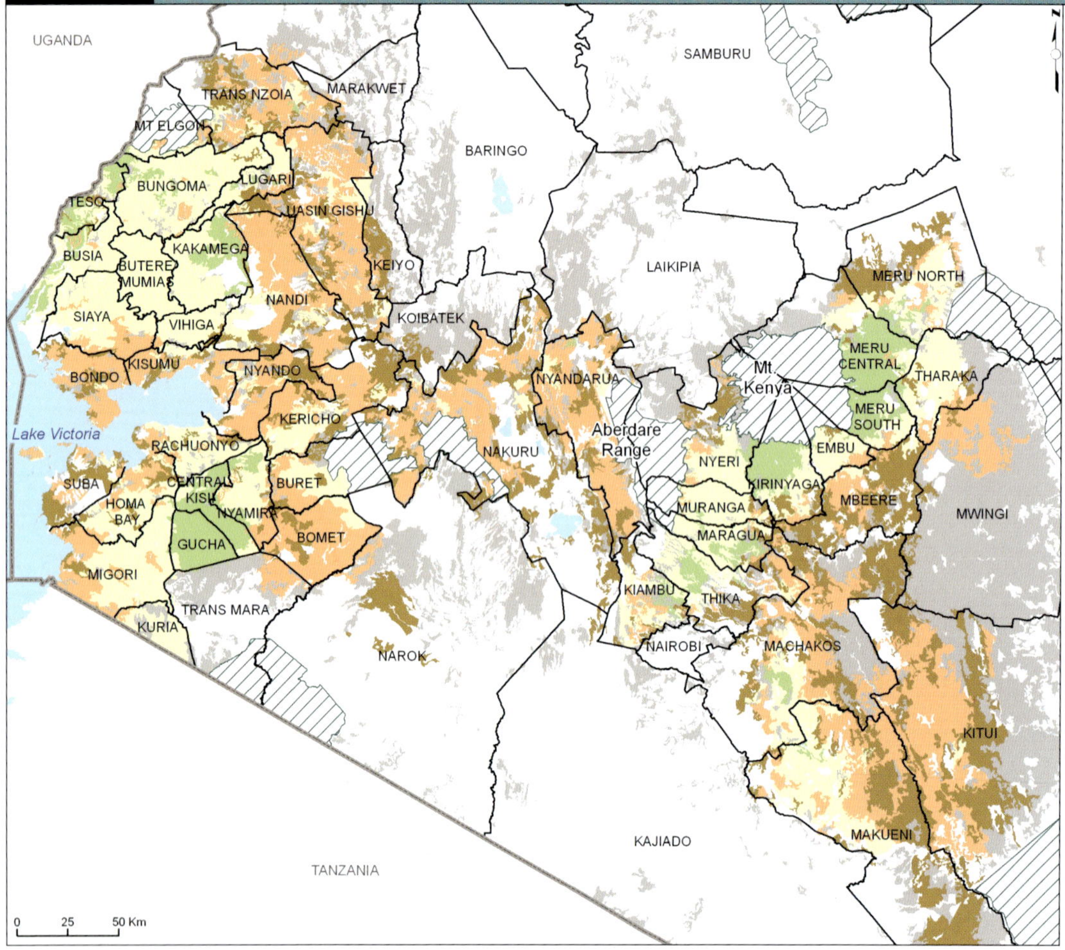

Sources: Administrative boundaries (CBS 2003), water bodies (FAO 2000), parks and reserves (IUCN and UNEP/WCMC 2006), cropland areas (FAO 2000), and number of crops grown in sampled croplands (WRI calculation based on ICRAF and DRSRS 2001).

Note: The map combines detailed crop information from 5,747 aerial photos for a growing season in 1997, each providing a sample point of detailed crop information. These samples are averaged to spatial units (polygons) of croplands from Kenya's most recent land cover map (FAO 2000). These averages represent conservative estimates. The raw data indicate that in some sample points farmers grow up to eight different crop species simultaneously.

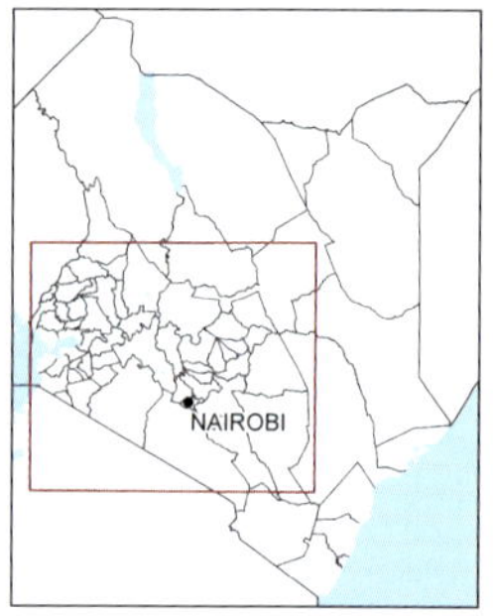

AGRICULTURAL CROPS IN SAMPLED CROPLAND
(average number)
- > 4
- 2 - 4
- 1 - 2
- <= 1
- Cropland not sampled

OTHER FEATURES
- District boundaries
- Major national parks and reserves (over 5,000 ha)
- Water bodies

On much of Kenya's most fertile cropland, farmers grow an average of two to four crops at the same time. Sites where farmers grow only one or two crops at a time typically are marginal farming areas with less rainfall, or highly productive areas where farmers grow a single cash crop, such as wheat-growing areas of Narok District or rice-growing areas in Mbeere District. Areas with the greatest number of crops grown at one time are concentrated in Gucha, Kirinyaga, and Meru Central Districts.

- The **average size of farmers' fields** (Map 5.7) is a key indicator for biodiversity because smaller fields have more edges and boundaries, often planted with diverse species of trees and shrubs. Such heterogeneous landscapes make for better habitat for native plants, birds, and small mammals than do large fields of a single crop, such as rice or wheat. On the other hand, small field size may also make agricultural production more labor-intensive and less efficient.

Simultaneously examining these three indicators can help identify priority areas for programs to further enhance biodiversity in agricultural landscapes, for instance through certified organic farming schemes, or programs to promote more planting of native tree species. Of course, these indicators do not account for other important factors that impact biodiversity on farmlands, such as pesticide and fertilizer use, soil conservation, and preservation of native vegetation on the banks of streams and other water bodies. However, ecologists currently understand the relationships between biodiversity and these factors only in very general terms. Farmers and other resource managers could benefit greatly from the availability of more specific guidance on these linkages.

Map 5.6 Extent of Tree Cover in Croplands of Central and Western Kenya, 2001

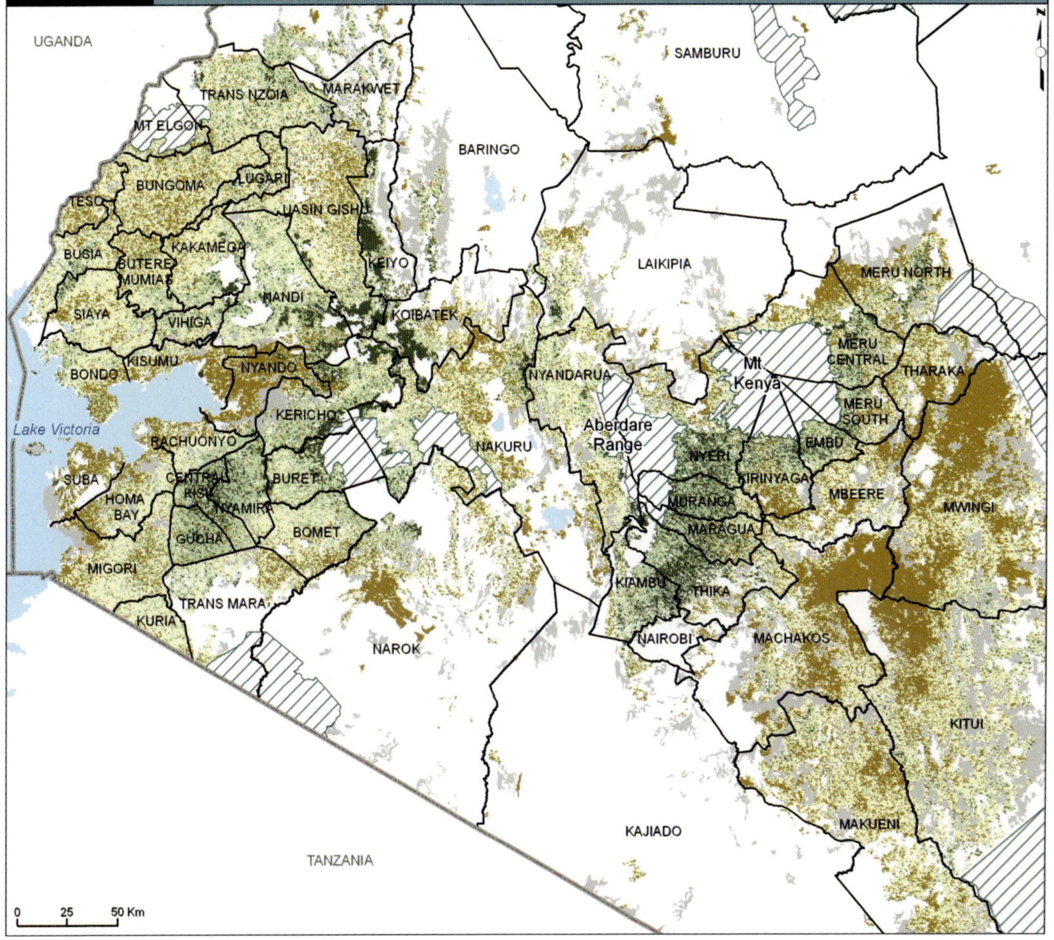

TREE COVER IN AREAS WITH MORE THAN 30% CROPLAND (percent)

- > 30
- 20 - 30
- 10 - 20
- <= 10
- Areas of < 30% cropland

OTHER FEATURES

- District boundaries
- Major national parks and reserves (over 5,000 ha)
- Water bodies

Sources: Administrative boundaries (CBS 2003), water bodies (FAO 2000), parks and reserves (IUCN and UNEP/WCMC 2006), areas with more and areas with less than 30 percent cropland (FAO 2000), and percent tree cover (Hansen et al. 2003).

Across most of the heavily cultivated landscapes of the highlands, farmers' fields contain 10 to 30 percent tree cover, consisting of fruit and other trees grown for fuel, forage, and building materials. Farmlands with relatively higher levels of tree cover are found in the eastern foothills of the Aberdare Range and on the southern slopes of Mount Kenya, as well as in Gucha, Central Kisii, and Nyamira Districts in the western highlands. Small belts of farmland with high levels of tree cover can also be seen in agricultural areas that border forest plantations and reserves in Kericho, Koibatek, and Keiyo Districts.

Map 5.7 Field Size in Croplands of Central and Western Kenya, 2000

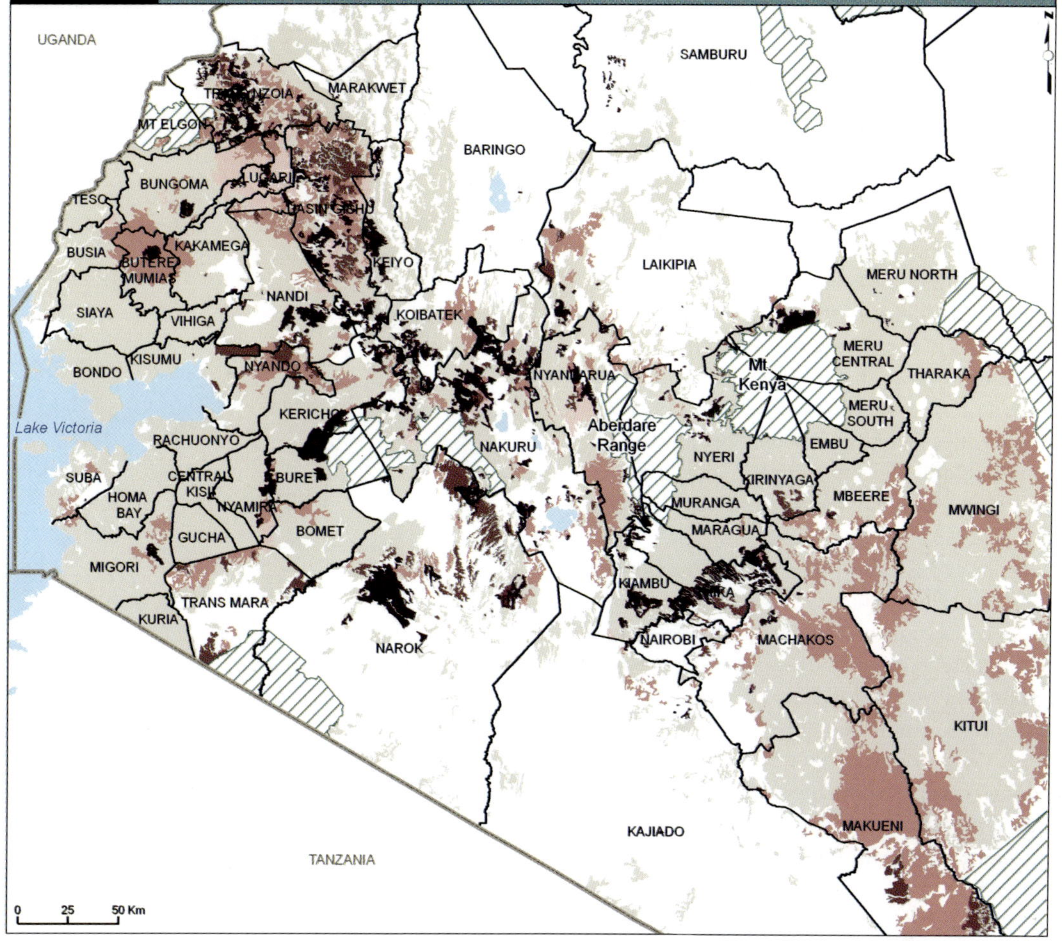

FIELD SIZE

- Large (> 5 ha)
- Large mixed with medium
- Medium (2 - 5 ha)
- Medium mixed with small
- Small (< 2 ha)

OTHER FEATURES

- District boundaries
- Major national parks and reserves (over 5,000 ha)
- Water bodies

Sources: Administrative boundaries (CBS 2003), water bodies (FAO 2000), parks and reserves (IUCN and UNEP/WCMC 2006), and field size in croplands (WRI extraction from FAO 2000).

Throughout Kenya's highlands, farmers tend to grow their crops in small fields. Sites where fields are large generally correspond to farming enterprises engaged in production of specific cash crops, such as wheat farming in Narok District or coffee growing in Kiambu and Thika Districts. Some of the large fields shown for Buret and Kericho Districts represent large fields of tea and tree plantations.

Impact of Rangeland Development on Biodiversity

Most of Kenya is too dry for rainfed agriculture. In the country's semi-arid and arid regions, people do little farming, but their use of the land for grazing livestock and for building towns, roads, water points, and other infrastructure can still affect biodiversity, although usually not as much as farming does. For instance, grazing of livestock by pastoral people near water points in northern Kenya "pushes" wildlife away from water, at least during the daytime (Williams 1998, de Leeuw et al. 2001). Recent studies in the southern Kenya rangelands of the Mara have shown that density of human settlement has an impact on wildlife densities. At lower human densities (less than 7 people per square kilometer) wildlife density increases, and at higher human densities wildlife density declines rapidly (Reid et al. 2003).

Maps 5.8 and 5.9 show how developing boreholes and other water points in northern Kenya impacts livestock and wildlife. In this region, rangelands dominate, consisting of savannas, grasslands, bushlands, and woodlands.

The species composition of livestock and wildlife herds varies considerably across this region, depending on vegetation type as well as availability of water. For instance, livestock herds in Turkana District are made up mostly of cattle, while herds in Samburu District typically include a mix of cattle and smaller livestock, such as goats and sheep. In the northeast, camels are more prevalent than elsewhere in the northern rangelands.

To compare the impacts of livestock on wildlife herds composed of different species, data on animal populations are converted to a common unit, known as a *tropical livestock unit* (TLU). Each TLU is equal to an animal weight of 250 kg; thus, one cow accounts for 0.7 TLU, one camel is counted as 1.8 TLUs, and it takes 14 goats or sheep to make up one TLU. For wildlife species, one elephant is equivalent to 7.0 TLUs, one buffalo counts as 2.5 TLUs, and one wildebeest accounts for 0.9 TLU. Meanwhile, it takes ten Thomson's gazelle to make one TLU.

Map 5.8 **Water Points and Livestock Density in the Northern Rangelands, 1994-96**

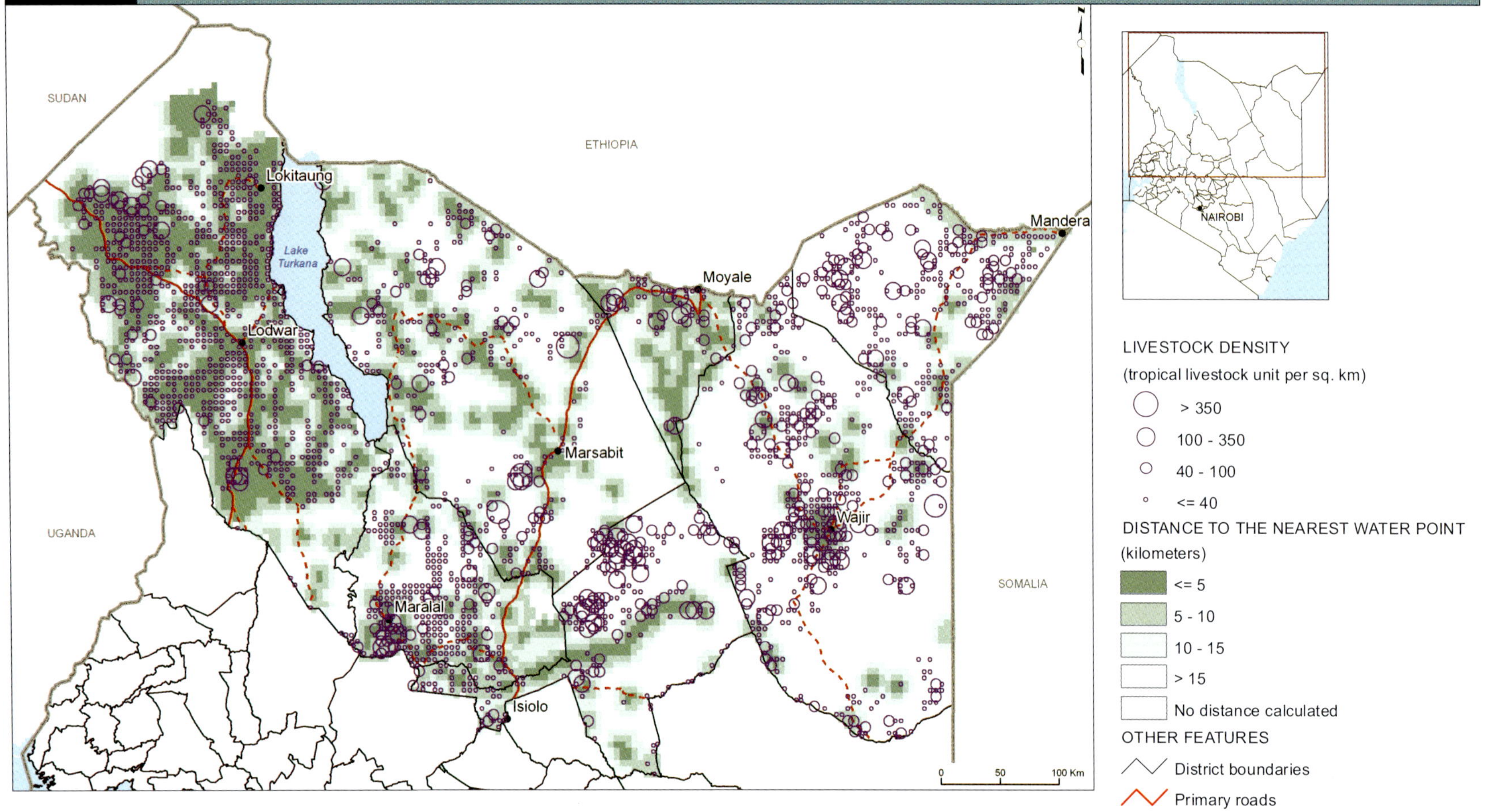

Sources: Administrative boundaries (CBS 2003), cities (SoK and ILRI 2000), water bodies (FAO 2000), parks and reserves (IUCN and UNEP/WCMC 2006), primary and secondary roads (SoK and ILRI 1997), water points (GTZ 1996), distance to water points (WRI calculation based on GTZ 1996), and 1994-96 livestock density (ILRI calculation based on DRSRS 2003, Grunblatt et al. 1995, Grunblatt et al. 1996).

Areas closest to water points (that is, zones less than 5 km from a water point, shown as dark-shaded map areas) tend to support the greatest density of livestock populations (shown as purple circles). For instance, livestock are prevalent throughout Turkana District (west of Lake Turkana), which has the densest network of boreholes and permanent water sources.

Note: Livestock numbers are aggregated (using TLUs) to squares of 5 km by 5 km, averaged by square kilometer, and then represented by a circle proportional to their density. Livestock includes cattle, sheep, goats, camels, and donkeys that were observed during low-altitude flights.

Map 5.9 Water Points and Wildlife Density in the Northern Rangelands, 1994-96

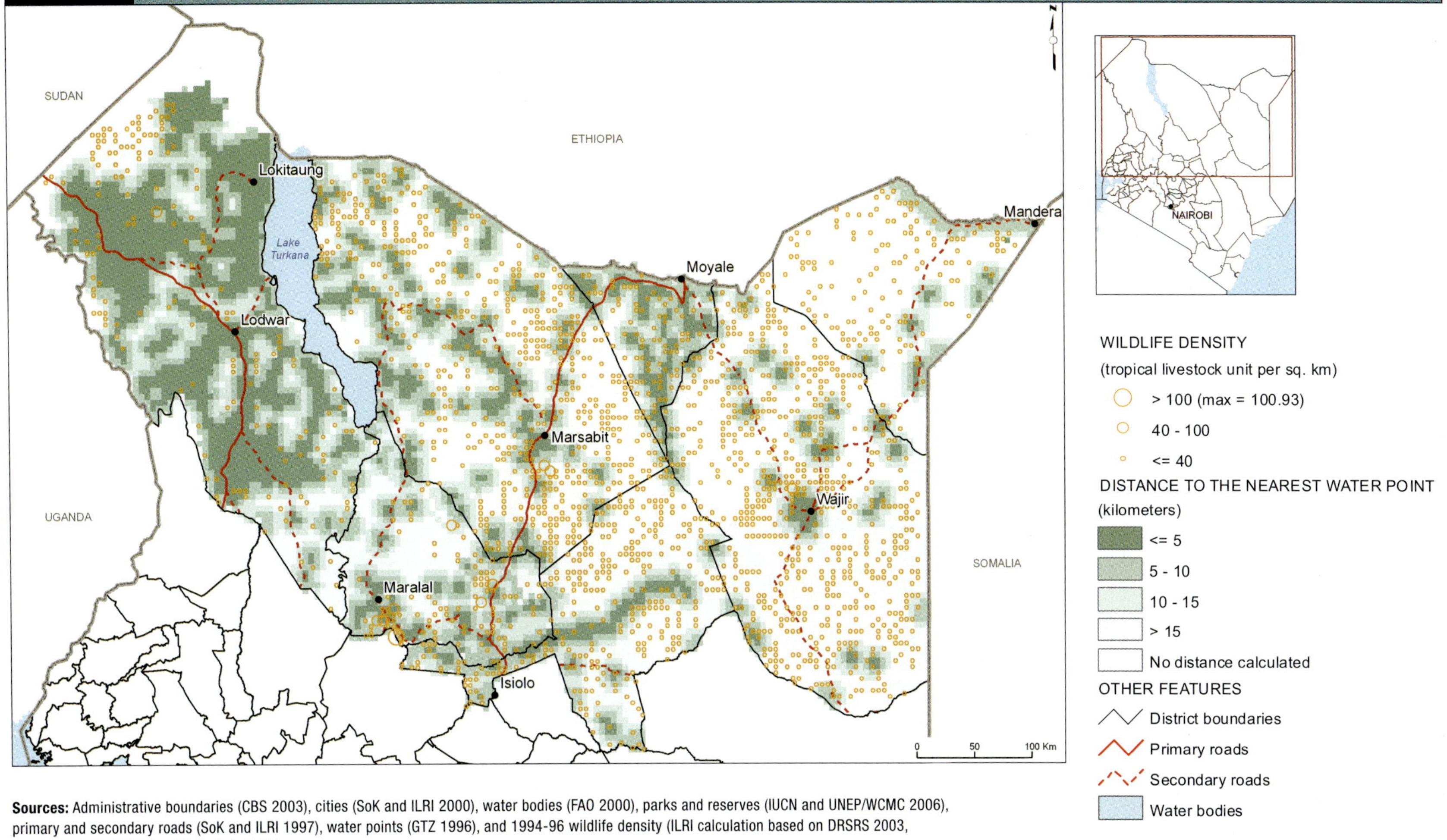

Sources: Administrative boundaries (CBS 2003), cities (SoK and ILRI 2000), water bodies (FAO 2000), parks and reserves (IUCN and UNEP/WCMC 2006), primary and secondary roads (SoK and ILRI 1997), water points (GTZ 1996), and 1994-96 wildlife density (ILRI calculation based on DRSRS 2003, Grunblatt et al. 1995, Grunblatt et al. 1996).

The greatest densities of wildlife (mapped as orange circles) are found in zones that are located at least 15 km from the nearest water point (shown as light-shaded areas). The drilling of boreholes in Turkana District, west of Lake Turkana, and elsewhere in the northern rangelands has permitted grazing of more livestock, but has pushed wildlife farther away from water sources. Note also that wildlife densities tend to be less (by weight) than densities of livestock in this region.

Note: Species numbers are aggregated (using TLUs) to squares of 5 km by 5 km, averaged by square kilometer, and then represented by a circle proportional to their density. The wildlife counts include 21 different large grazing animals that can be observed during low-altitude flights.

LOSING AND GAINING WILDLIFE

East Africa is one of the few places on Earth where people can watch the spectacle of migrating herds of millions of wildebeest, gazelle, antelopes, and other grazing animals. Several of Kenya's most celebrated natural areas—such as the savannas of the Mara, Athi-Kapiti, and Amboseli ecosystems—still support grand migrations of wildebeest and zebra. Elephants are another species that is on the move in large numbers in Kenya, in the Mara, Amboseli, Tsavo, and Laikipia-Samburu ecosystems.

In pastoral lands, humans and wildlife can coexist peacefully most of the time. However, conflicts with wildlife can erupt when people settle permanently and establish farms. Sometimes people kill wildlife purposely for food, for trophies, or to protect their crops and their lives. Most devastating to wildlife, however, is the loss of habitat that comes from competing human uses of the land for farms, towns, water points, or heavy grazing of domestic livestock.

The following section looks at Kenya's wildlife populations and how they have changed in recent decades. The maps and table presented here are based largely on data from aerial wildlife counts conducted periodically since the 1970s by the Department of Resource Surveys and Remote Sensing, as well as data from the Kenya Wildlife Service. Using this information, national-scale maps showing wildlife numbers over time for Kenya's rangelands as a whole can be constructed; finer-scale maps showing changes in the distribution of selected wildlife species in specific locations can also be made.

As these maps and the table demonstrate, Kenya has experienced severe contractions in the size and distribution of wildlife populations since the 1970s. For some species and in some areas, declining trends have been reversed and recovery has begun. In other cases, the losses continue and may even be accelerating. Assessments of Kenya's mammal populations, for example, show that 51 species (14 percent of the total number of species) are now threatened with extinction (IUCN 2006).

Data analysis and mapping indicate that wildlife populations have tended to fare better in or near Kenya's parks and game reserves. However, many species, especially the large grazing animals, spend

Map 5.10 Wildlife Density in the Rangelands, 1977-78

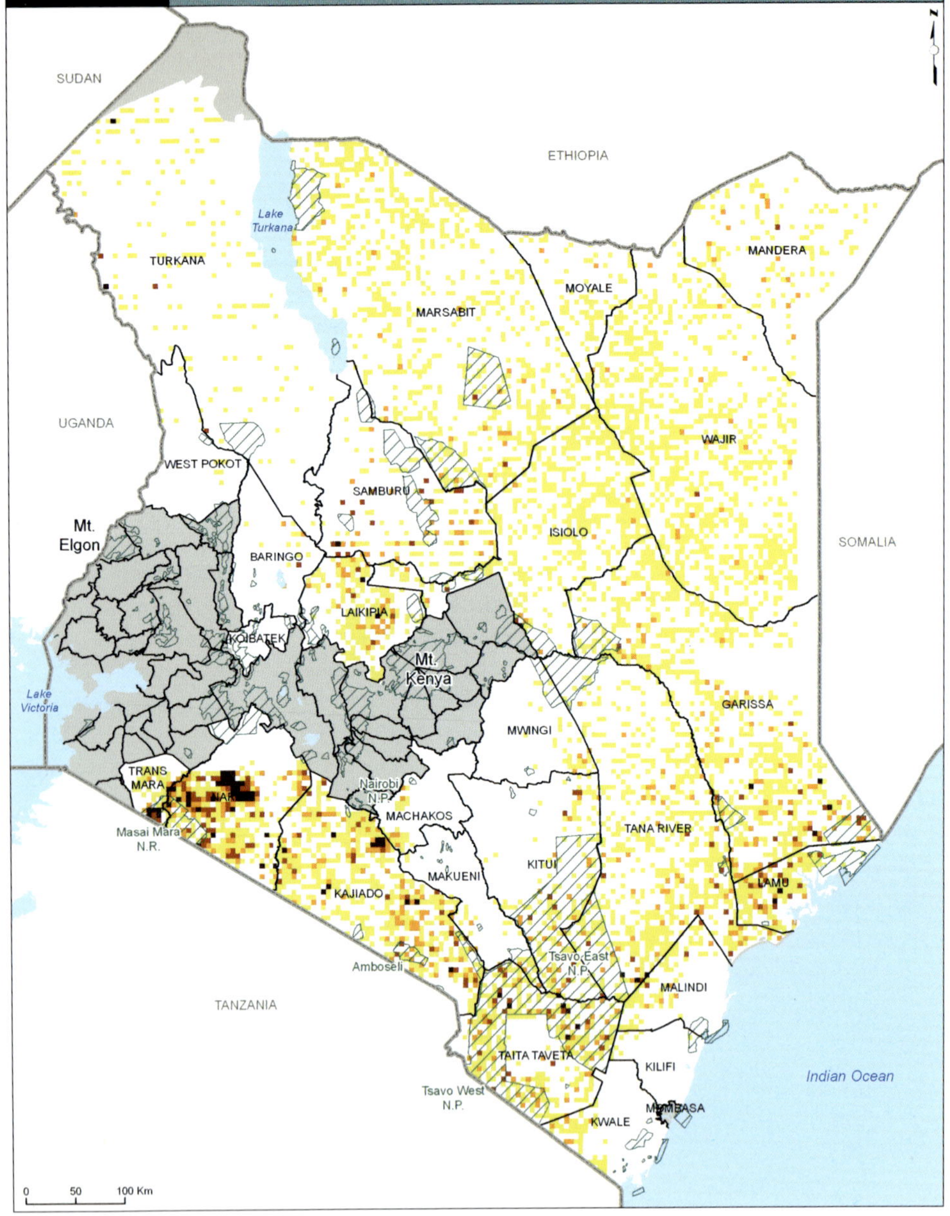

Sources: Administrative boundaries (CBS 2003), water bodies (FAO 2000), parks and reserves (IUCN and UNEP/WCMC 2006), and 1977-78 wildlife density (ILRI calculation based on DRSRS 2003, Grunblatt et al. 1995, Grunblatt et al. 1996).

Map 5.11 Wildlife Density in the Rangelands, 1994-96

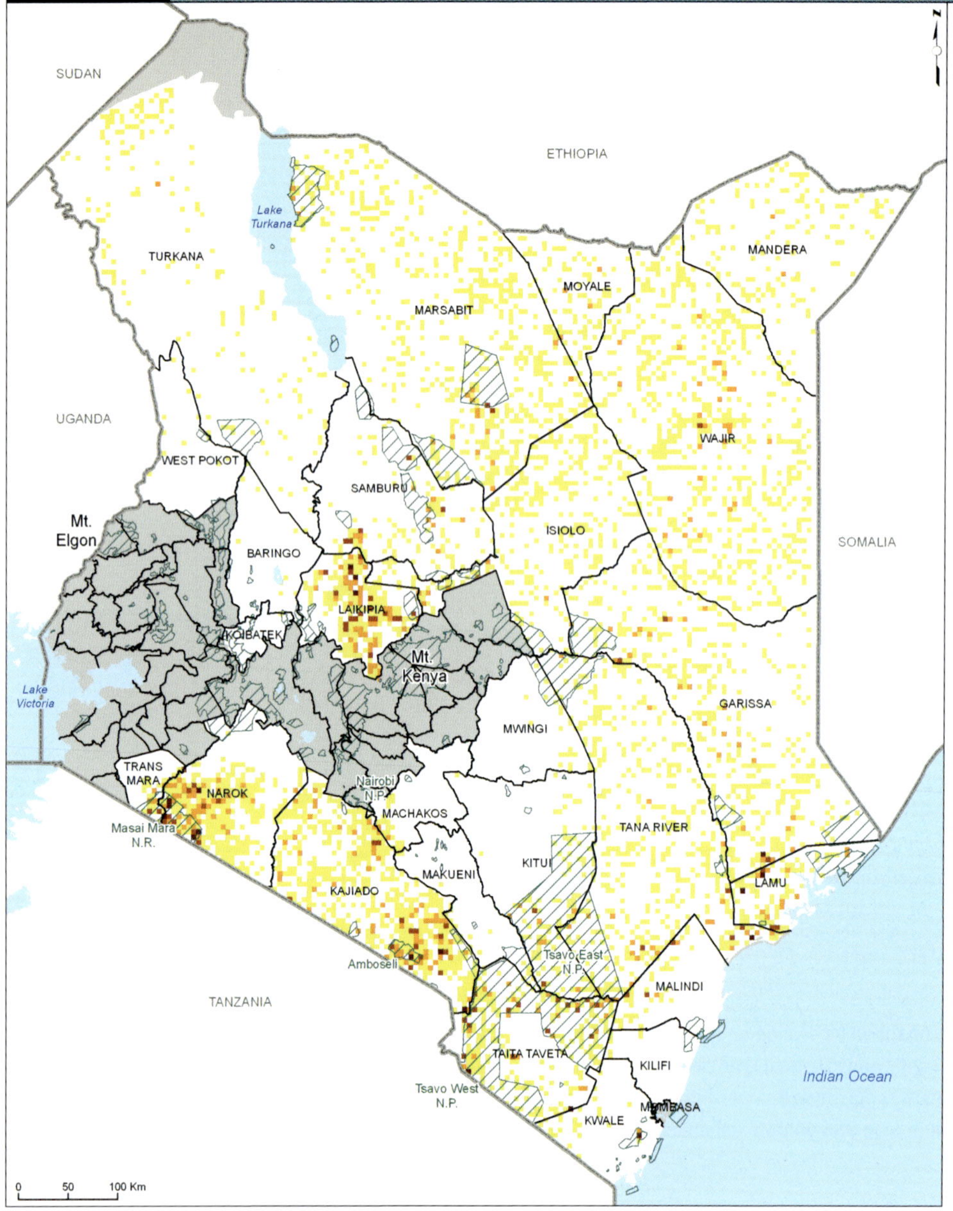

Areas of high wildlife density (dark brown-shaded areas) are less prevalent in the 1990s than in the 1970s. Especially striking is the disappearance of sites with dense wildlife populations in central Narok District, north of the Masai Mara Game Reserve. In general, declines in wildlife density have been steeper in communal and privately owned lands than in parks and reserves.

Note: Species numbers are aggregated (using TLUs) to squares of 5 km by 5 km and then averaged by square kilometer. The wildlife counts include 21 different large grazing animals that can be observed during low-altitude flights.

WILDLIFE DENSITY
(tropical livestock unit per sq. km)

- > 125
- 45 - 125
- 15 - 45
- 0 - 15
- No observation of wildlife
- No data

OTHER FEATURES

- District boundaries
- National parks and reserves
- Water bodies

Sources: Administrative boundaries (CBS 2003), water bodies (FAO 2000), parks and reserves (IUCN and UNEP/WCMC 2006), and 1994-96 wildlife density (ILRI calculation based on DRSRS 2003, Grunblatt et al. 1995, Grunblatt et al. 1996).

a significant amount of their life cycle outside the borders of these protected areas. The way people use private and communally held lands strongly affects Kenya's wildlife and will play an increasingly important role in ensuring the long-term survival of many species.

Trends in Spatial Distribution of Wildlife

Most of Kenya's rangelands contain wildlife, but the density of wildlife populations varies over time and space according to several factors. The most important factors are the availability of water and forage, as well as competition with human uses of the land for growing crops; raising domestic livestock; and building towns, roads, and water points.

In three Districts—Laikipia, Kajiado, and Narok—wildlife are especially abundant. Here, large herds of many different species congregate, especially in areas close to some of Kenya's best-known national parks and reserves, including Masai Mara National Reserve, Amboseli National Park, and Nairobi National Park. Wildlife by no means restrict their ranges to these protected areas; they also migrate across private and communally held lands and even across international borders.

Maps 5.10 and 5.11 depict the density of wildlife populations across Kenya's rangeland Districts. As indicated earlier, animal densities are converted to a common unit (known as a *tropical livestock unit* (TLU), which is equivalent to an animal weight of 250 kg) to represent the density of wildlife herds composed of different species.

Note that the maps depict *average* wildlife densities (in terms of TLU per sq km) over the course of a year, and do not show the significant differences in wildlife populations that occur between the rainy and dry seasons. These maps are most useful for pinpointing areas with the highest average wildlife densities and comparing these areas to other mapped features, such as the boundaries of parks and reserves, or extent of land under cultivation versus less modified ecosystems. For instance, in Map 5.10, areas of Narok District that showed high wildlife densities in the 1970s but not the 1990s correspond in large part to areas in which rangelands have been converted to croplands (see Maps 5.1 and 5.4).

Map 5.12 **Changes in Wildlife Density in the Rangelands, 1977-78 to 1994-96**

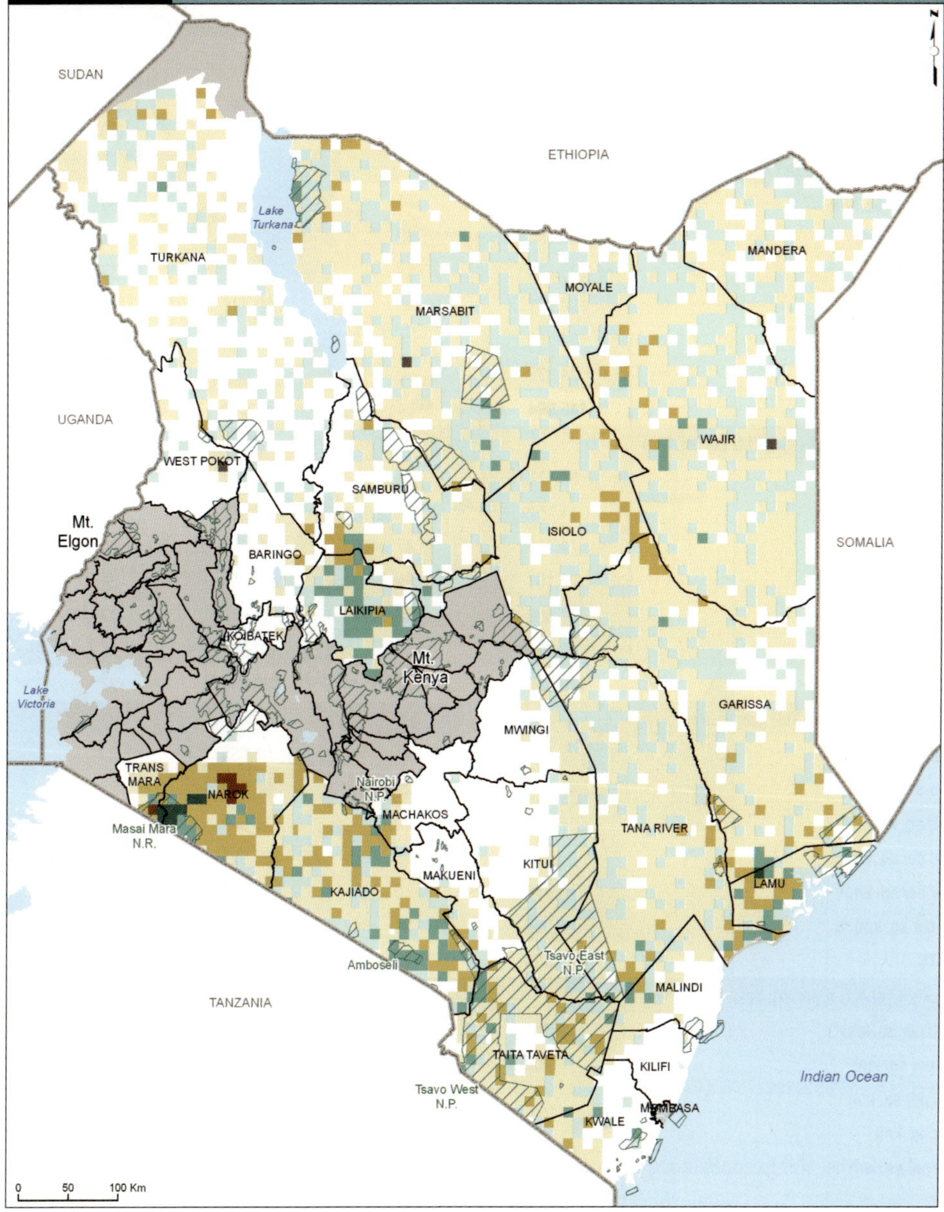

Sources: Administrative boundaries (CBS 2003), water bodies (FAO 2000), parks and reserves (IUCN and UNEP/WCMC 2006), and changes in wildlife density between 1977-78 to 1994-96 (ILRI calculation based on DRSRS 2003, Grunblatt et al. 1995, Grunblatt et al. 1996).

Many places in Kenya experienced gains in wildlife density between the 1970s and the 1990s, with sites of most rapid recovery (dark bluish-green areas) concentrated in southwest Narok District, near the Masai Mara Game Reserve; in Kajiado District, near Amboseli National Park; in Laikipia District, northwest of Mount Kenya National Park; and in selected areas near the coast in Lamu District. Sites with sharp declines in wildlife density (dark brown-shaded areas) are found throughout large parts of central Narok District, south of Nairobi in Kajiado District (see also detailed Map 5.15 of Kitengela dispersal area), northern Laikipia District, locations along the Samburu-Laikipia border, and in Isiolo and Garissa Districts near the Wajir border.

Note: To estimate changes in wildlife densities, species numbers are aggregated (using TLUs) to squares of 10 km by 10 km and then averaged by square kilometer for each reference period. The wildlife counts include 21 different large grazing animals that can be observed during low-altitude flights.

CHANGE IN WILDLIFE DENSITY

- High increase (> 10,000 TLU/ sq. km)
- No change
- High decline (> 28,000 TLU/ sq. km)
- No observation of wildlife at both dates
- No data

OTHER FEATURES

- District boundaries
- National parks and reserves
- Water bodies

Map 5.15 Spatial Distribution of Wildebeest Population in the Athi-Kapiti Plains, 1977-78 to 1994-96

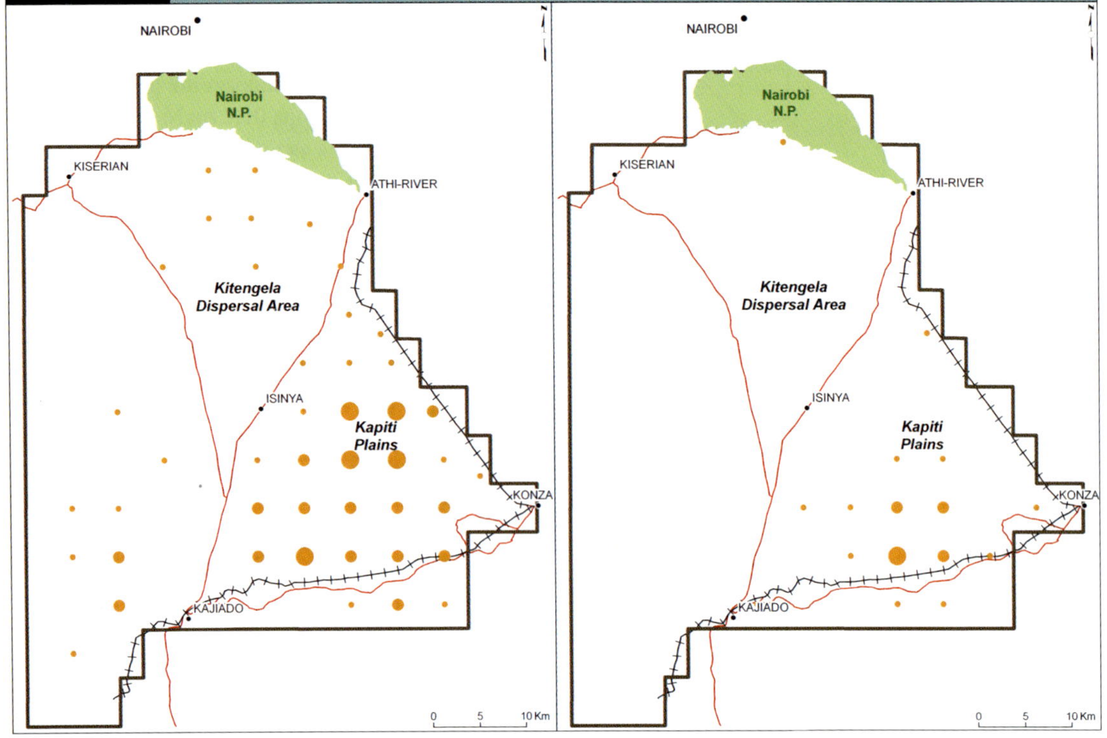

1977-78 **1994-96**

Sources: Towns and market centers (SoK and ILRI 2000), parks and reserves (IUCN and UNEP/WCMC 2006), and presence of wildebeest (Reid et al. 2006, Gichohi 1996).

PRESENCE OF WILDEBEEST (number observed)

- High (> 45)
- Medium (15 - 45)
- Low (<= 15)

OTHER FEATURES

- Study area boundary
- Railroads
- Roads
- National parks and reserves

The number and size of wildebeest populations in the Athi-Kapiti plains south of Nairobi National Park declined sharply between the 1970s and the 1990s. Historically, these plains provided migration corridors and rainy-season grazing grounds for vast herds of wildebeest, but land-use changes and fencing of private lands for domestic livestock ranching now threaten seasonal wildlife movements.

Note: Wildebeest observed during the wet season by low-altitude flights are aggregated to squares of 5 km by 5 km and then represented by a circle proportional to their numbers.

Figure 5.2 Trends in Wildebeest Populations in the Kitengela Dispersal Area, 1977–2002

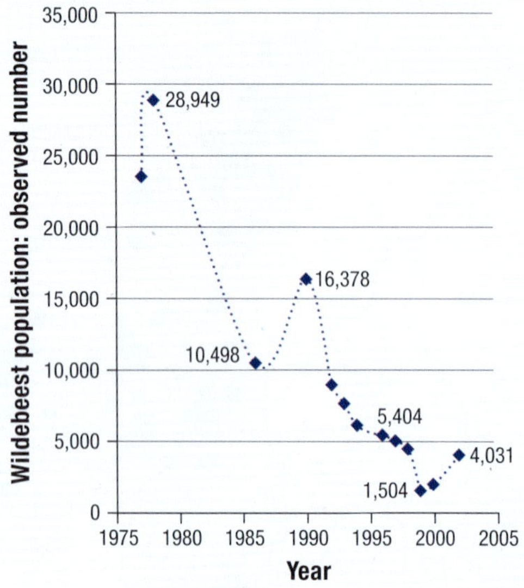

Source: Reid et al. 2006, Gichohi 1996.

The plight of Grevy's zebra exemplifies the broader problems of wildlife in the northern rangelands, including competition for land and water from humans and their domestic livestock, as well as illegal hunting. These problems also affect other wildlife species in northern rangelands, such as Hunter's hartebeest, lesser kudu, and giraffe (de Leeuw et al. 2001, Adanje and Ottichilo 1999).

Map 5.15 shows the distribution of wildebeest populations in the Athi-Kapiti plains of Kenya's southern rangelands. Historically, wildebeest herds numbering in the tens of thousands migrated through this area of southern Kenya, moving between their dry-season grazing grounds in what is now Nairobi National Park and their wet-season dispersal zone in the Athi-Kapiti plains. In recent years, subdivision of land and erection of fences by private landowners has blocked access to traditional wildlife migration routes and dispersal areas.

Wildebeest and zebras that once migrated without hindrance are finding it increasingly difficult to move between their rainy- and dry-season grazing grounds. As shown by Figure 5.2, wildebeest numbers in the Kitengela pastoral area plummeted from a peak of almost 29,000 in 1978 to just over 1,500 in 1999. Although these populations regained ground to reach more than 4,000 in 2002, managing human-wildlife conflicts in the privately owned lands of the Kitengela plains remains a major challenge (see Box 5.2). Wildebeest and other wildlife sometimes trample fences and crops during their attempts to follow their historical migration routes. Landowners also face hardships due to outbreaks of livestock diseases that are carried by migrating wildlife, as well as loss of livestock to lions and other predators that follow migratory herds out of the park and into private ranchland.

Maintaining viable migration corridors and dispersal areas outside of parks, game reserves, and other protected areas is critical to the future of many of Kenya's migratory species, including much of the wildlife that draws thousands of international visitors each year. Although wildlife numbers appear stable in Amboseli National Park, other parks and reserves, such as Masai Mara and Tsavo, are not faring so well (Hansen et al., submitted). The search for land-use management options and other measures to conserve wildlife while also protecting people, their livelihoods, and their aspirations is an urgent priority.

Box 5.2 Creating Economic Incentives to Maintain Wildlife Corridors on Private Lands

One of the biggest challenges facing wildlife conservation in Kenya is how to encourage private landowners to manage rangelands in ways that allow seasonal migration of grazing animals while also providing local income and livelihoods. The presence of wildlife often creates uncompensated financial losses for local people, who share in few if any tourist revenues or other wildlife-related benefits. This need not be the case, however.

The Kitengela Wildlife Conservation Lease Program is demonstrating that appropriate economic incentives can be highly effective in promoting peaceful coexistence of people, livestock, and wildlife. Established in April 2000, the program provides monetary compensation to landowners in the Kitengela area who agree to keep their fallow land unfenced; refrain from cultivating, building on, or selling the designated land; and actively manage their land for wildlife protection and sustainable livestock grazing. At Ksh 725 (about US$ 10.36) per hectare per year, program payments to participating households average Ksh 28,000 (US$ 400) to Ksh 56,000 (US$ 800) annually (Gichohi 2003), a figure close to the income that households earn from rearing livestock. Lease payments are made in three installments at the beginning of each school term to encourage families to use the revenue for school fees, the largest item in the household budgets of many local pastoralists.

Since the program's inception, the land area covered by conservation leases in the Kitengela has grown from 89 hectares in 2000 to more than 1,120 hectares in 2001 and, by July 2003, to about 3,500 hectares held by 115 participating families. Local landowners offering an additional 5,800 hectares for conservation leasing are waiting to join the program, pending availability of sufficient funding (Gichohi 2003). The program has relied on external funding (The Wildlife Trust, Friends of Nairobi National Park, and the Wildlife Foundation) for its initial phase. Plans are underway to raise US$ 1 million to include an additional 25,000 hectares under the program (Gichohi 2003).

A key element behind the success of the Kitengela conservation leasing program has been the partnership formed between the local community and the International Livestock Research Institute (ILRI). With a long history of custodianship of the wildlife that shares their land, local Maasai pastoralists asked ILRI to help them evaluate the economic returns of various land-use options, including conservation leasing. ILRI's analysis showed that the income available from livestock was low and unreliable, and the returns from cropping were even less profitable. ILRI also made available high-resolution maps of income levels throughout Kenya, helping the local Maasai to evaluate their opportunities in a national context (Dawson 2004).

Empowerment of the local Maasai community to understand their economic options and make informed decisions about their future has thus become one of the most important program outcomes. Moreover, education levels have risen significantly among local children, especially girls (Gichohi 2003)—a critically important development benefit widely considered to be the most powerful means of lifting communities out of extreme poverty.

In sum, the Kitengela Wildlife Conservation Lease Program has proved successful in enabling the local community to see wildlife in a more positive light and to share in the economic benefits that wildlife bring to Kenya as a whole. Its long-term sustainability will depend on securing sufficient funding for the cash transfers, particularly in light of rapidly changing land values in such close proximity to Nairobi.

Figure 5.3 Trends in Total Wildlife Population in Laikipia District, 1985–2005

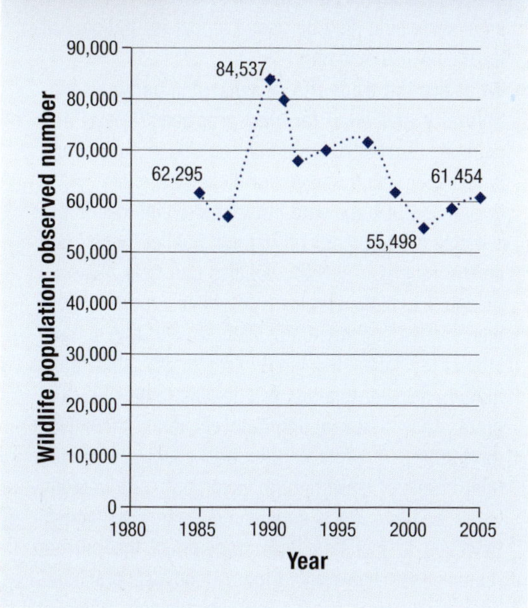

Sources: Grunblatt et al. 1995, Georgiadis and Ojwang' 1997, 1999, 2001, Georgiadis et al. 2003.

Figure 5.4 Trends in Elephant Population, 1973–2005

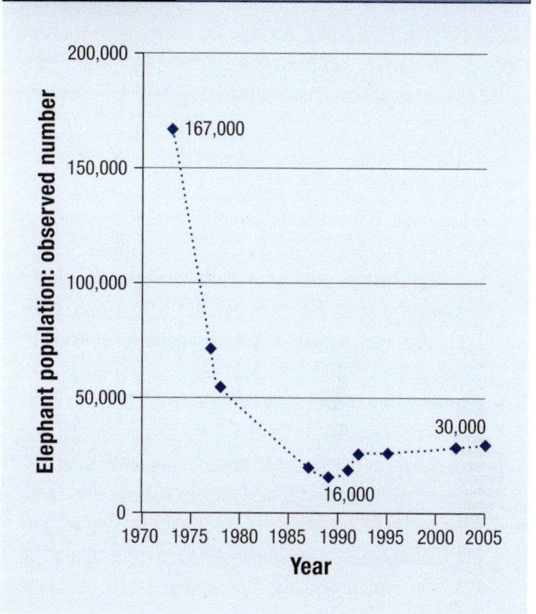

Sources: Said et al. 1995, Kenya Wildlife Service (2005 unpublished data).

Wildlife Areas and Species in Recovery

Although overall national trends show declining wildlife populations, not all the news about Kenya's wildlife is bad. As seen earlier in Map 5.12, local gains have been registered in several areas, with the strongest rises in wildlife populations occurring in Laikipia District, as well as in particular areas close to Masai Mara National Reserve and Amboseli National Park. Moreover, in some species, such as elephants and rhinos, the population crashes experienced in the 1980s and 1990s have bottomed out, and steady progress toward recovery has begun.

The factors driving these and other "success stories" vary from place to place and species to species. In general, many different organizations and initiatives play a role—from national-level policies to community-based efforts and changes in local land-use patterns. Success in maintaining Kenya's unique biological endowment demands an understanding of both national-level trends and a more localized, landscape-level perspective.

Figure 5.3 charts the recovery of wildlife populations in Laikipia District. Steady increases have been seen since 2001, following a decade of persistent drought and serious wildlife declines in the 1990s. Although the trend toward decline has been reversed, wildlife have not yet regained the numbers that prevailed prior to the population crashes of the 1990s. For the most part, the conservation efforts under way in Laikipia have been supported by private and communal landowners rather than through initiatives based in national parks or reserves. These landowners, many of which own large ranches, receive significant economic benefits from wildlife, especially through ecotourism, giving them a powerful incentive to preserve wildlife habitat and prevent poaching.

A second wildlife success story is the recovery of elephant populations (see Figure 5.4). Gains in elephant numbers are being recorded in several different parts of the country, including the rangelands north of Mount Kenya in Laikipia and other Districts, as well as southern rangeland areas near Amboseli and Tsavo National Parks. After suffering huge losses from poaching in the 1970s and 1980s, elephant populations have begun to recover, stabilizing in the 28,000-30,000 range. This figure is less than a fifth of the initial 1970s population, but well above the low points of 16,000-20,000 reached in 1987 to 1991. Crucial factors behind the recovery of Kenya's elephant populations have been the anti-poaching and community conservation efforts led by the Kenya Wildlife Service and others, as well as the international ban on trading in ivory and other elephant products.

Box 5.3 Mapping Biodiversity: Links to National Decision-Making

Below are a few of the questions prominent in current biodiversity-related policy debates in Kenya. For each of these questions, we highlight how additional research and geospatial analysis can help inform the policy development process.

- **What is the status of wildlife in Kenya's rangelands?** Spatial data on wildlife status have been collected systematically since 1977, enabling decision-makers to examine not only national trends but also District- and local-level changes in wildlife populations (see, for example, Map 5.12 and Table 5.1). Data can be examined for individual species or for groups of species, such as grazing animals.
- **Where do we target conservation efforts for rangeland species?** Using spatial information on wildlife status, resource planners and communities can decide where and how to target conservation efforts for selected species, such as elephants or Grevy's zebra. Analysts can combine mapped information on species population and distribution (such as Maps 5.13 and 5.14) with other spatial data, for example on cropping or water points, to identify potential conflict areas, such as areas where crops might be vulnerable to damage by wildlife or areas with competing water demands.
- **How are changing patterns of land use affecting rangeland species?** Satellite images, aerial photos, and map products derived from these sources are useful in identifying land cover and land use in a specific location (see, for example, Map 5.5) and detecting changes over time. By combining information on land cover and land use with data on wildlife distribution and migration, analysts can gain insights into the possible causes of changes in wildlife populations. This can inform policy and program responses, such as setting aside specific areas for wildlife-compatible land uses or targeting payment mechanisms to compensate farmers who use their land in ways that maintain wildlife migration corridors, as is now being done in the Kitengela dispersal area south of Nairobi National Park (see Box 5.2). Such maps and analyses would also be useful inputs for national or local land-use planning.
- **What are the tradeoffs or synergies between biodiversity and local farming practices?** Maps that combine spatial information on local agricultural landscapes (such as average field size, extent of interplanting of trees with other crops, and location of remaining fragments of natural habitat) with data on the ranges of wildlife species can help resource planners identify areas of potential importance for biodiversity conservation, even in densely settled, intensively cultivated lands. Armed with this information, decision-makers will be able to develop programs that create appropriate economic incentives for farmers to grow certain tree species, diversify their crops, or leave natural vegetation buffers along water bodies. Much additional research is needed, however, to increase understanding of the precise relationships between farmers' land-use practices, biodiversity conservation, and sustainable flows of ecosystem services.
- **How could local livelihoods be enhanced by changes in the delivery and valuation of ecosystem services such as wildlife, water, or forests?** Maps that compare biodiversity and related ecosystem services to the spatial distribution of livelihoods and poverty can help decision-makers better understand the relationships between poverty and natural resource use. For example, the expansion of cropping into forested or marginal lands often alters hydrological processes in ways that impact the livelihoods of downstream water users. (See maps in Chapter 8 for an illustration of how livestock keepers, wildlife, and protected areas are impacted by upstream development in the upper Tana River region.)

Box 5.4 Creating a Poverty and Demographic Profile for Rangeland Districts with Different Wildlife Densities

Table 5.2 draws on information about the density of wildlife populations in each of Kenya's 24 rangeland Districts and combines it with District-level data on poverty indicators and other demographic features. The wildlife averages rely on counts from low-altitude flights and include 21 different large grazing animals, such as elephants, giraffes, zebras, wildebeest, and impalas. (To permit comparison of animal densities across herds with different species mixes, wildlife counts are converted to a common unit, known as a *tropical livestock unit* (TLU), equivalent to 250 kg of animal weight.)

Using the table, analysts can look for relationships between a District's poverty indicators and the status of its wildlife populations. Districts are grouped according to the average density of their total wildlife populations in the 1990s, enabling comparison of the demographic and poverty characteristics of Districts with high, medium, and low levels of wildlife. For instance, one could examine whether Districts with high densities of wildlife tend to have higher or lower than average poverty rates, or whether Districts with similar wildlife densities have similar poverty rates.

What Does the Poverty Profile Show?

- The six Districts with the highest wildlife densities have a total population of 1.34 million, of whom 668,000 are poor, for a combined poverty rate of 50 percent.
- Meanwhile, the six Districts with the lowest wildlife densities are home to 1.1 million poor people, among a total population of 1.8 million people, making for an aggregate poverty rate of 61 percent. The other 12 Districts with mid-level wildlife densities have comparable high poverty rates. (Tana River is an exception with a poverty rate of 38 percent—CBS (2005) indicates that this rate, however, is associated with a higher standard error and underestimates the poor.)
- Kenya's top five Districts with the highest wildlife densities have lower poverty rates (just slightly better than the national rural average of 53 percent). Community conservation efforts targeting these Districts should recognize that these communities are slightly better off and choose appropriately tailored approaches and communication strategies. Wildlife interventions in the remaining rangeland Districts, however, are facing a double challenge: levels of well-being that are much below Kenya's rural average and fewer total numbers of animal species.
- Among the Districts with the highest wildlife densities, Laikipia and Kajiado Districts have the lowest poverty rates (39 and 44 percent, respectively). All other Districts in this group have poverty rates greater than 50 percent.
- Correlations between poverty rate and wildlife abundance are difficult to interpret and should not be seen as causal. It is important to bear in mind that data on District-wide averages can mask significant spatial variation. For instance, the low average poverty rate for Laikipia District is a composite of poverty rates for many diverse localities, ranging from relatively affluent areas of high rainfall and fertile pastureland near Mount Kenya and the Aberdare Range, to drier, poorer areas in the central and northern parts of the District. Local poverty rates are likely to diverge substantially between these areas, with much higher incidence of poverty in the latter than in the former.

Similar tables could be constructed comparing other wildlife and poverty indicators, including the indicators of human well-being presented in Chapter 2. For example, comparing poverty maps with areas that show changes in wildlife in more specific locations (below District-level) could help to pinpoint which poor communities could still benefit from wildlife viewing as a revenue source and which had foregone that option. Similarly, a profile that combines poverty and species range maps, such as elephants, could examine whether poor communities and their crops share a greater risk of potentially harmful wildlife interactions.

Table 5.2 People, Poverty, and Wildlife Density in the Rangeland Districts

DISTRICT (average wildlife density in *tropical livestock Units* per sq. km)	AREA (SQ. KM)	NUMBER OF PEOPLE (000)	AVERAGE POPULATION DENSITY (PERSON/ SQ. KM)	NUMBER OF POOR (000)	AVERAGE POVERTY RATE (%)	KSH NEEDED PER MONTH TO REACH POVERTY LINE[1] (MILLIONS)
Very High Average Wildlife Density (>1.8 TLU per sq. km)						
Trans Mara (9.41 TLU per sq. km)	2,848	159	55.8	93	59	26.5
Laikipia (9.13 TLU per sq. km)	9,480	246	25.9	97	39	15.6
Narok (7.03 TLU per sq. km)	15,104	325	21.5	168	52	39.5
Kajiado (5.48 TLU per sq. km)	21,905	306	14.0	136	44	27.0
Lamu (5.20 TLU per sq. km)	6,171	107	17.3	56	53	13.2
Taita Taveta (4.35 TLU per sq. km)	17,109	203	11.9	118	58	33.6
TOTAL 6 DISTRICTS	**72,617**	**1,346**	**18.5**	**668**	**50**	**155.4**
High Average Wildlife Density (0.9 – 1.8 TLU per sq. km)						
Tana River[2] (1.44 TLU per sq. km)	38,218	176	4.6	67	38	8.0
Malindi (1.44 TLU per sq. km)	7,754	214	27.6	140	65	43.4
Kwale (1.38 TLU per sq. km)	8,252	420	50.9	264	63	78.5
Samburu (1.24 TLU per sq. km)	21,074	109	5.2	50	46	10.5
Kitui (1.04 TLU per sq. km)	20,451	490	24.0	345	70	124.0
Garissa[2] (1.01 TLU per sq. km)	44,665	193	4.3	123	64	33.5
TOTAL 6 DISTRICTS	**140,414**	**1,602**	**11.4**	**989**	**62**	**297.9**
Medium Wildlife Density (0.4 – 0.9 TLU per sq. km)						
Machakos (0.88 TLU per sq. km)	6,227	810	130.1	485	60	132.2
Wajir[2] (0.71 TLU per sq. km)	56,702	276	4.9	181	65	49.3
Isiolo (0.66 TLU per sq. km)	25,353	67	2.6	35	52	8.2
Marsabit (0.55 TLU per sq. km)	61,426	100	1.6	53	53	12.5
Makueni (0.53 TLU per sq. km)	7,995	728	91.1	454	62	135.0
Moyale (0.49 TLU per sq. km)	9,589	38	4.0	27	71	9.4
TOTAL 6 DISTRICTS	**167,292**	**2,019**	**12.1**	**1,235**	**61**	**346.6**
Low Wildlife Density (< 0.4 TLU per sq. km)						
Mandera[2] (0.22 TLU per sq. km)	25,987	195	7.5	124	64	33.8
Mwingi (0.10 TLU per sq. km)	10,090	289	28.6	181	63	51.6
Turkana (0.07 TLU per sq. km)	68,380	332	4.9	207	62	66.7
Kilifi (0.07 TLU per sq. km)	4,778	462	96.7	332	72	127.5
Baringo (0.05 TLU per sq. km)	8,645	242	28.0	112	46	23.6
West Pokot (0.04 TLU per sq. km)	9,102	288	31.6	151	53	35.5
TOTAL 6 DISTRICTS	**126,982**	**1,808**	**14.2**	**1,107**	**61**	**338.7**
TOTAL 24 DISTRICTS	**507,305**	**6,775**	**13.4**	**3,999**	**59**	**1,139**

Sources: Poverty and demographic estimates (1999) are WRI/ILRI calculation based on CBS 2002, CBS 2003, CBS 2005. Average wildlife density (1994-96) is an ILRI calculation based on DRSRS 2003, Grunblatt et al. 1995, Grunblatt et al. 1996.

Note: [1] The poverty gap measures the average expenditure shortfall (gap) for the poor in a given administrative area relative to the poverty line. It is a crude estimate of the minimum amount of resources needed to eradicate poverty (see Chapter 2).
[2] Poverty data are by Constituency level and have a higher standard error (see Chapter 2).

SUMMING UP

- Biodiversity—the full variety of plants, animals, and microorganisms found on Earth—is the source of many benefits crucial to human well-being. It provides the underlying conditions necessary for the delivery of ecosystem services.
- With over 6,500 plant species, more than 1,000 bird species, and over 350 species of mammals, Kenya is second in Africa in species richness for these species groups. Bushland, woodland, savanna, and grassland ecosystems together cover 75 percent of Kenya. Agroecosystems extend over 19 percent of its area.
- Of the 60 Important Bird Areas set up to ensure the survival of local and migratory bird species, half were in decline, about a quarter were improving, and eight were stable, as indicated in a 2003-04 assessment.
- Throughout large parts of Kenya's agroecosystems, farmers' fields are interspersed with patches of forests, woodlands, and other vegetation types. This suggests that farmers could manage their lands in ways that support biodiversity. Average field size, extent of tree cover in croplands, and average number of crops grown represent important components of agrobiodiversity in a landscape. Maps of these three indicators show the following: Throughout central and western Kenya, field sizes are small (less than 2 hectares). Croplands with high levels of tree cover are east of the Aberdares, south of Mount Kenya, as well as in Gucha, Central Kisii, and Nyamira Districts. Kirinyaga, Meru Central, and Gucha are the Districts where farmers grow the greatest number of crops at one time.
- Kenya's rangelands support primarily livestock and grazing mammals such as gazelle, wildebeest, zebras, and other wildlife species—an important source of tourism revenues. In 1994-96, livestock numbers dominated the rangelands, representing about 84 percent of all the grazing animals in Kenya's rangelands.
- The total population of large grazing wildlife species in the rangelands declined by 61 percent between 1977-78 and 1994-96. Central parts of Narok District, areas in northern Kajiado District, locations along the Samburu-Laikipia District border, and parts of Isiolo and Garissa Districts experienced the sharpest declines. Competition for land and water from humans and their livestock, as well as illegal hunting, have been behind these declines. For example, maps of water sources, wildlife, and livestock distribution in the northern rangelands show that livestock near water points is "pushing" wildlife away from water.
- Trends for particular rangeland species parallel these aggregated declines. Grevy's zebra, a species unique in the northern rangelands, numbered less than 2,000 in 2004, down from about 13,000 in 1977. Wildebeest in the Kitengela pastoral area south of Nairobi plummeted from almost 29,000 in 1978 to just over 1,500 in 1999.
- Despite these overall and local declines of large grazing mammals, their densities have increased in some areas between 1977-78 and 1994-96. Such gains were near the Masai Mara Game Reserve and Amboseli National Park, as well as in Lamu and Laikipia Districts. In the latter District, private and communal landowners have been a major contributor to this trend reversal, rather than initiatives based on new government policies.
- After suffering huge losses from poaching in the 1970s and 1980s, elephant populations have begun to recover, stabilizing around 28,000-30,000. Antipoaching and community conservation efforts, as well as the international ban on trading in elephant products, have been the crucial factors behind this recovery.

In Chapter 6

CONTENTS

LIST OF MAPS

AUTHORS AND CONTRIBUTORS

Sandra van Dijk (ILRI)
Robin Reid (ILRI)
Michael Arunga (ILRI)
Moses Ikiara (KIPPRA)
Janet Nackoney (WRI)
Florence Landsberg (WRI)
Karen Holmes (consultant)
Greg Mock (consultant)
Norbert Henninger (WRI)
Fred Kaigua (KATO)
Sammy Towett (KWS)
Alex Odero (KWS)
Mohammed Said (ILRI)
Ololtisatti Ole Kamuaro (ILRI)
Dan Omolo (ILRI)
Ville Vuorio (ILRI)
Dickson Ole Kaelo (IRLI)
Dan Tunstall (WRI)
Hyacinth Billings (WRI)
Carolina de Rosas (WRI)
Jamie Worms (WRI)

WHAT THIS CHAPTER SHOWS

To demonstrate that Kenya's tourism economy depends on a foundation of healthy ecosystems, this chapter highlights key ecosystem components and their uses. The first section presents ecosystem assets important for nature-based tourism: maps show the system of protected areas, areas rich in birdlife and bird biodiversity, the locations where wildlife with high 'viewing value' concentrate, and a more detailed view of ecosystem assets along the Indian Ocean coast. The second section looks at the patterns of use of these ecosystem assets. A brief overview of tourism infrastructure is followed by a series of graphs summarizing recent trends in numbers, revenue, and distribution of visitors among the main tourist attractions. The chapter concludes with a more detailed examination of visitor and revenue patterns for Kenya's protected areas.

6

Tourism

Regarded by many as the "jewel of East Africa," Kenya is one of the world's foremost tourist destinations. Tourism in Kenya is based primarily on the country's stunning natural attractions, including magnificent wildlife in their native habitat as well as some of Africa's finest beaches. This unique natural endowment has turned Kenya's tourism industry into a leading economic sector, generating revenues of almost Ksh 49 billion (US$ 700 million) in 2005 and directly employing 176,000 people—about 10 percent of all jobs in the formal sector (CBS 2006).

LINKS BETWEEN ECOSYSTEM SERVICES AND TOURISM IN KENYA

About 70 percent of visitors to Kenya come to enjoy the country's natural beauty and engage in nature-based activities, such as wildlife viewing; hiking; and enjoying sun, sand, and surf on its beautiful beaches (see Figure 6.1). A common factor linking these activities and places is their dependence on healthy ecosystems and the services they provide, including clean air and water, scenic landscapes and vistas, and diverse assemblages of animal and plant species.

One of the paradoxes of such nature-based tourism is that, in the absence of thoughtful, forward-looking management, the relentless pressure of human visitors can degrade the very ecosystem assets that attract tourists in the first place. Over-concentration of tourist activities and infrastructure, notably along some of Kenya's coastal beaches as well as in certain national parks and game reserves, has led to environmental damage as well as a decline in the quality of the tourism experience. Along the coast, beaches have been seriously degraded and polluted, coral reefs and mangrove forests have been substantially damaged or destroyed, and marine species have been harmed. In some game parks, vegetation has been degraded, wildlife behavior has been disrupted, and resources have been overused (Ikiara and Okech 2002).

These troubling trends have helped to erode Kenya's tourist appeal and contributed to the challenges facing the country's tourism industry. In the late 1990s, Kenya experienced steep declines in the tourism sector, with revenues falling about 20 percent annually between 1996 and 1998 (Ikiara and Okech 2002). Domestic instability, combined with widespread fear of global terrorism, depressed tourism activity and earnings well into the first decade of the 21st Century (Ikiara 2001; Belau 2003).

Building a Sustainable Tourism Industry

More recently, the downward slide of the tourism industry has been at least partially reversed, with international arrivals rising by about 40 percent from 2002 to 2005 (see Figure 6.1), and annual tourism earnings more than doubling during the same period (CBS 2004; CBS 2006). However, the industry's future is far from assured, as it confronts strong competition from other wildlife tourism destinations (such as Botswana, South Africa, and Tanzania), as well as ongoing domestic challenges, including electricity and water shortages, environmental degradation, and declining wildlife populations (Ikiara 2001).

It is incumbent upon decision-makers in Kenya's public and private sectors to find the right mix of policies and investments that can foster the growth of sustainable tourism. Tourism marketing continues to focus on traditional attractions thereby perpetuating over-concentration at some sites (Ikiara and Okech 2002). New approaches that can help attract and allocate investment in underutilized areas are needed, while simultaneously protecting the unique landscapes, wildlife, and other ecosystem assets that draw higher-spending tourists. Finding ways to direct a larger share of tourism proceeds to benefit local people and communities is also critically important.

This chapter highlights the role of Kenya's ecosystems in supporting a vibrant tourism sector. It takes a look at the range of ecosystem assets that are important for the industry, including Kenya's network of parks and protected areas, as well as the spatial distribution of selected wildlife species with high 'viewing value.' Later sections focus on patterns of human use, investment, and revenue generation.

KEY ECOSYSTEM ASSETS FOR THE TOURISM SECTOR

Tourism has a long and notable history in Kenya, pre-dating independence. As early as the 1930s, large numbers of overseas visitors had begun traveling to Kenya on big-game hunting expeditions (United Nations 2002). Today, the typical international visitor still comes to Kenya in search of big game—this time, armed with binoculars and a digital camera. Most overseas tourists spend a night in Nairobi on arrival, embark on a two-day or so safari to view wildlife, and devote the rest of their holiday to a longer stay on the coastal beaches (Ikiara and Okech 2002).

Thus, Kenya's tourism potential is inextricably linked to its natural assets. From the white sand beaches and teeming coral reefs of the Indian Ocean coast to the summits of its majestic mountains, Kenya has been endowed with diverse landscapes of dramatic natural beauty. Running through the country is the most spectacular stretch of the Great Rift Valley, with its stunning geology and its alkaline and freshwater lakes alive with birdlife. The savannas of southern Kenya are home to national parks and game reserves, such as Amboseli, Masai Mara, and Tsavo, that provide unparalleled opportunities for viewing wildlife.

More than 80 of Kenya's top 120 tourist destinations are national parks and wildlife reserves, which encompass some 45,000 square kilometers, or about 8 percent of Kenya's total land area (GoK 1995). Most parks and wildlife reserves are located in rangeland ecosystems (see Map 6.1), which tend to be the least modified, wildest places in Kenya. Dotting the mountain slopes and foothills of Kenya's highland landscapes are several forest reserves, mostly surrounded by more densely settled agricultural lands.

Kenya also contains colorful, diverse birdlife, and bird watching is a small but growing segment of the tourist industry. Some 60 Important Bird Areas (IBAs), covering 5.7 million hectares (10 percent of the country's land area), have been designated,

Map 6.1 **Topography, Protected Areas, and Important Bird Areas**

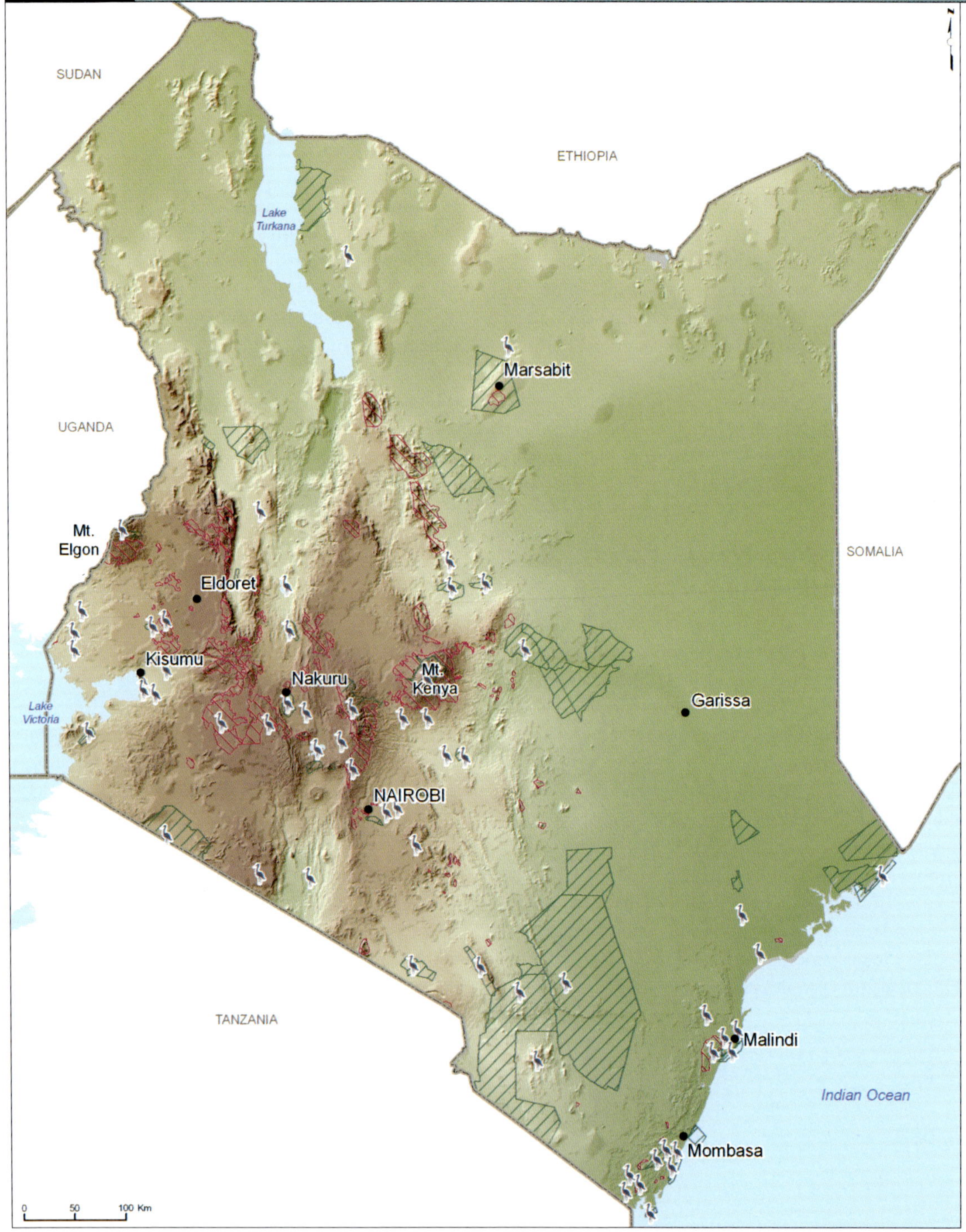

Sources: Cities (SoK and ILRI 2000), water bodies (FAO 2000), parks and reserves (IUCN and UNEP/WCMC 2006), 250-meter Digital Elevation Model (SoK, JICA, and ILRI 1996), and centroid of Important Bird Areas (Fishpool and Evans 2001).

The topography of Kenya encompasses dramatic landscapes and magnificent scenery, from the Great Rift Valley to Mount Kenya and the central highlands to the wide, flat vistas of the southern savannas. To safeguard these landscapes and other natural assets, Kenya has invested in a network of protected areas, including national parks and game reserves throughout the country (green hatched areas), as well as forest reserves, located mostly in the central highlands (red hatched areas). Concentrated along the southern coast and in the highlands are Kenya's 60 Important Bird Areas (indicated by blue bird symbols), which are prime spots for bird watching and are globally important for bird conservation.

Note: The map depicts each Important Bird Area by a point in the center of its associated area. Some IBAs are much smaller than the point shown in this national map and others cover a much larger area, such as IBAs associated with the large protected areas of Masai Mara or the two Tsavo National Parks. IBAs range from 1 hectare to more than 1 million hectares in size (Bennun and Njoroge 1999).

Important bird areas

OTHER FEATURES

National parks and reserves

Forest reserves

Water bodies

indicating sites of international significance for the presence of threatened species, irreplaceable bird populations, or exceptionally large numbers of migratory birds (Bennun and Njoroge 1999).

Some kinds of tourism are more closely linked to ecosystem services than others. Different kinds of tourism place different demands on different types of services. On one end of the spectrum is the tourist who is specifically seeking a 'wilderness experience;' at the other is, for example, the tourist who enjoys being part of a crowd at the beach. Thus, the type of tourism determines the demand for ecosystem services. It also determines the number and density of tourists who can enjoy the recreational, spiritual, and aesthetic services provided by a given ecosystem without compromising these services (Scholes and Biggs 2004).

Spatial Distribution of Wildlife with High Tourism Value

Viewing wildlife in its natural habitat is the primary motivation for about 80 percent of international visitors to Kenya (Filion et al. 1994; Ikiara and Okech 2002). Different ecosystems support different wildlife species (see Map 6.2), and well-informed tourists can choose their destinations accordingly.

For instance, the open savanna and bush woodland of Tsavo National Park support elephants, buffaloes, lions, antelopes, gazelles, giraffes, zebras, and a few rhinos; crocodiles, hippos, and a wealth of birdlife also make their homes there. Visitors to densely wooded mountain slopes can see forest-dwelling species, including the black leopards and the black and white colobus monkeys that inhabit the lower slopes of Mount Kenya. Still other species are found near Kenya's mountain lakes, such as the giant flocks of flamingoes at Lake Nakuru or Lake Bogoria, and the egrets, herons, and fish eagles of Lake Baringo (iExplore 2006).

To a large extent, wildlife tourism in Kenya is driven by the 'big five' species: lions, leopards, elephants, rhinoceros, and buffalo. The emphasis on this small group of highly 'charismatic' species originated in the days of big game hunting, when they were considered especially dangerous and

Map 6.2 **Spatial Distribution of Selected Species with High 'Viewing Value,' 1994-96**

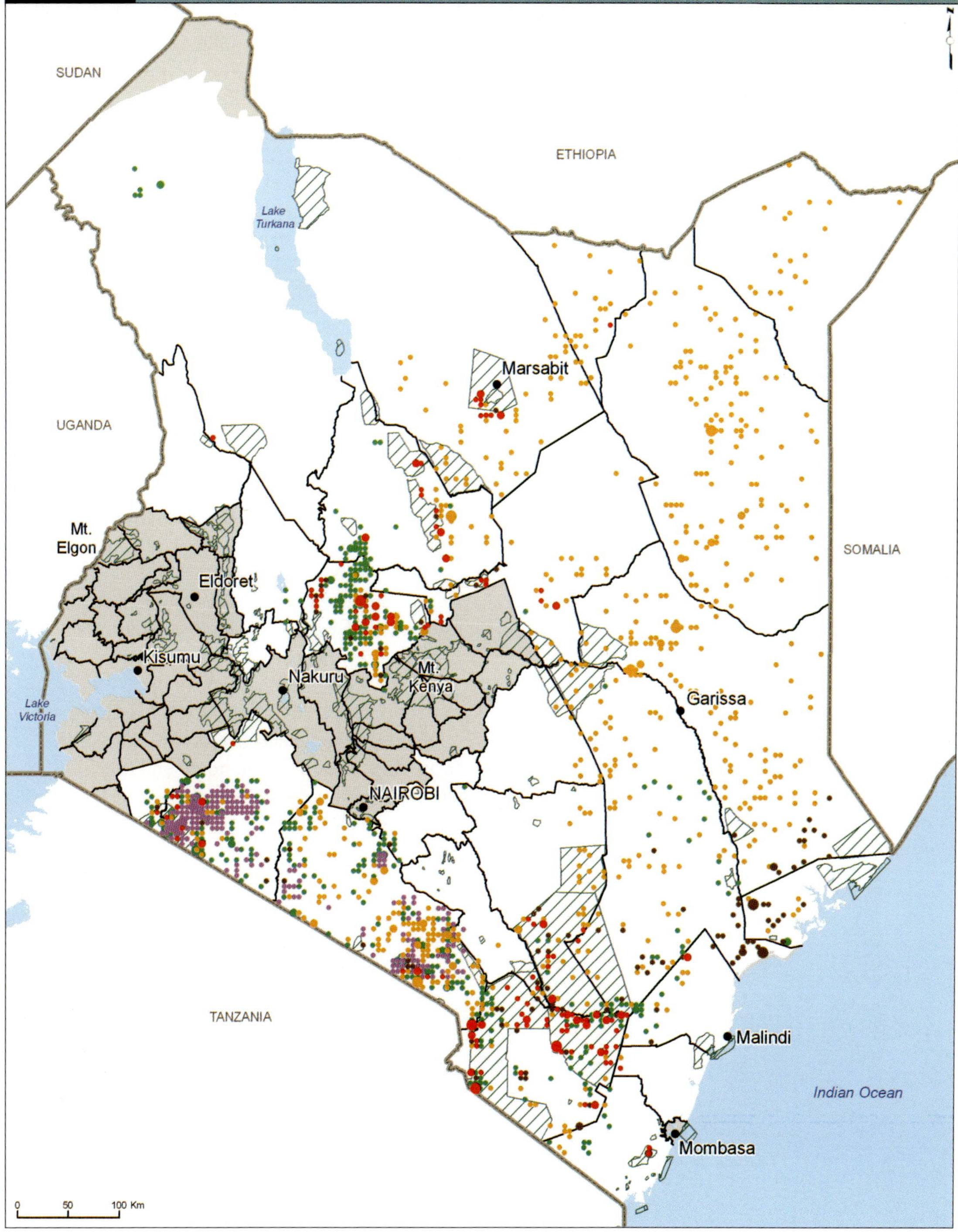

Sources: Administrative boundaries (CBS 2003), cities (SoK and ILRI 2000), water bodies (FAO 2000), parks and reserves (IUCN and UNEP/WCMC 2006), and 1994-96 wildlife numbers (ILRI calculation based on DRSRS 2003, Grunblatt et al. 1995, 1996).

Wildlife is broadly distributed across Kenya, but particular species often exhibit a specific pattern of spatial distribution. For instance, giraffe populations (indicated by golden dots) are found throughout Kenya's rangeland Districts, while elephants are found in the rangelands of Laikipia District as well as Amboseli, Marsabit, and Tsavo National Parks (indicated by red dots). Note that the distribution of some highly charismatic species is not shown, as data on animals that are nocturnal (e.g., lions and leopards) or extremely rare (e.g., rhinos) are not easily collected by aerial survey.

Note: The wildlife counts came from a rangeland census using low-altitude flights. Animals are aggregated to squares of 5 kilometers by 5 kilometers.

ELEPHANT (number)
- > 2,000
- 700 - 2,000
- 90 - 700

BUFFALO (number)
- > 5,000
- 2,000 - 5,000
- 40 - 2,000

GIRAFFE (number)
- > 1,000
- 600 - 1,000
- 50 - 600

WILDEBEEST (number)
- > 20,000
- 10,000 - 20,000
- 14 - 10,000

ZEBRA (number)
- > 3,000
- 800 - 3,000
- 15 - 800

No observation of any of these species

Not sampled

OTHER FEATURES

District boundaries

National parks and reserves, forest reserves

Water bodies

Map 6.3 **Migration of Wildebeest and Zebra in the Mara-Serengeti Ecosystem**

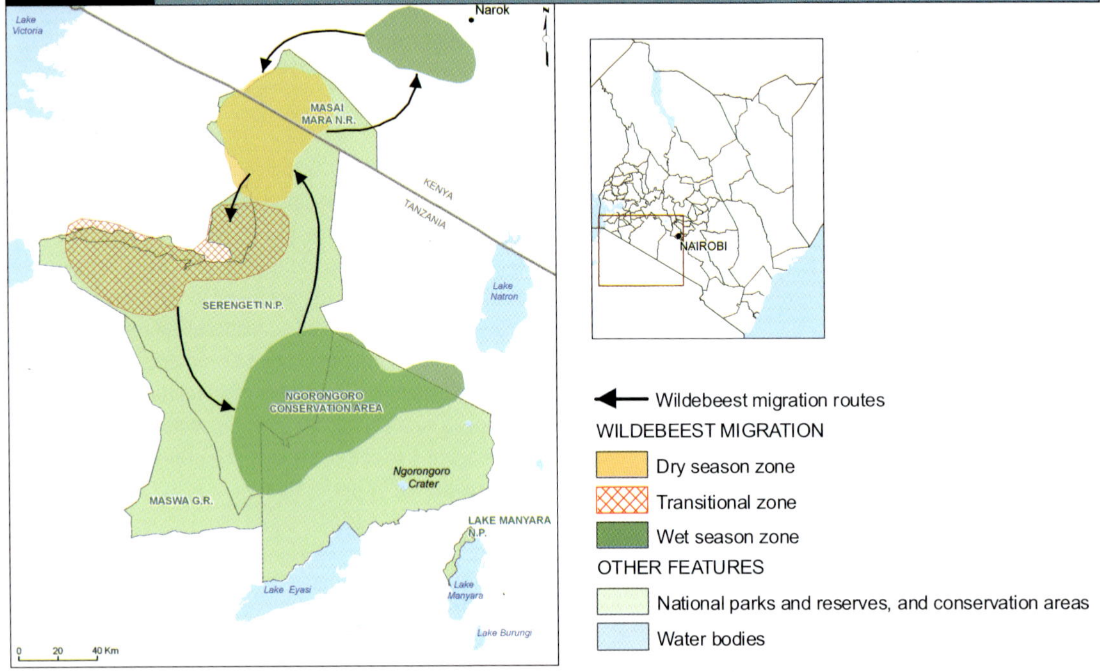

Sources: Water bodies (FAO 2000), parks and reserves (IUCN and UNEP/WCMC 2006), and wildebeest migration areas and routes (ILRI digitization based on Serneels and Lambin 2001).

Wildebeest and zebra follow seasonal rainfall patterns as they migrate between the Serengeti plains of Tanzania and the rangelands of Kenya's Narok District. Masai Mara National Reserve provides a source of forage and water for these animals during the dry season (gold-shaded area), while rangelands north of the reserve (dark green-shaded area), near Narok Town, serve as a wet-season grazing area. However, conversion of these rangelands to cropland is disrupting migration patterns, leading to declining wildlife populations.

thus highly prized as the hunter's quarry (Scholes and Biggs 2004). Today, their popularity is perpetuated by marketing. However, promoting a select group of Kenya's wildlife contributes to over-concentration of tourists in a few locations, leading to an erosion in the quality of the tourism experience as well as endangering wildlife and ecosystem integrity (Ikiara and Okech 2002). Meanwhile, other parks and protected areas, richly endowed with different but equally fascinating species, remain underutilized.

A second major wildlife attraction for tourists is the annual migration of wildebeest and zebra in the Mara-Serengeti ecosystem, when thousands of animals risk their lives crossing the Mara River in search of lush green grass. Unfortunately, land conversion north of Masai Mara National Reserve, from open range to wheat farms, is interfering with the northern loop of this migration (see Map 6.3). As a result, wildlife numbers are on the decline, with wildebeest populations in the Masai Mara ecosystem falling from 120,000 in 1977 to 31,000 in 2002 (Ottichilo et al. 2001; Ojwang et al. 2006). While Kenya has gained in food production, changing land use patterns have come at a price: undermining one of the area's principal tourist attractions.

Coastal Ecosystems that Support Tourism

Soon after independence, Kenya shifted the focus of its investments in hotels and tourist infrastructure from big game hunting to beach tourism. Along Kenya's 530 kilometers of Indian Ocean coastline are ecosystems containing a diverse array of assets that are important for tourism, such as sandy beaches and coral reefs—all rich in marine life and supporting a large population of seabirds (Maps 6.4 and 6.5).

Traditionally, Kenya has targeted high-density, mass-market beach tourism that relies on a relatively limited set of ecosystem services—primarily sand, sea, and sun (Ikiara and Okech 2002). Although the range of required ecosystem services may be small, the magnitude of the environmental pressures resulting from high-volume, low-yield coastal tourism can be great.

To date, development of coastal tourism in Kenya has proceeded without much regard for environmental limits or the carrying capacity of coastal ecosystems. Tourism-related impacts have been aggravated by over-concentration of tourism infrastructure and activities in particular areas, notably the beaches of the North Coast (i.e., from Mombasa to Kilifi) and Diani Beach on the South Coast (NEMA 2003).

However, some types of coastal tourism require lower visitor densities and a broader, more diverse set of ecosystem services. For instance, dive tourism, a lucrative segment of the global tourism industry, requires clean water, intact reefs, and diverse, colorful species of fish and marine invertebrates.

Visitors to Kenya's coast can enjoy a wide range of lower-density activities, such as snorkeling, scuba diving, deep sea fishing, and dhow trips for watching dolphins and dugongs (an herbivorous marine mammal related to the manatee). To protect the ecological integrity of Kenya's coral reefs, the government has designated six marine reserves—Kisite, Kiunga, Malindi, Mombasa, Mpunguti, and Watamu—encompassing a significant portion of the reef and its surrounding waters.

Kenya's coastal ecosystems also contain sites offering fine opportunities for wildlife viewing, such as the remnants of coastal forests that once covered much of East Africa's Indian Ocean shoreline. These areas are extremely important ecologically, and some have untapped potential for development of low-density, ecologically sensitive tourism. For example, in the Arabuko-Sokoke Forest, less than 10 kilometers inland from Malindi, over 260 species of birds have been recorded, including 6 globally threatened species (Arabuko-Sokoke Forest Management Team 2002).

Also a short distance inland from the coast are areas that provide habitat for species with high viewing value. For instance, the Shimba Hills Reserve, about 15–20 kilometers inland from the coast, is famous for its sable antelope, the last remaining breeding population of these animals in the country. The reserve also contains a sizeable leopard population (Kenya.com 2006; iExplore 2006).

Map 6.4 Northern Coast: Ecosystem Assets and Infrastructure Important for Tourism

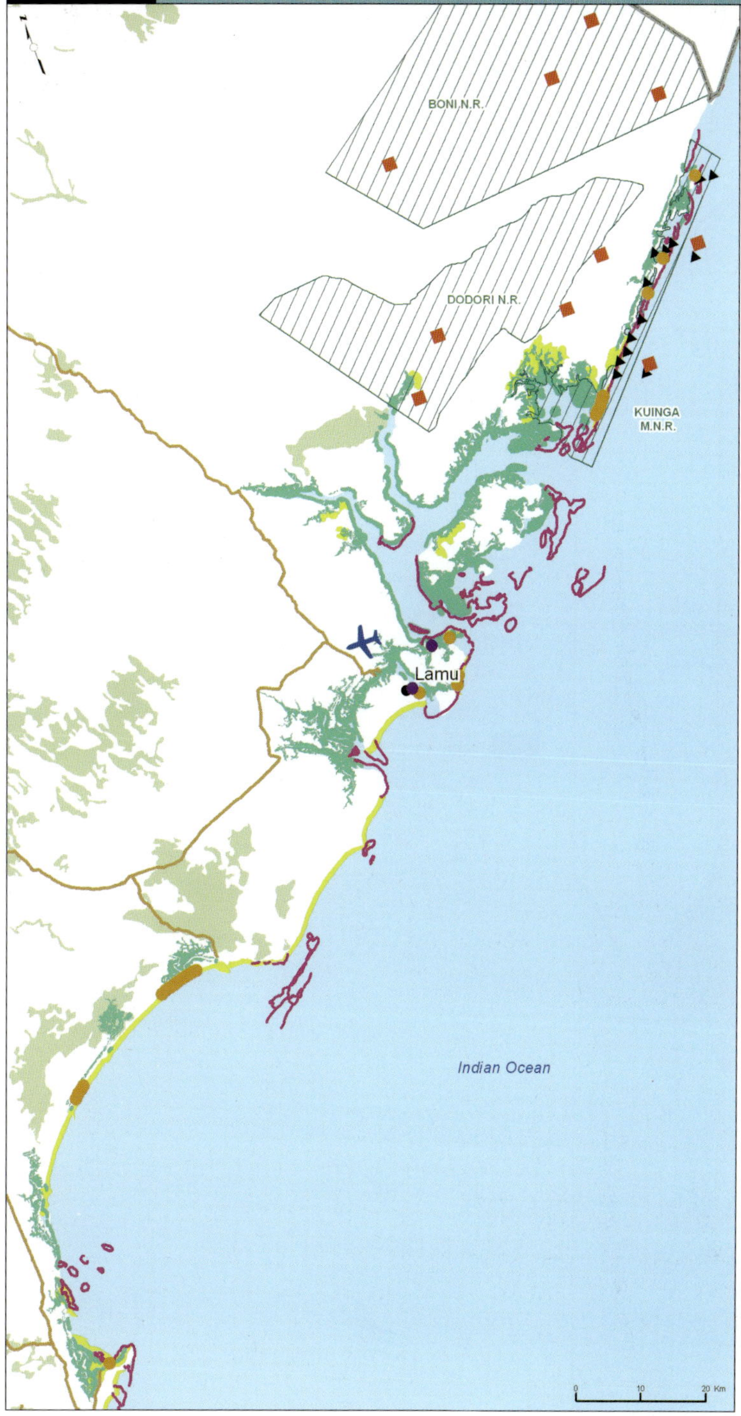

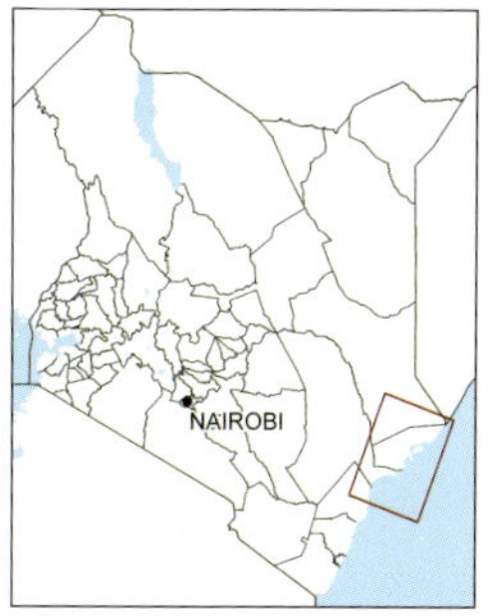

Sources: Cities (SoK and ILRI 2000); water bodies, closed forests, and sand beaches (FAO 2000); parks and reserves (IUCN and UNEP/WCMC 2006); major airfields (NIMA 1997); sable antelope sites, marine mammal sites, and location of hotels (UNEP 1998); major roads, coral reefs, mangroves, and turtle nesting sites (UNDP et al. 2006); and number of hotel beds (ILRI/WRI calculation based on RoK 2003, UNEP 1998).

Map 6.5 Southern Coast: Ecosystem Assets and Infrastructure Important for Tourism

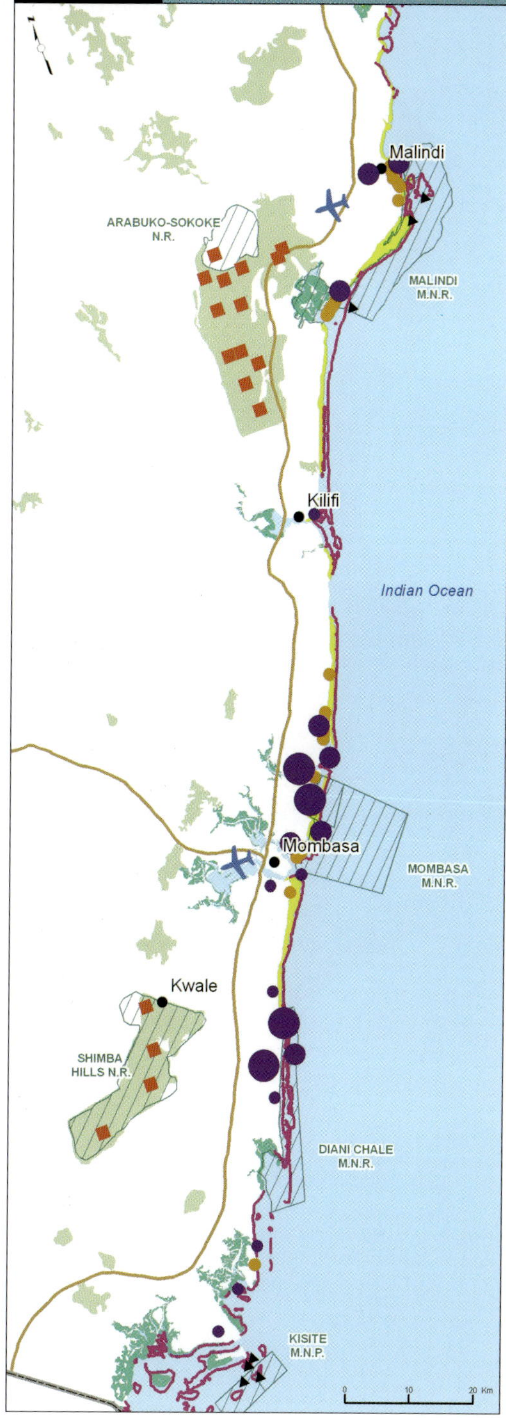

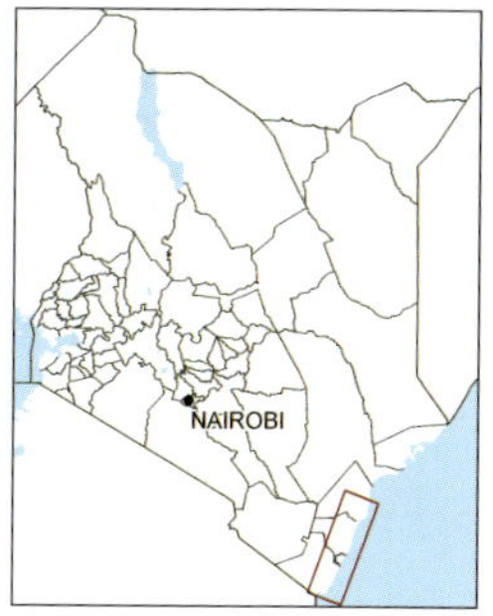

Sources: Cities (SoK and ILRI 2000); water bodies, closed forests, and sand beaches (FAO 2000); parks and reserves (IUCN and UNEP/WCMC 2006); major airfields (NIMA 1997); sable antelope sites, marine mammal sites, and location of hotels (UNEP 1998); major roads, coral reefs, mangroves, and turtle nesting sites (UNDP et al. 2006); and number of hotel beds (ILRI/WRI calculation based on RoK 2003, UNEP 1998).

Kenya's coast contains numerous ecosystem assets that attract tourists, including sandy beaches (yellow-shaded areas) and coral reefs (in purple). The coast also offers opportunities for wildlife viewing, including trips to visit turtle nesting sites (gold dots) and watch dolphins (black triangles), as well as inland visits to nearby forested areas (light green areas) that are home to the rare sable antelope (orange squares). Infrastructure for tourist accommodation (purple dots) is concentrated in and around Mombasa, the Diani Beach area, and Malindi.

TOURIST ATTRACTIONS
- Turtle nesting sites
- Sable antelope sites
- Dugong sites and dolphin schools
- Sand beaches
- Coral reefs

HOTEL ACCOMMODATION
(number of beds)
- > 1,500
- 500 - 1,500
- 20 - 500

FORESTED AREAS
- Closed forests
- Mangroves

OTHER FEATURES
- Major airfields
- Major roads
- National parks and reserves

NATURE-BASED TOURISM: INFRASTRUCTURE, VISITOR AND REVENUE TRENDS, AND SPATIAL DIVERSIFICATION

Abundant wildlife, spectacular landscapes, and beautiful beaches are not enough to sustain a vibrant tourism sector. Tourism infrastructure is crucial as well. Investments are needed to develop and maintain a wide variety of services, including transport systems; water treatment and distribution facilities; communications services; tourist accommodations; and a system of parks, game reserves, and other protected areas.

Fortunately, many of Kenya's parks and reserves have well-developed infrastructure, including the roads leading to the park as well as roads and accommodations located inside the park. Several popular parks are within a day's drive of Nairobi, including Lake Nakuru, Hell's Gate, Lake Naivasha, the Aberdare, and Mount Kenya National Park (Map 6.6). The highlands, where most of Kenya's population resides, has a good network of roads and airstrips serving most major tourist destinations. More distant attractions, such as Masai Mara National Reserve, Amboseli and Tsavo National Parks, and coastal destinations near Mombasa or Malindi are also quite accessible by air or road.

On the other hand, parks requiring significant travel time by car and with a less developed tourism infrastructure capture only a small share of Kenya's visitors (see Table 6.1). This includes Marsabit National Park and Reserve in the northern rangelands, Central Island National Park in Lake Turkana, and Mount Elgon National Park close to Uganda.

The type and location of tourism infrastructure is to a large extent a legacy of Kenya's past investment decisions. To date, these investments have resulted in over-concentration of tourists in certain areas of the country (Ikiara and Okech 2002). Unfortunately, crowding tourists into a few parks and reserves diminishes the quality of the tourism experience and lessens Kenya's appeal for international visitors. It also concentrates the costs and benefits of

Map 6.6 National Parks, Reserves, and Other Tourism Infrastructure

Sources: Cities (SoK and ILRI 2000), water bodies (FAO 2000), parks and reserves (IUCN and UNEP/WCMC 2006), major roads (SoK and ILRI 1997), and campsites, tented camps, hotels, and lodges (approximately placed by ILRI/WRI based on MacMillan Education 1993, UNEP 1998, RoK 2003).

Key components of tourism infrastructure, such as roads, airstrips, and lodging, are well developed in certain parts of Kenya, including the highlands, sections of the Indian Ocean coast, and near popular parks and reserves.

Note: The sites showing tourist accommodations are a rough approximation based on readily available publications. The paucity of spatially referenced data may have resulted in omission of sites. In addition, a single symbol underrepresents the greater number of hotels and bed capacity in certain areas such as Nairobi and the coastal region, which together captured about 75 percent of total hotel occupancy in 2005 (CBS 2006).

TOURIST ACCOMMODATION

- Camp sites and tented camps
- Hotels and lodges

OTHER FEATURES

- Major roads
- National parks and reserves, forest reserves
- National parks and reserves with visitor data
- Water bodies

tourism development in limited areas of the country, which can entrench existing social and economic inequities. Spatial diversification of infrastructure investment can help to protect wildlife and ecosystems from damage by too many visitors, while at the same time helping to strengthen the economic performance of the tourism sector.

Trends in the Tourism Economy and Visitor Distribution

Travel and tourism are leading economic activities in Kenya. Tourism contributes to the economy not only through direct earnings (hotel revenues, park entrance fees, etc.) but also through indirect economic effects, such as increased demand for goods and services in other economic sectors, such as agriculture, transport, entertainment, and textiles. These indirect contributions greatly magnify tourism's economic impact. Overall, the tourism sector accounted for 8.7 percent of Kenya's gross domestic product (GDP) and ranked as the third largest foreign exchange earner in 2002 (Ikiara and Okech 2002). Moreover, tourism is identified in Kenya's *Economic Recovery Strategy* (GoK 2003) as a potentially important contributor to poverty reduction (see Box 6.2).

However, Kenya's tourism earnings have been somewhat volatile in recent years. Since 1980, the number of international visitors has increased dramatically, from about 400,000 in 1980 to almost 1.5 million in 2005 (Figure 6.1). However, the growth curve has not always been smooth. Tourism earnings grew rapidly in the early 1990s, but fell steeply in the latter half of the decade (Figure 6.2). Particularly in the late 1990s, Kenya's tourism industry faced downward trends in per capita spending, average length of stay, hotel occupancy rates, and quality of service (Ikiara 2001; Ikiara and Okech 2002). Another downturn hit the industry in the early years of the current decade, when concerns about global terrorism depressed worldwide demand for international travel (Belau 2003).

In more recent years, the tourism economy has improved significantly, with a growing number

Figure 6.2 **Tourism Earnings, 1980–2005**

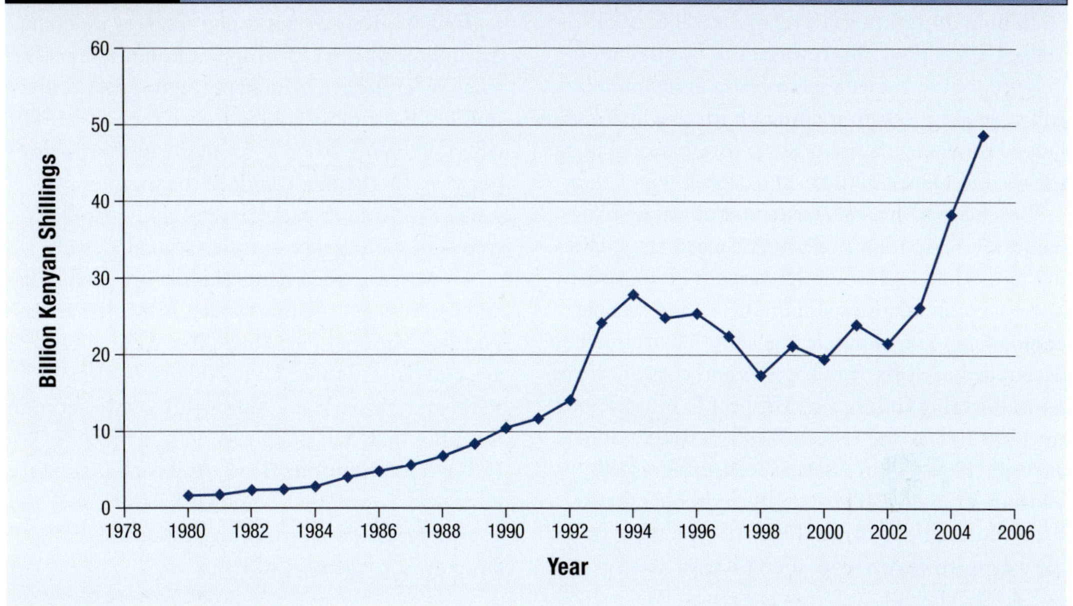

Sources: Ikiara 2001, Ikiara and Okech 2002, CBS 2004, 2006.

Figure 6.1 **International Arrivals by Purpose, 1980–2005**

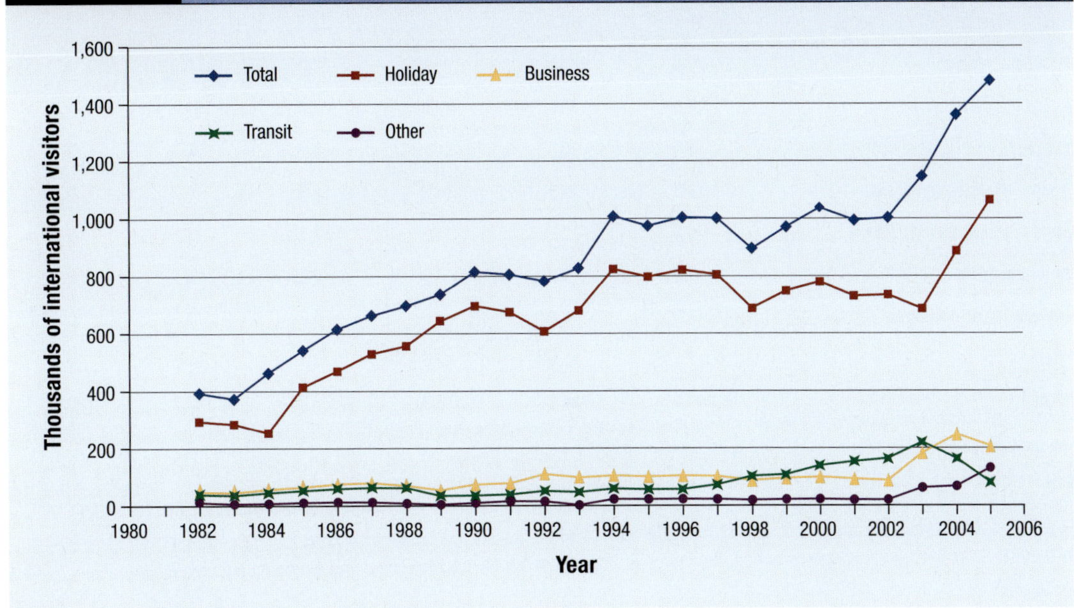

Sources: Ikiara 2001, Ikiara and Okech 2002, CBS 2004, 2006.

Figure 6.3 **Distribution of Occupied Bed-Nights Among Tourist Attractions, 1980–2005**

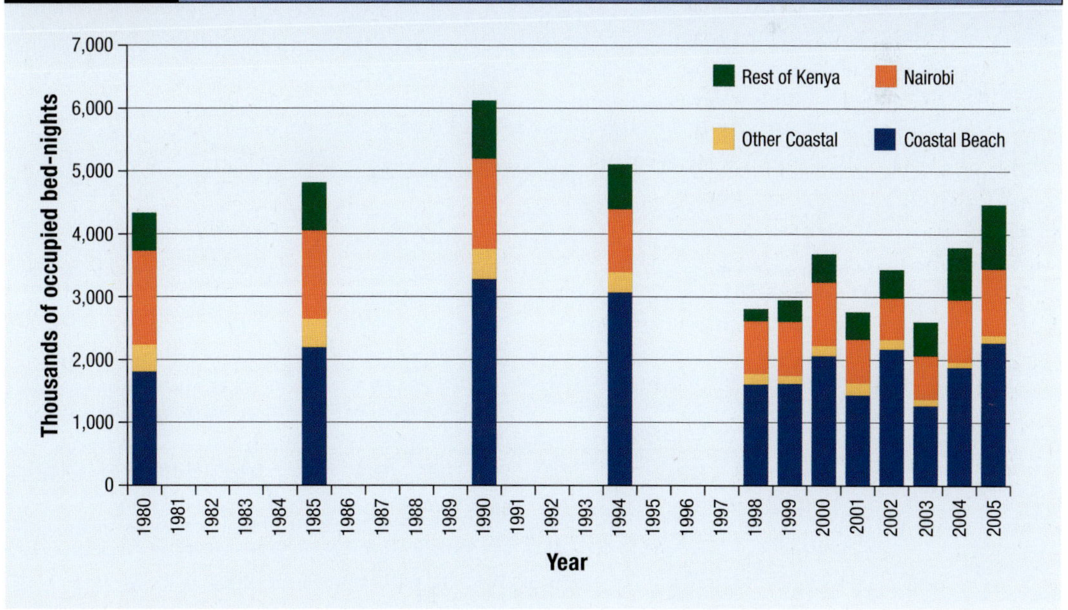

Sources: Ikiara 2001, Ikiara and Okech 2002, CBS 2004, 2006.

of international visitors and higher earnings. For instance, 2005 tourism revenues totaled almost Ksh 50 billion, up 125 percent relative to 2002 (CBS 2004; CBS 2006). This reversal can be attributed in large part to Kenya's increased political stability and stronger marketing efforts, both of which have helped to create a more positive international image (Ikiara and Okech 2002).

Beaches and coastal ecosystems continue to account for a large share of tourism earnings, including more than half of all nights spent by tourists in hotel accommodations (Figure 6.3). However, an emphasis on mass tourism has led to environmental deterioration of Kenya's beaches and coastal ecosystems, lowering the country's appeal to international travelers (Ikiara and Okech 2002). Stronger efforts to protect ecosystem assets as well as increased investment in new, high-quality, less concentrated tourism development will likely be needed in order to sustain strong earnings along Kenya's coast.

Use of Protected Areas: Visitor Trends and Revenue Generation

In 2005, Kenya's parks and reserves welcomed 2.1 million visitors, the highest number registered since records have been kept (Figure 6.4). This number has almost doubled since the early 1980s, when the figure stood at around 1 million visitors per year. Over time, trends in the number of visitors to Kenya's parks have roughly paralleled trends in the number of international arrivals.

However, a small handful of Kenya's 84 parks and reserves get the most visits. Just three areas—Nairobi National Park (including Animal Orphanage and Safari Walk), Lake Nakuru National Park, and Masai Mara National Reserve—account for more than half of all visitors (see Table 6.1). If Tsavo East National Park, Amboseli National Park, and Tsavo West National Park are also considered, then six parks are responsible for close to 72 percent of all visits.

Figure 6.4 **Visitors to National Parks and Game Reserves, 1980–2005**

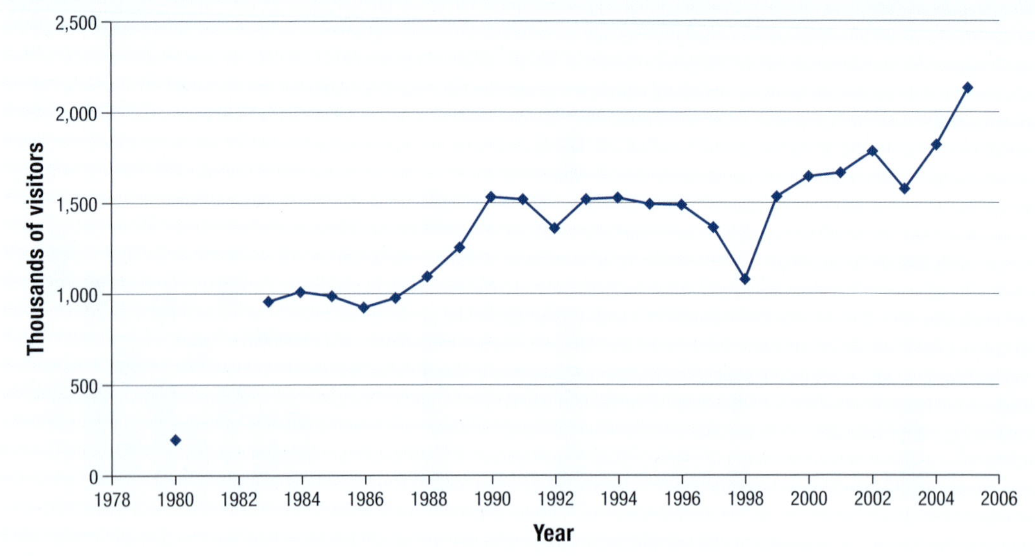

Sources: Ikiara 2001, Kahata and Imbanga 2002, Ikiara and Okech 2002, CBS 2004, 2006.

Table 6.1 **Number of Visitors to Parks and Game Reserves, 2001-05**

	VISITORS (000)					
	2001	2002	2003	2004	2005[1]	2005 (PERCENT)
Nairobi TOTAL	366.2	459.3	342.9	419.9	485.2	22.7
Nairobi Animal Orphanage	*151.1*	*254.5*	*205.3*	*239.4*	*257.8*	*12.1*
Nairobi Safari Walk	*113.5*	*114.4*	*66.3*	*88.0*	*127.5*	*6.0*
Nairobi National Park	*101.6*	*90.4*	*71.3*	*92.5*	*99.9*	*4.7*
Lake Nakuru National Park	209.4	229.8	216.7	257.0	344.6	16.2
Masai Mara National Reserve	207.2	231.1	233.0	240.0	285.2	13.4
Tsavo East National Park	132.7	152.8	119.2	158.5	180.1	8.4
Amboseli National Park	91.5	92.0	54.7	101.6	126.2	5.9
Tsavo West National Park	78.7	76.3	62.6	92.7	105.7	5.0
Haller Park	87.2	87.0	99.9	101.2	100.8	4.7
Kisumu Impala Sanctuary	96.9	117.7	69.6	63.3	87.9	4.1
Lake Bogoria National Reserve	59.6	18.7	64.7	64.7	65.7	3.1
Kisite Marine N.P./Mpunguti Marine N.R.	45.7	47.1	35.9	51.7	59.2	2.8
Aberdare National Park	40.5	41.5	30.3	44.0	48.3	2.3
Mount Kenya National Park	26.3	27.9	25.5	27.7	39.5	1.9
Mombasa Marine National Park	29.1	30.5	31.4	32.3	36.2	1.7
Hell's Gate National Park	73.0	60.9	75.1	38.9	35.6	1.7
Malindi Marine National Park	26.5	29.8	22.8	27.5	32.8	1.5
Watamu Marine National Park	30.0	29.3	21.1	28.4	32.4	1.5
Shimba Hills National Reserve	18.3	14.4	16.2	18.7	17.3	0.8
Mount Longonot National Park	13.8	12.8	12.2	9.5	11.5	0.5
Meru National Park	7.8	8.2	5.7	6.4	8.9	0.4
Samburu National Reserve	6.3	6.0	6.0	6.2	7.3	0.3
Other[2]	17.4	11.0	30.5	30.3	22.5	1.1
TOTAL	**1,664.1**	**1,784.1**	**1,575.9**	**1,820.5**	**2,132.9**	**100.0**

Source: CBS 2006.

Note: [1] Provisional
[2] Others include Arabuko Sokoke, Ol-Donyo Sabuk, Marsabit, Saiwa Swamp, Ruma National Park, Mwea National Reserve, Central Island National Park, Kiunga, Mount Elgon, Nasolot, Ndere, and Kakamega National Reserve.

Table 6.2 Annual Average Visitors and Revenues for Selected Protected Areas, 2000-04

	NUMBER (000)	SHARE OF TOTAL VISITORS TO PARK (PERCENT)	REVENUES		SHARE OF TOTAL REVENUES FOR PARK (PERCENT)
			KSH (MILLION)	US$ (000)	
Kenyans					
Nairobi TOTAL	365.4	85	32.1	421	34
Nairobi National Park	*49.2*	*52*	*4.9*	*65*	*8*
Nairobi Animal Orphanage	*224.4*	*96*	*18.0*	*236*	*78*
Nairobi Safari Walk	*91.7*	*93*	*9.2*	*121*	*71*
Lake Nakuru National Park	113.0	53	11.3	149	5
Tsavo East National Park	38.0	27	3.8	50	2
Amboseli National Park	21.6	25	2.2	28	1
Tsavo West National Park	27.1	34	2.7	36	3
Aberdare National Park	7.9	19	0.8	10	1
International Residents					
Nairobi TOTAL	25.6	6	12.1	159	13
Nairobi National Park	*21.1*	*22*	*10.5*	*139*	*18*
Nairobi Animal Orphanage	*2.4*	*1*	*0.5*	*6*	*2*
Nairobi Safari Walk	*2.1*	*2*	*1.1*	*14*	*8*
Lake Nakuru National Park	12.9	6	6.4	85	3
Tsavo East National Park	4.2	3	2.1	28	1
Amboseli National Park	4.6	5	2.3	31	2
Tsavo West National Park	4.6	6	2.3	30	2
Aberdare National Park	4.5	11	2.2	29	3
International Visitors					
Nairobi TOTAL	36.8	9	50.8	667	53
Nairobi National Park	*24.9*	*26*	*43.5*	*572*	*74*
Nairobi Animal Orphanage	*7.6*	*3*	*4.7*	*61*	*20*
Nairobi Safari Walk	*4.3*	*4*	*2.6*	*35*	*20*
Lake Nakuru National Park	87.5	41	199.6	2,624	92
Tsavo East National Park	97.5	70	200.3	2,633	97
Amboseli National Park	61.5	70	140.3	1,845	97
Tsavo West National Park	46.9	60	96.4	1,267	95
Aberdare National Park	28.1	69	64.0	842	95
TOTAL					
Nairobi TOTAL	427.7	100	94.9	1,247	100
Nairobi National Park	*95.2*	*100*	*59.0*	*775*	*100*
Nairobi Animal Orphanage	*234.4*	*100*	*23.1*	*303*	*100*
Nairobi Safari Walk	*98.2*	*100*	*12.9*	*169*	*100*
Lake Nakuru National Park	213.4	100	217.4	2,857	100
Tsavo East National Park	139.7	100	206.2	2,710	100
Amboseli National Park	87.7	100	144.8	1,903	100
Tsavo West National Park	78.6	100	101.4	1,333	100
Aberdare National Park	40.4	100	67.0	881	100

Source: KWS 2005.

Note: Visitor data from the KWS Tourism Section on citizens, residents, and nonresidents was averaged for the years 2000 to 2004. The average number of visitors per year was multiplied with the respective entry fees, using adult rates (http://www.kws.org/tariffs.html) and the average exchange rate of 17 February 2005 (http://www.oanda.com/convert/classic). Data are rounded to nearest thousand, million, or percent.

The most popular parks generally get between 100,000 and 350,000 visits per year. Meanwhile, other sites with rich wildlife resources and striking scenery, such as Meru and Samburu National Parks, receive fewer than 10,000 visits annually.

The distribution of visitors varies among Kenya's parks and reserves. For parks near urban centers, Kenyans typically make up the majority of visitors. For instance, more than 90 percent of visitors to the Nairobi Animal Orphanage and Safari Walk are Kenyans (see Table 6.2). At greater distances from urban centers, most park visitors are international tourists. About 70 percent of visitors to the Aberdare, Amboseli, and Tsavo East National Parks are overseas tourists.

The distribution of park revenues follows a different pattern. Because entrance fees are higher for nonresidents, international tourists are responsible for most of the revenues generated at Kenya's parks and reserves. For example, overseas visitors account for more than 90 percent of revenues to all national parks listed in Table 6.2, with the exception of Nairobi area parks. At the Nairobi Animal Orphanage and Safari Walk, Kenyans account for more than 70 percent of all revenues collected.

Box 6.1 Mapping the Role of Ecosystems in Tourism: Links to National Decision-Making

Kenya's *Economic Recovery Strategy for Wealth and Employment Creation, 2003–2007* (GoK 2003) identifies tourism as a key sector for poverty reduction and employment creation. Besides the direct contribution of tourism to GDP, foreign exchange earnings, and employment creation, the sector also exerts strong multiplier effects by encouraging economic activity and expansion in additional sectors—transport, agriculture, and entertainment, among others.

Another key dimension of Kenya's national decision-making on tourism is spatial diversification. Overconcentration of tourists in a handful of parks, reserves, and coastal beaches encourages ecosystem degradation through intensive use. Spatial diversification of tourism could help protect ecosystems, while also promoting more equitable distribution of tourism's benefits and costs among local communities.

Below are examples of how mapping and analysis of ecosystem services and related indicators could contribute to national decision-making on tourism, sustainable development, and poverty reduction.

Spreading tourism impacts and benefits. Various maps presented in this atlas—such as the distribution of wildlife species and wildlife density, as well as the location of threatened or endangered species—could help policymakers identify new areas that have the potential to attract significant tourist interest. These maps could be combined with maps of existing infrastructure to pinpoint additional investments needed to expand tourism in underutilized areas. Some areas where such investment might be targeted are:

- *Lamu hinterlands.* The area surrounding Lamu is rich in potential tourist attractions, such as beautiful beaches, coral reefs, mangrove forests, and wildlife viewing (including the endangered sable antelope). Investment in transport and other tourism-related infrastructure could help this area capture a greater share of the tourism market.
- *Samburu National Park and surrounding Laikipia ecosystem, including the northern slopes of Mount Kenya.* Samburu is among the least visited of Kenya's national parks in spite of the fact that the area contains a great diversity of wildlife viewing opportunities (see wildlife maps in this and the biodiversity chapter). For example, visitors can encounter the largest elephant population outside of the Tsavo National Parks; half of Kenya's rhino population; and the only herd of Jackson's hartebeest, a threatened antelope (Laikipia Wildlife Forum 2006). Tour operators, private ranches, community-owned lodges, and wildlife conservancies have begun to market the Samburu-Laikipia ecosystem as an alternative destination and a leader in ecotourism in Kenya.

In all cases, great care should be taken to ensure that development of tourism infrastructure does not undermine the integrity of ecosystems, and that stakeholders in each area are consulted and potential resource conflicts are avoided.

Tourism marketing and promotion. Maps of ecosystem assets could be used to promote tourism by showing the accessibility and spatial distribution of popular tourist destinations.

Increasing community involvement in tourism development. Maps can be used to display data from spatial analysis aimed at understanding which tourist destinations actually benefit local communities. Mapping can also play a role in efforts to minimize human-wildlife conflicts in the areas surrounding parks and protected areas—an increasingly important part of tourism strategies in the area.

Expanding the role of ecotourism. Mapping can be an important part of efforts to make ecotourism a larger component of the Kenyan tourism sector. Detailed studies are needed to assess the impacts of ecotourism, including surveys of how many visitors choose ecotourism as well as evaluations of how much ecotourism is benefiting local communities. Information from the recent National Inventory of Ecotourism Projects in Kenya (ESOK 2005) could be combined with map information to help identify areas with high ecotourism potential.

Assessing the impact of infrastructure quality. Maps can help to examine the relationships between declining tourism and problems with the quality of local accommodations. Random spot checks of hotel quality could be carried out and the results mapped to reveal areas with systematic problems. This map could then be overlaid with mapped areas of declining tourism to determine if there is any spatial correlation.

Creation of tourism information systems and tools. In recent years, tourism planning in Kenya has often called for improved access to information systems and technologies, yet many of these recommendations have yet to be implemented. Cooperative efforts between the Ministry of Tourism and Wildlife and the Central Bureau of Statistics to strengthen data collection and establish a comprehensive database for tourism statistics could help to interpret trends in visitor numbers and demographics for key parks and other tourist destinations. Such a database could also form the basis for sector analysis tools or a tourism forecasting model. Once tourism data and statistics are available in a database format, it will be easier to map this information and undertake spatial analyses.

Upgrading of security. Maps could be used to depict the availability and effectiveness of police units by tourist destination. Such maps could prove especially helpful in pinpointing the need for investments to upgrade security in areas of high tourism potential that are currently unsafe.

Promotion of domestic tourism. Since domestic tourism is the most significant income source for several parks, it is critically important to continue promoting these attractions to Kenyans and using Kenya's ecosystem assets for educational purposes. Mapping can help identify attractions of particular interest to Kenyan citizens.

Box 6.2 Creating a Poverty and Demographic Profile of Communities Neighboring the Most Visited Protected Areas

Map 6.7 Areas within 25 Kilometers of Most Visited Protected Areas

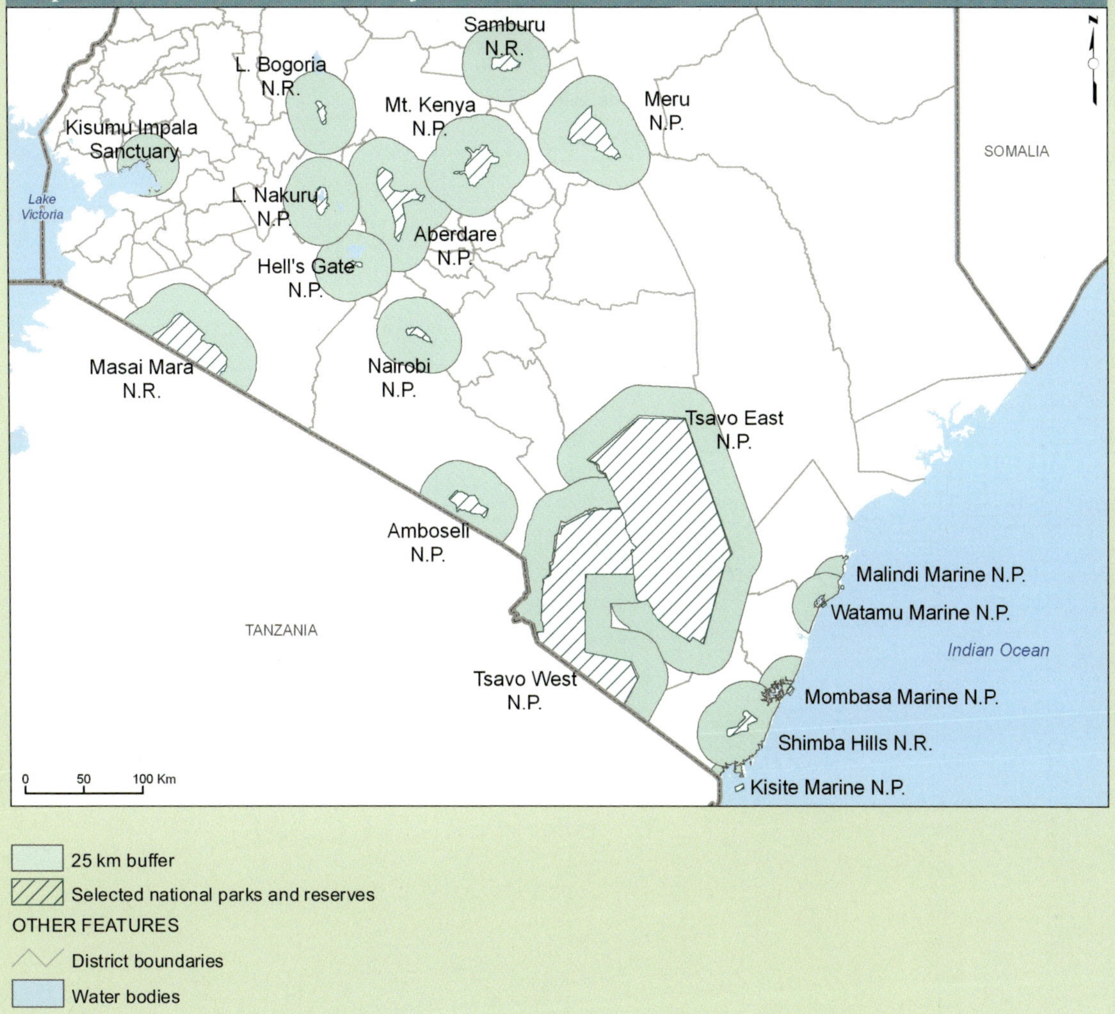

Parks and reserves are important contributors to local economies as well as to overall national income. Table 6.3 presents information on the socioeconomic attributes of populations living near Kenya's most visited parks and protected areas. It is based on data for communities within a 25-kilometer radius of the boundaries of each protected area (Map 6.7).

Such information enables comparison of the demographic and poverty characteristics of communities surrounding parks with high, medium, and low levels of visitation. These comparisons can in turn help identify relationships between park visitation and the economic status of nearby communities.

What Do the Map and Poverty Profile Show?

- Predictably, the largest numbers of poor people live in the vicinity of parks near urban and other densely populated areas. For instance, more than 970,000 live near Nairobi area parks, and almost 305,000 live near Mombasa Marine Park. In addition, large numbers of the poor live near protected areas in the densely populated highlands, including Aberdare National Park (about 324,000), Mount Kenya National Park (250,000), and Lake Nakuru National Park (over 245,000). Of the parks mentioned above, some are extremely popular (Nairobi area parks and Lake Nakuru National Park), while others are among the less-visited parks (Mount Kenya National Park and Mombasa Marine Park).
- For two of the parks with the greatest number of visitors (Masai Mara and Amboseli), the number of poor people in surrounding communities is quite small (69,000 and 16,000, respectively), reflecting the low population densities in these areas.
- Patterns regarding poverty *rates* are quite distinct from patterns involving the absolute number of poor people. While the number of poor people living near Masai Mara is quite low, the average poverty rate among these communities is 63 percent, which is among the higher rates for all parks shown in the table. Other parks with very high poverty rates (55–69 percent) in the surrounding communities include both parks with many visitors (Tsavo East and West, for instance) and parks with relatively few visitors (such as Meru and Watamu Marine). Parks with lowest poverty rates (34–38 percent) in nearby communities tend to be located in relatively better off central parts of the country (for example, Aberdare and Hell's Gate National Parks).
- The size of the poverty gap in communities surrounding popular parks and reserves varies enormously, from more than Ksh 400 million (US$ 5.7 million) per month for the densely populated communities near the Nairobi area parks, to only about 4–6 million Ksh (US$ 57,000–85,000) per month for the communities in less densely populated areas, such as those near Amboseli and Samburu National Parks. The poverty gap is the amount of money that would be required to raise the income of every poor person to just reach the poverty line (shown in the right-hand column in Table 6.3).
- These patterns suggest that poverty rates are not associated with the level of visitation to the selected national parks, but with other factors. In fact, the poverty rates of communities within a 25-kilometer buffer in general are closer to Kenya's rural average rate of 53 percent; when they are lower than this average rate, they tend to reflect countrywide spatial patterns (e.g., rates of 38 percent or lower for the Aberdare, Hell's Gate, and Mount Kenya National Parks). A comparison with poverty rates further away and a more detailed local analysis could provide additional explanations for these spatial patterns.

Similar tables could be constructed using different tourism statistics or poverty indicators. For instance, one could compare the revenue levels at particular parks to the magnitude of investment needed to close the poverty gap in nearby communities. (See Chapter 2 for examples of various indicators of human well-being in Kenya.)

Continued

Table 6.3 People, Poverty, and Communities within 25 Kilometers of the Most Visited Protected Areas

PROTECTED AREAS RANKED BY SHARE OF VISITORS TO ALL PARKS AND RESERVES IN KENYA	TERRESTRIAL AREA WITHIN 25 KILOMETERS OF PARK BOUNDARY (SQ. KM)	NUMBER OF PEOPLE (000)	AVERAGE POPULATION DENSITY (NUMBER OF PEOPLE PER SQ. KM)	NUMBER OF POOR (000)	AVERAGE POVERTY RATE (PERCENT)	KSH NEEDED PER MONTH TO REACH POVERTY LINE [1] (MILLION)
HIGH SHARE OF VISITORS (13.4–22.7% Country Total of All Visitors)						
Nairobi TOTAL[2] (22.7%)	3,359	2,434	725	970	40	414.7
Lake Nakuru National Park (16.2%)	3,438	616	179	245	40	61.1
Masai Mara National Reserve (13.4%)	3,669	108	30	69	63	23.1
TOTAL 3 AREAS	**10,466**	**3,158**	**302**	**1,284**	**41**	**498.9**
MEDIUM-HIGH SHARE OF VISITORS (5.0–8.4% of Country Total of All Visitors)						
Tsavo East National Park (8.4%)	14,358	229	16	143	62	44.9
Amboseli National Park (5.9%)	3,000	30	10	16	54	3.8
Tsavo West National Park (5.0%)	10,383	247	24	135	55	36.8
TOTAL 3 AREAS	**27,741**	**506**	**18**	**294**	**58**	**85.5**
MEDIUM SHARE OF VISITORS (2.3–4.1% of Country Total of All Visitors)						
Kisumu Impala Sanctuary (4.1%)	1,563	715	457	430	60	163.1
Lake Bogoria National Reserve (3.1%)	3,141	183	58	77	42	15.5
Kisite Marine N.P./Mpunguti Marine N.R. (2.8%)	284	27	95	15	54	3.6
Aberdare National Park (2.3%)	6,178	963	156	324	34	43.9
TOTAL 4 AREAS	**11,166**	**1,888**	**169**	**846**	**45**	**226.1**
LOW SHARE OF VISITORS (0.3–1.9% of Country Total of All Visitors)						
Mount Kenya National Park (1.9%)	4,959	682	138	250	37	40.1
Mombasa Marine National Park (1.7%)	945	604	639	305	51	118.8
Hell's Gate National Park (1.7%)	2,945	205	70	79	38	12.7
Malindi Marine National Park (1.5%)	767	117	152	78	66	25.5
Watamu Marine National Park (1.5%)	1,103	143	129	99	69	35.3
Shimba Hills National Reserve (0.8%)	3,160	393	124	221	56	75.1
Meru National Park (0.4%)	5,433	451	83	255	57	67.2
Samburu National Reserve (0.3%)	3,572	54	15	27	50	6.2
TOTAL 8 AREAS	**22,884**	**2,649**	**116**	**1,314**	**50**	**380.9**

Sources: Visitor data CBS 2006. Area estimate based on a 25-kilometer buffer (see Map 6.7) surrounding protected areas (IUCN and UNEP/WCMC 2006). Poverty and demographic estimates (1999) are WRI/ILRI calculation based on CBS 2002, 2003.

Note: [1] The poverty gap measures the average expenditure shortfall (gap) for the poor in a given administrative area relative to the poverty line. It is a crude estimate of the minimum amount of resources needed to eradicate poverty (see Chapter 2).

[2] Includes Nairobi National Park (4.7% of all visitors to Kenya's parks and reserves), Nairobi Animal Orphanage (12.1% of all visitors to Kenya's parks and reserves), and Nairobi Safari Walk (6.0% of all visitors to Kenya's parks and reserves). Table does not include Haller Park, a private park, which received 4.7% of all visitors to Kenya's parks and reserves. The park, a restored ecosystem in a former cement quarry, is 12 kilometers north of Mombasa at Bamburi Beach and overlaps significantly with the 25-kilometer buffer surrounding Mombasa Marine National Park. The 25-kilometer buffer around Mount Longonot National Park (0.5 percent of all visitors to Kenya's parks and reserves) overlaps with the one for Hell's Gate National Park and is therefore not included in this table.

SUMMING UP

- Tourism in Kenya relies on the country's natural attractions, including wildlife in its native habitat, as well as some of Africa's finest beaches and other coastal ecosystem assets. It ranges from low-density tourism focused on a 'wilderness experience' in less modified ecosystems, to high-density beach tourism requiring a relatively limited set of ecosystem services—primarily sand, sea, and sun.
- In 2005, the tourism industry generated revenues of almost Ksh 49 billion (US$ 700 million) and directly employed 176,000 people (about 10 percent of all jobs in the formal sector). About 70 percent of the visitors to Kenya came to see places of natural beauty and engage in nature-based activities.
- Kenya has invested in a network of protected areas to safeguard its natural heritage; support nature-based tourism; and achieve biodiversity, watershed protection, and other environmental objectives. More than 80 of Kenya's top 120 tourist destinations are national parks and wildlife reserves (about 8 percent of Kenya's total land area).
- Viewing wildlife in its natural habitat is the primary objective for about 80 percent of the international visitors who come to Kenya for holidays. Wildlife is broadly distributed across Kenya, but particular species with high 'viewing value' exhibit specific patterns of spatial distribution: For example, the rangelands of Laikipia District as well as Amboseli, Marsabit, and Tsavo National Parks all have high elephant numbers; the massive annual migration of wildebeest and zebra occurs in the plains of Kajiado District close to the Mara-Serengeti ecosystem. Declining wildlife numbers are undermining one of Kenya's principal tourist attractions (see Chapter 5). For instance, the wildebeest population in the Masai Mara ecosystem has fallen from 120,000 in 1977 to 31,000 in 2002.
- Beaches and coastal ecosystems continue to account for a large share of tourism earnings, including more than half of all nights spent by tourists in hotel accommodations in 2005. Coastal tourism includes both high-density beach tourism in and around Mombasa and tourism requiring lower visitor densities and a diverse set of ecosystem services. This includes snorkeling, diving, deep sea fishing, bird watching, and wildlife viewing—all taking advantage of Kenya's unique coastal ecosystem assets. For example, in the Arabuko-Sokoke Forest, less than 10 kilometers inland from Malindi, over 260 species of birds have been recorded, including 6 globally threatened species. Shimba Hills Reserve, about 15–20 kilometers inland from the coast, is famous for its sable antelope, the last remaining breeding population of these animals in the country. The government has designated six marine reserves—Kisite, Kiunga, Malindi, Mombasa, Mpunguti, and Watamu—encompassing a significant portion of the reef and its surrounding waters.
- In 2005, Kenya's protected areas welcomed 2.1 million visitors, the highest number ever registered. Of Kenya's 84 parks and reserves, Nairobi National Park (including the Animal Orphanage and Safari Walk), Lake Nakuru National Park, and Masai Mara National Reserve, together accounted for more than half of all visitors. More than 90 percent of the visitors to the Nairobi Animal Orphanage and Safari Walk, and more than 50 percent of the visitors to Nairobi and Nakuru National Parks were Kenyans. About 70 percent of the visitors to the Aberdare, Amboseli, and Tsavo East National Parks were from overseas. International tourists accounted for more than 90 percent of revenues for all national parks where such revenue data are available. Kenyans account for more than 70 percent of all revenues collected at the Nairobi Animal Orphanage and Safari Walk.
- To protect wildlife and ecosystems from serious damage caused by overly high visitor densities, tourism planners need to promote underutilized areas and spread visitor numbers more widely across destinations. This would also help to distribute tourism-related costs and benefits more evenly across the country. Improved spatial diversification of visitors will require increased and sustained investments in the transport system, safe water supplies, communications services, tourist accommodations, protected areas, and targeted marketing efforts. It will also require greater control and participation of local communities in wildlife management and tourism enterprises.

In Chapter 7

CONTENTS

LIST OF MAPS

AUTHORS AND CONTRIBUTORS

Norbert Henninger (WRI)
Stephen Mutimba (ESDA)
Evans Kituyi (University of Nairobi)
Linda Cotton (consultant)
Janet Nackoney (WRI)
Florence Landsberg (WRI)
Daniel Prager (WRI)
Greg Mock (consultant)
Dan Tunstall (WRI)
Hyacinth Billings (WRI)
Carolina de Rosas (WRI)

WHAT THIS CHAPTER SHOWS

This chapter gives a brief overview of the ecosystems that provide Kenya with wood and how Kenyans use this wood. The first section discusses the location of wood supply areas, each reflecting different levels of tree cover or woody biomass supply. The maps show the location of different types of forests, the percentage of tree cover in a landscape, the location of plantations, and areas where farmers are planting woodlots on their farmland. The second section focuses on the two most significant uses of wood in Kenya—as firewood and for making charcoal. The maps distinguish the areas that are most likely supplying firewood and charcoal from those used for other activities. The section also highlights the economic importance of charcoal production and shows where charcoal making and firewood collection contribute significantly to cash income. Finally, the section presents maps of annual biomass growth (outside of croplands) in order to estimate sustainable harvest levels for biomass energy.

7

Wood

Kenya's tree-covered landscapes fall under various classes such as forests, woodlands, bushlands, and wooded grasslands—each reflecting different tree densities and vegetation communities. They also include agroecosystems, where farmers grow both agricultural crops and trees. Forested areas are the source of an array of ecosystem services, providing soil and water conservation, a home for indigenous peoples (e.g., the Ogiek people), a grazing refuge during drought, or a site for cultural and religious ceremonies.

Trees are linked to hydrological and other water-related services, as tree cover influences runoff and water infiltration patterns. The remnants of multilayered forest habitats also contribute to Kenya's biodiversity, inasmuch as they provide a home for some of the country's rare bird species (African Bird Club 2006). The various tree-covered landscapes are also a source of products such as medicines, honey, meat, fruits, vegetables, fiber, nuts, and tubers. The wood from these ecosystems is used for firewood, charcoal, timber, posts, and poles, and is vital to Kenya's economy and the livelihoods of its people.

From an economic point of view, the forest sector officially contributes about Ksh 9.9 billion (US$ 141 million) to the national economy per year—about 1.3 percent of Kenya's gross domestic product (CBS 2004). However, this number does not fully reflect the forest sector's economic contribution. It omits some significant contributions, such as the value of energy produced from wood, and the value of various nontimber forest products.

For example, 60,000 full-time wood carvers use about 15,000 cubic meters of wood per year (Choge et al. 2002). Although wood sculptures consume less than 1 percent of Kenya's annual wood harvest (FAO 2005), they generate export earnings of around Ksh 1.6 billion (US$ 23 million) per year and financially support an estimated 400,000 dependents. Nonetheless, this revenue is not included in economic analyses of the forest sector.

Revenues from the charcoal market are at least ten times greater than those from wood carvings, and charcoal production is a voracious consumer of Kenya's trees. Yet, it is also not counted in the official forest sector statistics. Estimates of the economic value of Kenya's charcoal production range from Ksh 17.5 to Ksh 32 billion per year (depending on volume and price)—about US$ 250 to US$ 457 million (MoE 2002; ESDA 2005a).

Most Kenyans rely on wooded ecosystems to provide them with either firewood or charcoal. As Table 7.1 indicates, biomass (firewood, wood for charcoal, industrial wood, wood wastes, and farm residues) is Kenya's dominant fuel, accounting for over 80 percent of total energy consumption in 2000. In comparison, only 1.4 percent of the total energy consumed came from electricity, primarily used by commerce and industry and by urban households. Imported petroleum's share in Kenya's total energy consumption is about 18 percent, used mostly for commerce, industry, and transport (MoE 2002). Of all the wood supplied by the nation's ecosystems, Kenyans use some 80–90 percent for energy purposes (1995 estimate from 1994 *Kenya Forestry Master Plan* cited in Holding Anyonge and Roshetko 2003; FAO 2005). They use the remaining 10 to 20 percent for timber, posts, and poles.

Table 7.1 **Kenya's Total Energy Consumption by Sector and Fuel Type, 2000**

	SHARE IN KENYA'S TOTAL ENERGY CONSUMPTION (percent)								
	Firewood	Charcoal	Industrial Wood	Wood Wastes	Farm Residue	SUBTOTAL BIOMASS	Electricity	Petroleum	TOTAL ENERGY
Households: Rural	32.5	17.6	0.0	0.3	5.3	*55.7*	0.0	1.0	56.7
Households: Urban	0.8	13.9	0.0	0.2	0.0	*14.9*	0.4	1.0	16.3
Cottage Industry	3.0	6.6	0.0	0.0	0.0	*9.6*	0.2	0.1	9.9
Agriculture	0.0	0.0	0.0	0.0	0.0	*0.0*	0.0	0.6	0.6
Transportation	0.0	0.0	0.0	0.0	0.0	*0.0*	0.0	6.6	6.6
Commerce and Industry	0.0	0.0	0.3	0.0	0.0	*0.3*	0.8	8.8	9.9
TOTAL	**36.3**	**38.1**	**0.3**	**0.5**	**5.3**	***80.5***	**1.4**	**18.1**	**100.0**

Source: MoE 2002.

OVERALL DISTRIBUTION OF WOODY BIOMASS

Maps 7.1 to 7.3 delineate forests, tree cover, plantations, and areas where farmers have planted woodlots on farmland. Areas where the vegetation consists of densely spaced trees are generally designated as forests. Most of Kenya's closed forests (those where tree crowns cover a high proportion of land surface) fall under government jurisdiction (i.e., as gazetted forest reserves). Extraction of forest products from these reserves is highly regulated or illegal.

The inventory of the Kenya Indigenous Forest Conservation Programme (Wass 1995) estimated Kenya's 1995 closed forest cover to be 1.4 million hectares (about 2.5 percent of the total land area). It included 1.24 million hectares of indigenous closed canopy forest (1.06 million hectares in gazetted forests and 0.18 million hectares outside these forest reserves) and 0.16 million hectares in plantations. Other natural woody vegetation covers approximately 37.3 million hectares with 2.1 million hectares of woodlands, 24.6 million hectares of bushlands, and 10.6 million hectares of wooded grasslands (MoE 2002).

A different assessment of Kenya's forests—one that relied on satellite imagery and used a different definition for closed forests—estimated Kenya's 1995 closed forest area to be 984,000 hectares, representing 1.7 percent of the country's total land area (UNEP 2001). Media reports and local observations indicate tremendous pressure on Kenya's closed forest estate and suggest that the amount of closed forest area is now lower than indicated in the last forest inventory. Both legal conversion (e.g., the excision of land parcels from the gazetted forest reserve in the 1990s (Matiru 1999)) and illegal conversion of forests (extraction of timber, production of charcoal, growing of crops or marijuana) have contributed to this decline in forest area.

More recently, high-resolution aerial surveys of selected forests in the Aberdare Range, Mount Kenya, Mount Elgon, and the Mau Escarpment confirm that some of these trends are taking place on a more local scale, pinpointing significant unplanned forest exploitation and degradation

Map 7.1 Forest Areas, 2000

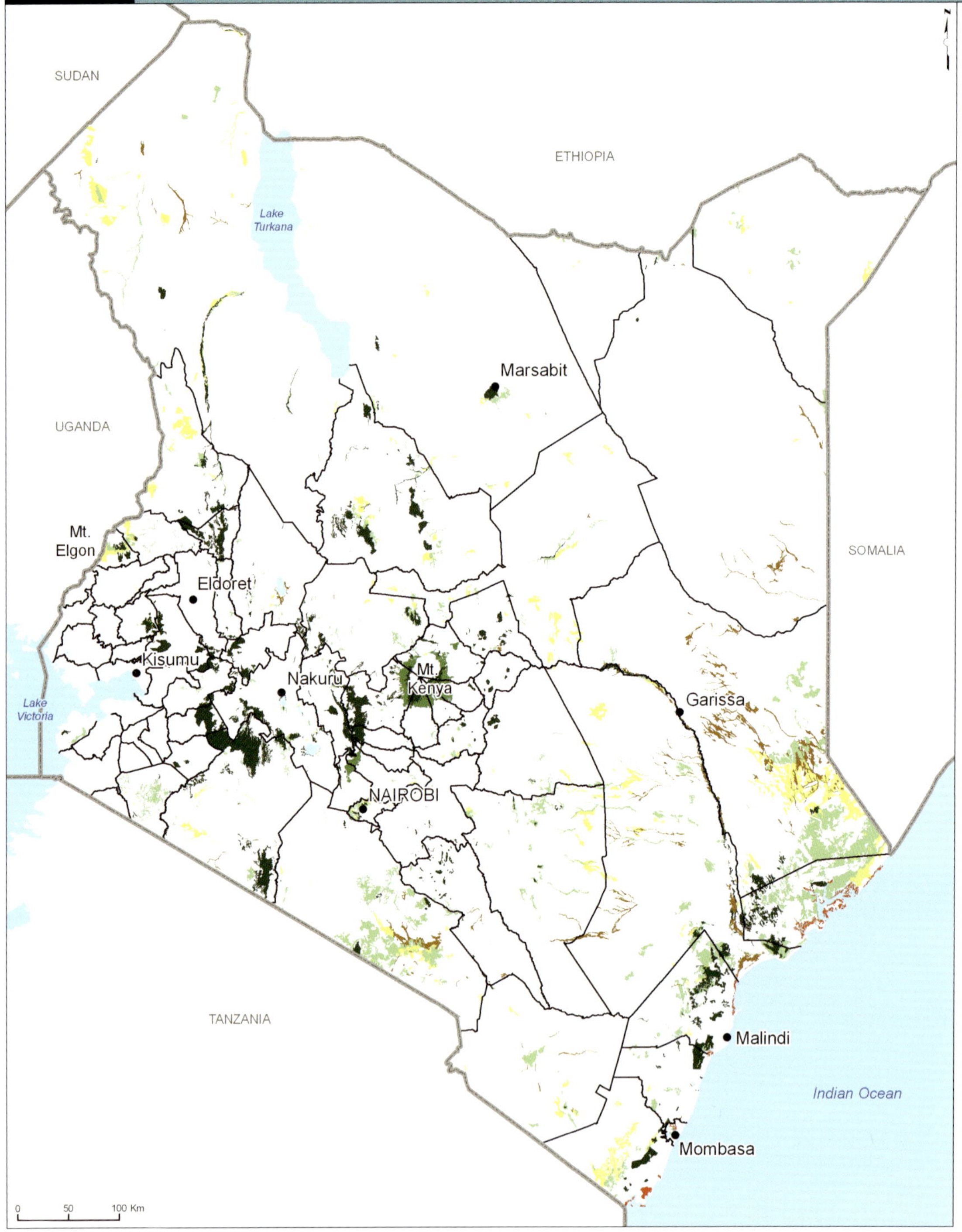

Sources: Administrative boundaries (CBS 2003), cities (SoK and ILRI 2000) and water bodies and forest types (FAO 2000).

Most of Kenya's closed canopy forests are concentrated in the highlands, surrounded by areas with high population densities and intensive agricultural production. Within rangeland areas, closed forests grow primarily in mountain ranges and along permanent and seasonal rivers. The largest closed natural forest areas are west of Nakuru (on the slopes of the Mau Escarpment), north of Nairobi (Aberdare Range), and Mount Kenya. The Districts between Garissa and the Indian Ocean coast include large tracts of open forest types. Significant gallery forests follow permanent rivers, for example southwest of Lake Turkana (Turkwell River) and south of Garissa (Tana River). Large tree-covered seasonal wetlands are prominent in the southeastern rangelands.

FOREST TYPES
- Closed trees
- Multilayered trees (broadleaved evergreen)
- Open trees (65-40% crown cover)
- Very open trees (40-15% crown cover)
- Closed trees on temporarily flooded land
- Mangrove trees

OTHER FEATURES
- District boundaries
- Water bodies

(Gathaara 1999; Lambrechts et al. 2003; Akotsi and Gachanja 2004). On the positive side, however, these surveys have led to a change in policies and institutional responsibilities for the forests of Mount Kenya, resulting in a slowdown of forest decline (Vanleeuwe et al. 2003; Akotsi and Gachanja 2004).

Kenyan authorities have not recently carried out a detailed national assessment on the changes in woodland and bushland. Significant land use changes are occurring, however, in Kenya's rangelands, such as in Narok and Trans Mara Districts (Serneels and Lambin 2001; Lamprey and Reid 2004; and Norton-Griffiths et al. in press). In many areas, landowners have found it profitable to have charcoal burners clear all the trees and then sell off or lease the land for crop production.

Forests and Tree Cover

In classifying different vegetation communities, experts consider the density of tree cover; the occurrence of different types of woody vegetation such as bushes, shrubs, and trees; and the presence of plants growing below the woody vegetation as groundcover, such as grass. This section explores two different approaches to mapping woody vegetation.

Map 7.1 identifies vegetation communities that are classified as forests using *Africover* categories on a national scale for the year 2000 (FAO 2000). Forests, by definition, have the greatest density of trees and the highest volume of wood per square kilometer (i.e., forests must exceed a certain threshold in tree cover and a minimum height in the woody vegetation). The map highlights natural and seminatural forested landscapes such as closed canopy forests and other forest types with more open crown cover; it does not include forest plantations or trees on cultivated landscapes.

Map 7.2 displays tree cover density on a continuous scale from 0 to 100 percent. It is derived from satellite imagery estimating woody vegetation within grid cells of 500 meters by 500 meters. In these grid cells, a tree is defined as mature vegetation greater than five meters in height. Such an approach

Map 7.2 **Tree Cover, 2001**

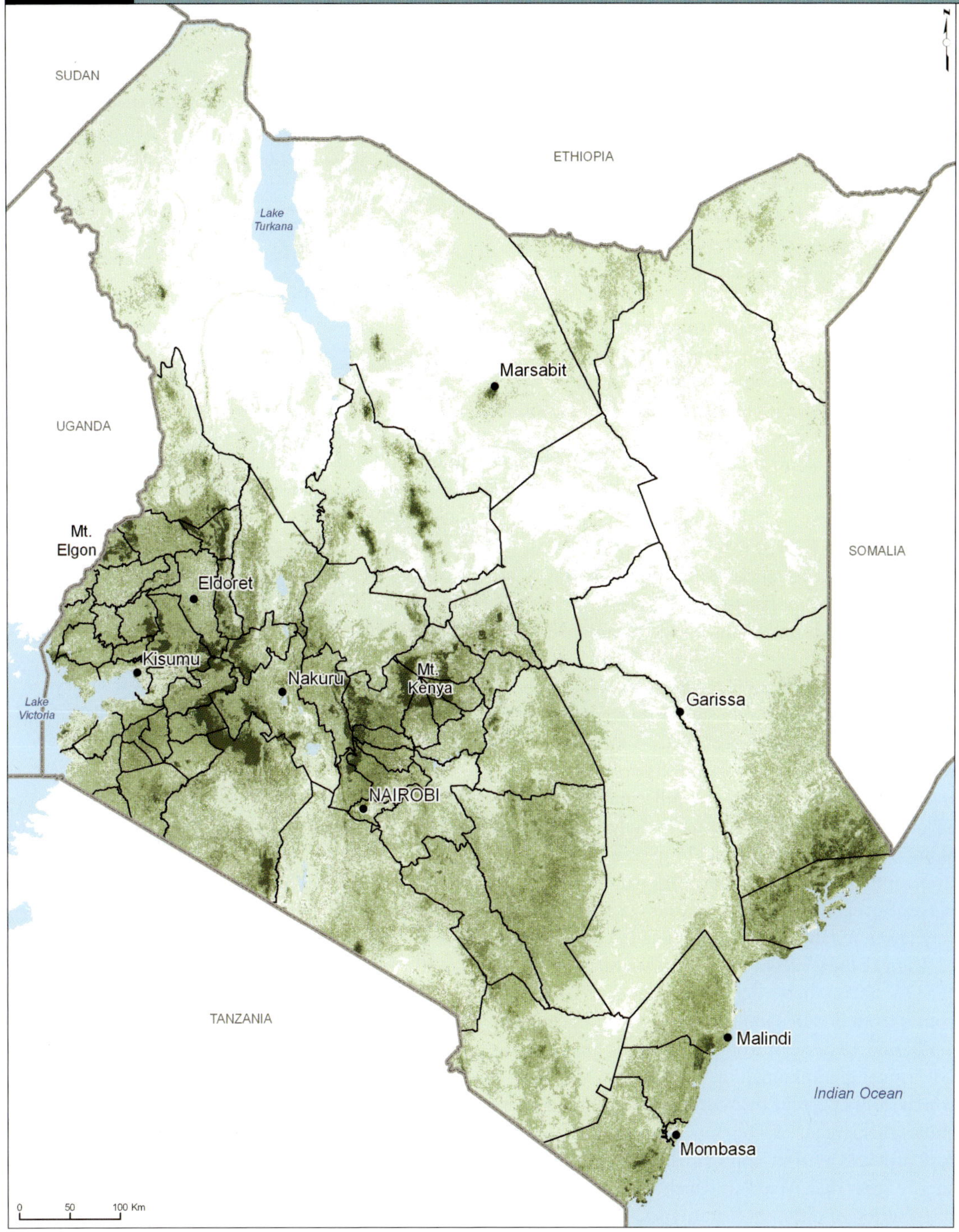

Sources: Administrative boundaries (CBS 2003), cities (SoK and ILRI 2000), water bodies (FAO 2000), and percent tree cover (Hansen et al. 2003)

Higher density tree coverage occurs in the high rainfall zones: in mountain ranges such as Mount Kenya, Aberdares, Mount Elgon, and Mau Escarpment in the country's interior, and areas close to the coast. Because of their relatively large area, lands classified as woodlands, bushlands, and wooded grasslands (see Map 1.3) together contain most of Kenya's woody biomass, albeit at much lower tree density and volume per area than the small remnants of closed forests. Agricultural land can have a high percentage of tree cover as reflected in the varying tree density in high-rainfall croplands, for example in Central Province.

TREE COVER
(percent)
- > 50
- 30 - 50
- 20 - 30
- 10 - 20
- 0 - 10
- No tree cover

OTHER FEATURES
- District boundaries
- Water bodies

avoids the problem presented by the traditional classification scheme in Map 7.1 (i.e., closed versus open canopy forest), which set a threshold of tree cover for each forest class. Map 7.2 can therefore highlight the importance of trees that fall below the minimum tree or canopy cover thresholds (i.e., trees outside forests). It includes both trees on cultivated and managed landscapes (croplands) but also on natural and seminatural landscapes (i.e., woodlands, bushlands, and wooded grasslands). This approach is, therefore, a more detailed and accurate representation of vegetation cover for our purposes.

Generally, most of the closed and open forest areas of Map 7.1 coincide with higher tree densities in Map 7.2. The coarse resolution of the satellite data used for Map 7.2 results in a map with fewer small features (e.g., trees in wetlands) or linear features (e.g., forests along riverbanks). Only Map 7.2, however, can reflect the varying tree density in high-rainfall croplands, such as the highlands. Considered together, the maps indicate that, surprisingly, closed canopy forests do not contain most of Kenya's woody biomass; woodlands, bushlands, and wooded grasslands together have a higher total volume of woody biomass, due to their vast size. Selected studies and anecdotal evidence suggest that closed canopy forests are only a minor contributor of woodfuel at a national level (MoE 2002; ESDA 2005a). However, it should be noted that forest reserves or patches of dense forest can be quite significant sources of woodfuel on a local scale (for example, when a government forest reserve is degazetted), and that illegal logging and charcoal production is taking place within forest reserves.

Plantations and Woodlots

Reliable statistics on the exact sources of Kenya's wood supply for energy and other uses are not routinely available. However, the *Kenya Forestry Master Plan* estimated that of the 1995 national wood supply, 9 percent came from indigenous forests, 49 percent from woodlands and bushlands, 33 percent from farmlands and settlements, and 9 percent from forest plantations (1994 *Kenya Forestry Master Plan* cited in Holding Anyonge and Roshetko 2003). It predicted that by 2020, the supply from farmlands and settlements would more than double, increasing its share to 54 percent.

As Table 7.2 indicates, burning firewood and charcoal account for roughly equal percentages of total wood consumption—about 45 percent each (MoE 2002). Together they use up 80–90 percent of Kenya's wood supply (1994 *Kenya Forestry Master Plan* cited in Holding Anyonge and Roshetko 2003; FAO 2005).

A more recent household survey conducted by the Ministry of Energy (MoE 2002) found that at the household level, about 8 percent of firewood supplies came from Trust Land (land held by County Councils on behalf of local communities, groups, families, and individuals) and another 8 percent from gazetted forests (government land). The remaining 84 percent were supplied by agroforestry systems and on-farm sources. This consisted of firewood purchased in the market (20 percent)—mostly from small private farms—and other more specific agroforestry sources. The latter included vegetation along boundaries and fences (25 percent), vegetation within croplands (13 percent), woodlots (8 percent), vegetation along roadsides (5 percent), and vegetation obtained from neighbors (13 percent).

Kenya's most recent *National Charcoal Survey* (ESDA 2005a) shows that 82 percent of charcoal comes from private land (either farmland or rangelands) and 18 percent from public lands (including government, communal, or Trust Land). Map 7.2, which shows percent tree cover and Map 7.3, which shows the percent of woodlots in croplands therefore provides a better approximation of woodfuel supply areas than Map 7.1, which displays the distribution of different forest types. While only 34 percent of rural and 82 percent of urban households in Kenya regularly use charcoal, rural households together consume more charcoal than urban households (MoE 2002). Of the total national charcoal production, rural households consume 47 percent (it is usually the more affluent families that can afford this fuel); urban households consume 36 percent; and cottage industries use 17 percent (most of it in towns and larger urban centers) (see Table 7.2).

Table 7.2 **Kenya's Total Biomass Energy Consumption by Sector and Fuel Type, 2000**

	SHARE IN KENYA'S TOTAL BIOMASS ENERGY CONSUMPTION (percent)					
	Firewood	**Charcoal**	**Industrial Wood**	**Wood Wastes**	**Farm Residue**	**TOTAL BIOMASS**
Households: Rural	40.0	22.0	0.0	0.4	7.0	69.4
Households: Urban	1.0	17.0	0.0	0.2	0.0	18.2
Cottage Industry	4.0	8.0	0.0	0.0	0.0	12.0
Agriculture	0.0	0.0	0.0	0.0	0.0	0.0
Transportation	0.0	0.0	0.0	0.0	0.0	0.0
Commerce and Industry	0.0	0.0	0.3	0.0	0.0	0.3
TOTAL	**45.0**	**47.0**	**0.3**	**0.6**	**7.0**	**100.0**

Source: MoE 2002

Based on the household and charcoal surveys, it is likely that at least 30–50 percent of Kenya's wood supply now comes from farms and settlements and is mainly used for energy purposes. Map 7.3 highlights where farm forestry and the associated woodlots are located. The map also shows plantations, which are a minor supplier of wood for energy (Wass 2000). The majority of wood harvested from plantations is for timber and poles, but some is also used to meet energy needs (Wass 2000; FAO 2005).

To delineate areas important for farm forestry, Map 7.3 relies on a sample of detailed aerial photos from 1997 for the agricultural areas in the central and western parts of the country (ICRAF and DRSRS 2001). The photo interpreters could clearly identify the extent of woodlots within the sampled cropland. Depending on the tree species and the age of the trees in the woodlot, the wood may be destined for biomass energy (either used directly as firewood in the immediate proximity or converted to charcoal and transported to urban markets) or for construction purposes (e.g., poles or timber).

The forest plantations shown in Map 7.3 are over-represented. All land intended to be forest plantations are shown on the map as plantations, even if significant areas were not replanted with trees. The total plantation area on Map 7.3 is 127,000 ha—close to the estimate that should be under forest plantations according to the 1994 *Kenya Forestry Master Plan*.

A 1999 assessment indicated that of the 120,000 hectares that are supposed to be used as forest plantations (the numbers used in the *Kenya Forestry Master Plan*), only 78,000 hectares were sufficiently stocked with trees. This is the result of a very limited annual replanting program. About 6,000 hectares per year are cleared and about 3,000 hectares per year are planted, leading to 40,000 hectares of unstocked plantations (World Bank and GoK 1999; Mbugua 2000; Wass 2000). Increased rates of replanting in plantations could ease the demand for wood in other areas.

Map 7.3 Central and Western Kenya: Woodlots on Croplands, 1997, and Plantations, 2000

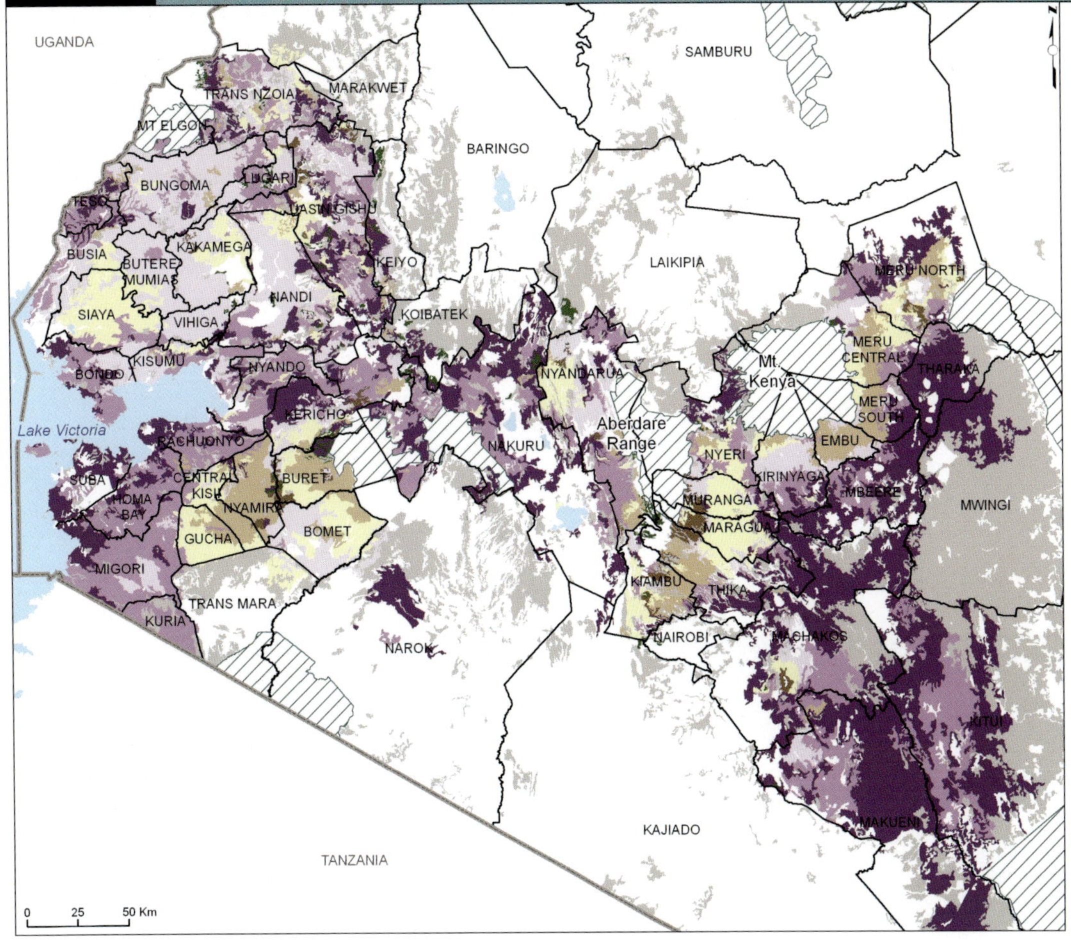

Sources: Administrative boundaries (CBS 2003), water bodies and cropland areas (FAO 2000), parks and reserves (IUCN and UNEP/WCMC 2006), and share of woodlots in croplands (WRI calculation based on ICRAF and DRSRS 2001).

Farms and settlements produce at least 30-50 percent of Kenya's wood supply, mainly for energy purposes. Agroforestry is the primary source of firewood. Private lands, either farmland or rangeland, are the major source of wood for charcoal.

This map shows the proportion of croplands covered by woodlots. Areas with higher percentages of woodlots cluster more extensively in the foothills of the Aberdare Range and Mount Kenya, and in most communities of Central Kisii, Nyamira, and Buret Districts. A relatively large area of the upper parts of Maragua and Muranga Districts is covered by cropland where woodlots cover more than 12 percent of the land. Close proximity to densely settled rural and urban areas, as well as other centers of high wood demand (for example, tea production) are among the factors behind these spatial patterns.

The share of woodlots is much lower in the western parts of the country. Farmers also do not plant woodlots in the more marginal cropping areas with lower rainfall, such as Makueni, Kitui, Mbeere, or Tharaka Districts. Note that these farmers may still plant trees for other purposes (see Map 7.2) and that woodlots are only one of many sources for firewood (other sources include vegetation used to demarcate boundaries, or vegetation on cropland).

Plantations (shown in dark green) cover only a very small percentage of the map area. The majority of them are government owned and most of the wood is used for timber. Major plantations are in the Rift Valley (e.g., Uasin Gishu, Keiyo, Koibatek and Nakuru Districts) and in the central part of the country (e.g., where Thika, Kiambu, and Nyandarua Districts border each other).

Note: The map combines detailed crop information (including the presence of woodlots) from 5,747 aerial photos for a growing season in 1997, each providing a sample point of detailed crop information. These samples are averaged to spatial units (polygons) of croplands from Kenya's most recent land cover map (FAO 2000).

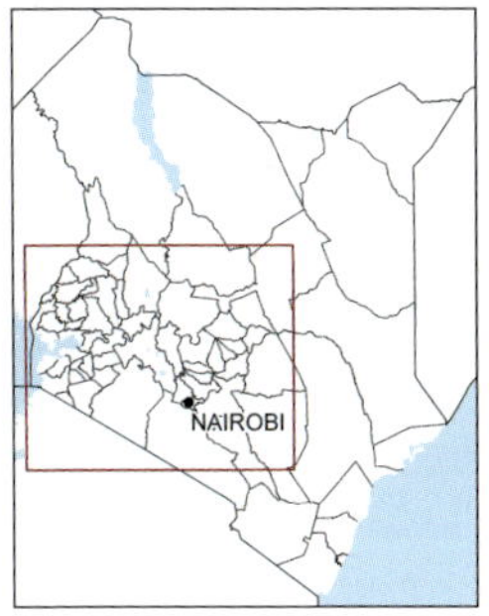

Tree plantations

WOODLOTS IN CROPLAND
(percent of sampled cropland)
- > 12
- 6 - 12
- 4 - 6
- 2 - 4
- 0 - 2
- No woodlots in sampled cropland
- Cropland not sampled

OTHER FEATURES
- District boundaries
- Major national parks and reserves (over 5,000 ha)
- Water bodies

KEY SUPPLY AREAS FOR FIREWOOD AND CHARCOAL

Woodfuel supply areas are difficult to map because of the ubiquitous use of the resource, the local scale of the firewood supply chain, and the limited availability of spatially disaggregated production data. Charcoal—an important fuel for urban households—is a special challenge because a 1986 Presidential directive banned the production and transport of charcoal (although it did not prohibit selling, buying, or using charcoal), and forced the charcoal market underground (Matiru and Mutimba 2002).

Firewood Collection and Charcoal Making

Data from recent studies make it possible to map several important variables related to firewood and charcoal in Kenya, including where firewood and charcoal appear to be important sources of income (Map 7.4), and which sources of wood are used for charcoal production (Map 7.5) (MoE 2002; ESDA 2005a; ALRMP et al. 2006). While the underlying data and the resulting maps still have significant gaps in coverage, together they provide an initial picture of the spatial patterns of charcoal and firewood production in Kenya.

Firewood is the dominant energy source for rural households, with 89 percent of rural Kenyans relying on firewood for their energy needs (MoE 2002). Typically, the firewood is used close to the source of extraction. More than 80 percent of households obtain their firewood within a 5-kilometer radius of their home (MoE 2002). The average length of time spent on collection is about two hours per day (MoE 2002)—a task that falls disproportionately on women and girls and takes time away from other productive activities.

About 82 percent of urban households and 34 percent of rural households (MoE 2002) use charcoal. Traders transport charcoal over great distances, primarily to urban markets. Since the 1980s, the proportion of rural households relying on this source of energy has slowly increased (MoE 2002).

Kenya's *National Charcoal Survey* estimates that 200,000 people produce charcoal, half of whom work full-time and half part-time (ESDA 2005a). About 300,000 people are involved in transportation and vending. These 500,000 people support approximately two million dependents. As Figure 7.1 shows, charcoal production provides significant employment, comparable to other important sectors in the economy (ESDA 2005b).

The *National Charcoal Survey* estimates gross revenue from charcoal production at Ksh 32 billion per year (US$ 457 million) (ESDA 2005a). A separate study (MoE 2002) provides a lower, but still significant, estimate of Ksh 17.5 billion per year (US$ 250 million). Charcoal revenues are calculated to be significantly higher than the returns from sugarcane, coffee, maize, and other cereals. Depending on the average retail price for charcoal and the estimated volume of national production, the gross revenue from charcoal lies between that of horticulture exports and that of livestock and related products (Figure 7.2).

Unlike other commodities, the government does not receive any tax revenues from the charcoal sector due to the 1986 Presidential directive that made the production and transport of charcoal illegal. Assuming a valued-added tax of 16 percent, the annual loss in tax revenues could be as high as Ksh 5.1 billion (US$ 72.9 million) per year (ESDA 2005b). While the aggregated revenues from the charcoal industry represent a significant amount, charcoal production remains a poorly remunerated occupation. The average monthly income is Ksh 4,496 (US$ 64) for a producer, Ksh 7,503 (US$ 107) for a vendor, and Ksh 11,298 (US$ 161) for a transporter (ESDA 2005a).

Figure 7.1 **Employment by the Charcoal Industry Compared to Other Formal Sectors, 2004**

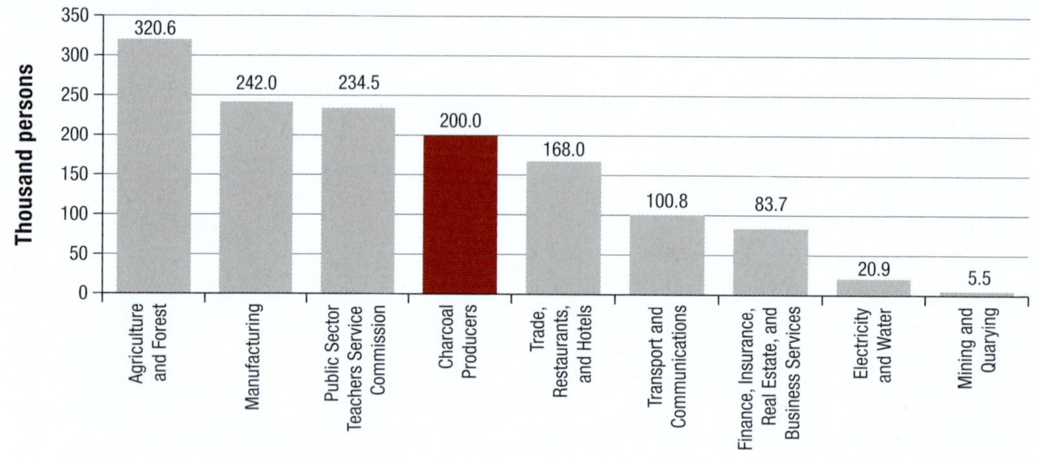

Sources: ESDA 2005b, CBS 2006.

Figure 7.2 **Gross Revenue from Charcoal Compared to Marketed Production of Agricultural Products, 2004**

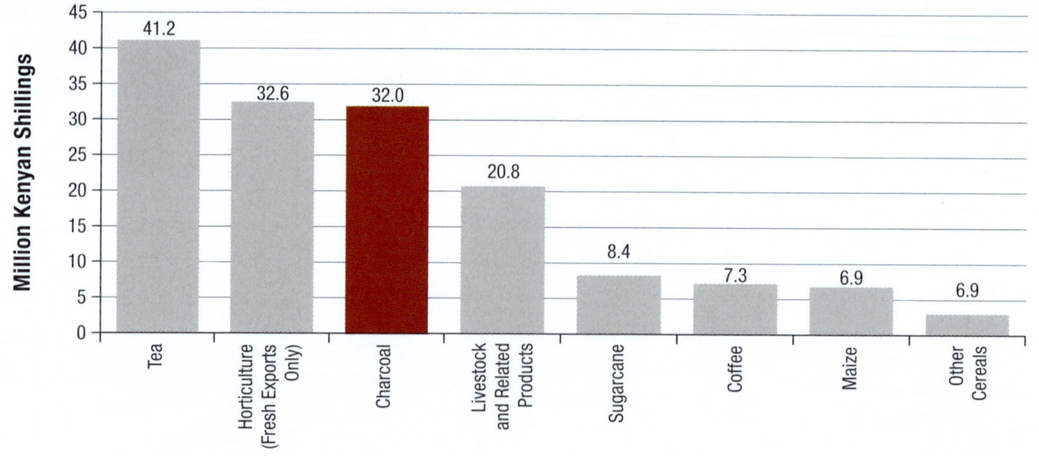

Sources: ESDA 2005b, CBS 2006.

Note: The economic value of agricultural products only reflects the quantities that were officially recorded in the market of 2004 (using average prices). The statistics do not count production for self-consumption or quantities traded informally. For example, millions of smallholder farms produce some maize for home consumption, but maize that reaches the commercial market comes mostly from large-scale farms (Jayne et al. 2000).

Map 7.4 Cash Income From Firewood Collection and Charcoal Making, 2003-05

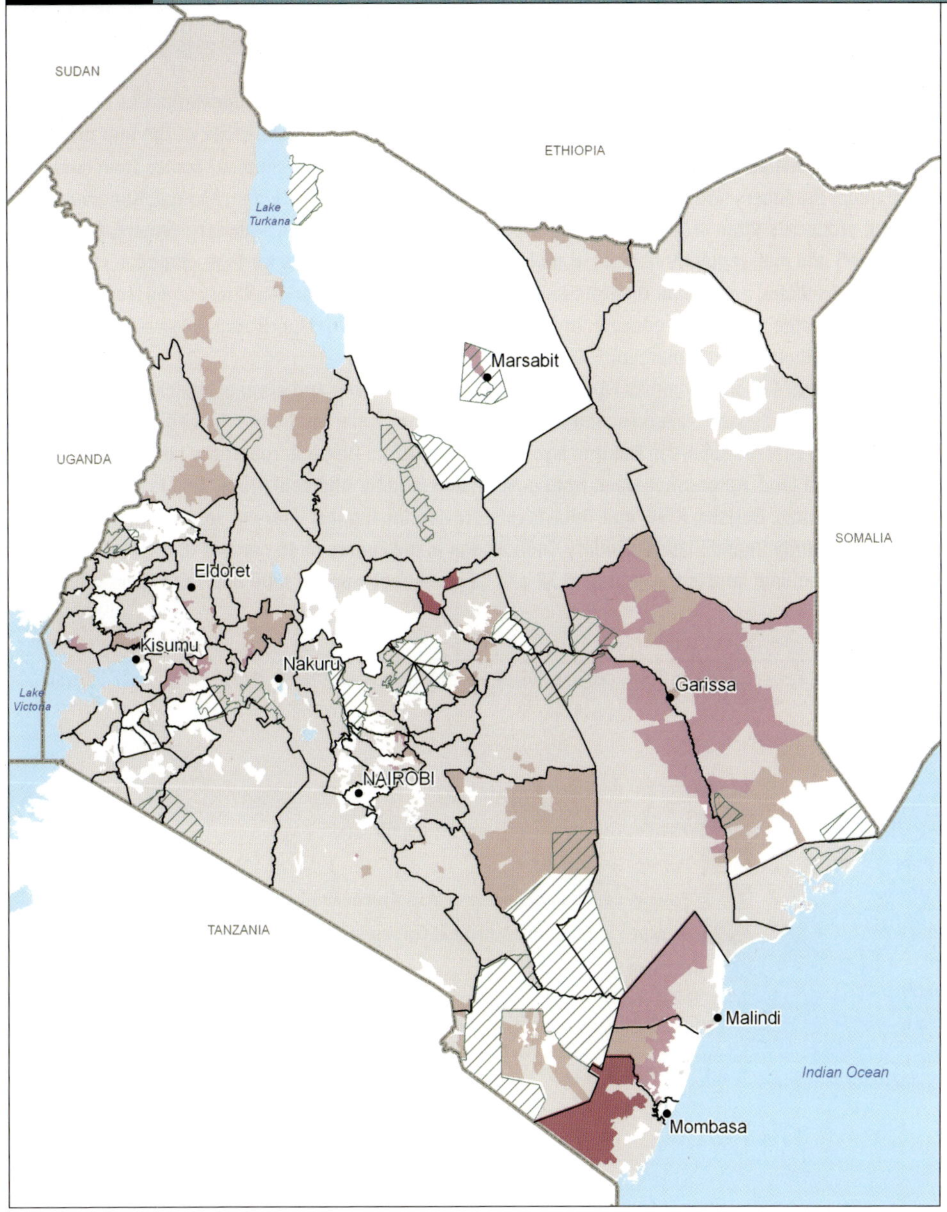

Sources: Administrative boundaries (CBS 2003), cities (SoK and ILRI 2000), water bodies (FAO 2000), and share of cash income from firewood collection and charcoal making (ALRMP et al. 2006).

Charcoal production and firewood collection is an important economic activity in Kenya. The sector contributes to income in most areas, except the more remote locations that have very little woody vegetation (e.g., parts of Marsabit District). These activities are also not a significant source of income in selected communities in the central part of the country and directly along the Indian Ocean (although households may still collect firewood or produce charcoal for their own use).

The majority of households in communities located about 50 kilometers inland from Mombasa (in Kwale District) obtain more than 20 percent of their cash from firewood and charcoal. Income from firewood and charcoal ranges between 10 and 20 percent of total income in the coastal hinterlands close to Malindi. Communities in the west (slightly inland from Lake Victoria) and along the Tana River (close to Garissa) show similarly high percentages. Charcoal from *mathenge* (*Prosopis juliflora*, also known as mesquite), an invasive shrub that is cleared from the land to save pasture, is the main source for this cash in Garissa District.

Note: Data are based on questionnaires sent to key food security experts in all Districts (generally about 6-10 people) to obtain information on predominant livelihood characteristics important for food security planning. In some cases where further clarification was necessary, questionnaires were sent to experts below District level (Division). This group of experts classified each of Kenya's 6,632 Sublocations by their predominant livelihood strategy and other livelihood characteristics including different sources of cash income.

CASH INCOME FROM FIREWOOD AND CHARCOAL
(percent of total income)

- > 20
- 10 - 20
- 5 - 10
- 0 - 5
- No cash income from charcoal and firewood

OTHER FEATURES

- District boundaries
- Major national parks and reserves (over 5,000 ha)
- Water bodies

Map 7.5 Sources of Wood for Charcoal in Selected Districts, 2004

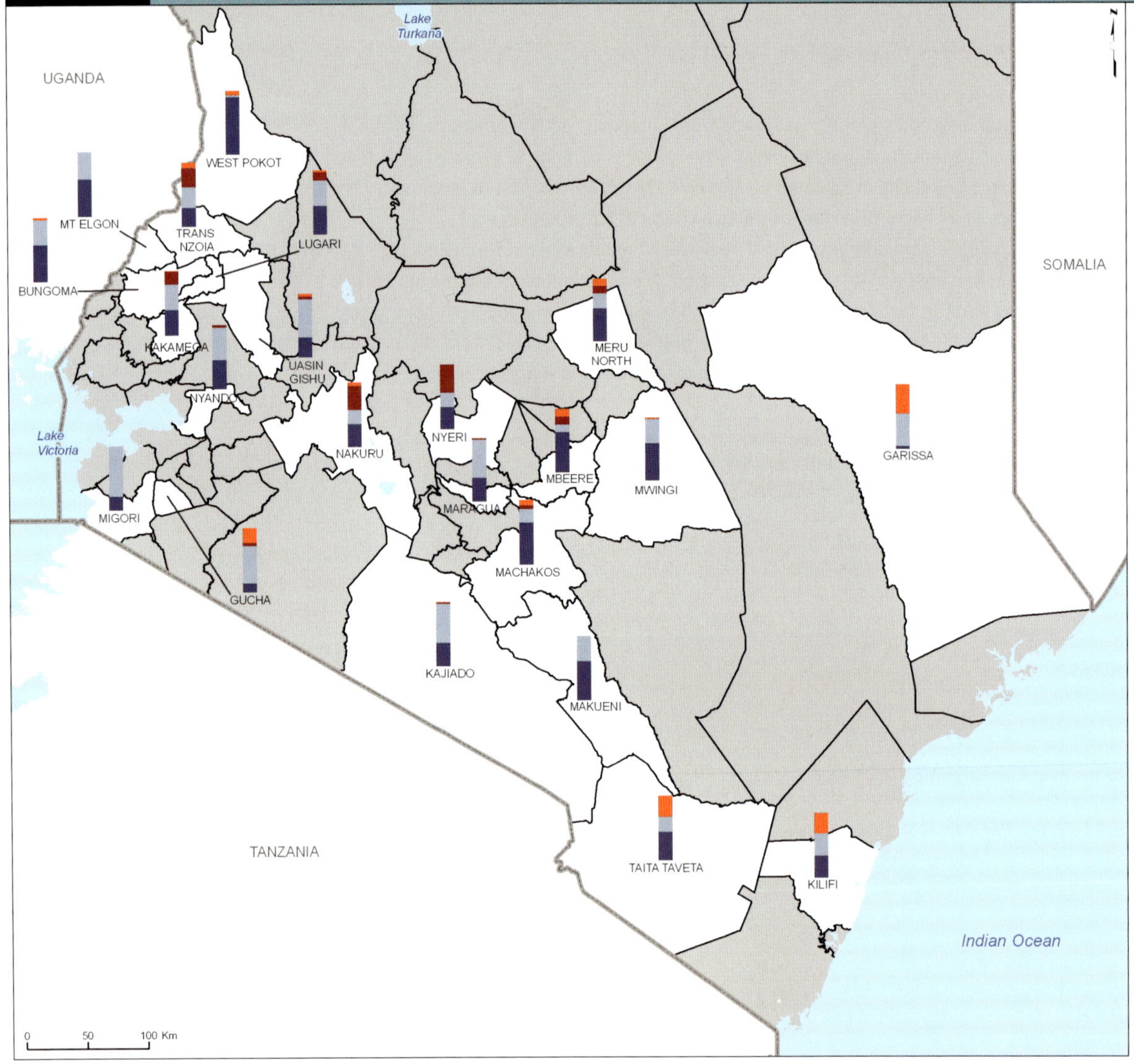

Sources: Administrative boundaries (CBS 2003), water bodies (FAO 2000), and sources of wood for charcoal (ESDA 2005a).

Eighty-two percent of Kenya's charcoal comes from private land (either farmland or rangelands) as highlighted by the dominance of blue shading in the stacked bars representing the relative share of the four principal wood sources. In eight of the 22 surveyed Districts, more than half of the wood for charcoal comes from land owned by the charcoal producers (West Pokot, Machakos, Makueni, Mbeere, Mount Elgon, Mwingi, Bungoma, and Meru North). These producers often grow trees for other purposes (e.g., fruit, shade, boundary demarcation, or construction material) and may regularly harvest branches or rely on tree remnants for their charcoal. In Migori, Kajiado, Maragua, Uasin Gishu, and Gucha Districts more than half of the wood for charcoal comes from private land that is not owned by the charcoal producers. In many parts of these Districts, private landowners hire labor to remove vegetation on their land for charcoal.

Only 18 percent of Kenya's charcoal comes from public lands (shown in red and orange), which include government land (e.g., national parks, game reserves, and forest reserves) and other land either owned communally or by a County Council. Charcoal producers in Nakuru, Nyeri, and Trans Nzoia Districts report the largest proportion of wood from government land. Removal of wood from government land for charcoal production is illegal. Among the sampled Districts, Garissa, Kilifi, and Taita Taveta provide the highest share of wood from other public lands (communal and County Council lands). County Council land is the source of 45 percent of the wood in Garissa District, and communal land is the source of 32 and 33 percent of wood in Kilifi and Taita Taveta Districts, respectively.

Note: Land in Kenya can be owned by government, County Councils, groups, and individuals (Kameri-Mbote 2005).

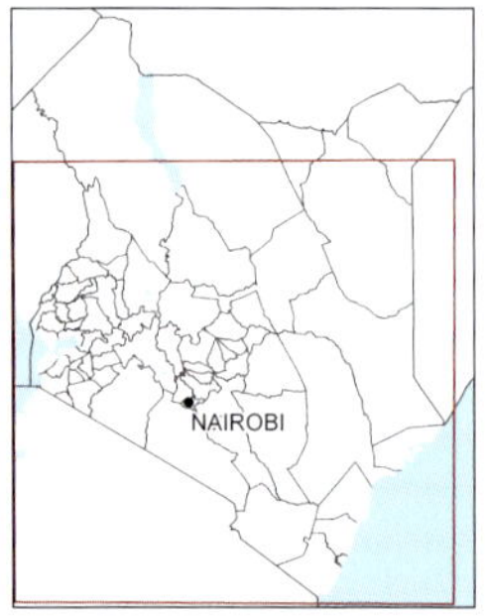

SOURCES OF WOOD FOR CHARCOAL

- 'Public': County Council and communal land
- Public: Government
- Private: Land not owned by charcoal producer
- Private: Land owned by charcoal producer
- No data

OTHER FEATURES

- District boundaries
- Water bodies

Growth of Biomass and Potential Harvest

Wood for charcoal and firewood can come from less modified ecosystems such as forests, woodlands, bushlands, and wooded grasslands. It can also come from managed landscapes such as fuelwood plantations, woodlots growing native and exotic trees on farmland, or trees and shrubs growing along the boundaries of cropland or roads. With care, wood can be harvested in a sustainable manner, with the harvest rate no higher than the annual biomass growth. Examples include removing only dead branches and any annual regrowth, or planting new trees to replace those that have been cut. Of course, wood can also be harvested in an unsustainable manner, leading to a decline in the stock of woody biomass. This results from harvesting more wood than grows back every year or clearing the land completely of all vegetation, either because of very high local energy demand or demand for land for new settlements, pasture, or croplands.

Assessments of supply and demand for fuelwood typically rely on studies estimating the annual growth of biomass. This growth rate depends on many factors, including rainfall, soil type, and the age of the vegetation community. The following maps draw upon detailed data from Kenya's most recent study examining energy supply and demand (MoE 2002). These maps represent the first attempt at a spatial representation of these data.

Map 7.6 is the result of applying the mean annual woody productivity rates of different types of vegetation in various agroecological zones to Kenya's most recent and detailed land cover maps (FAO 2000). It reflects the amount of wood that, in theory, could be sustainably harvested annually from vegetation outside croplands without depleting the biomass stock.

Map 7.6 Annual Growth of Biomass Outside of Croplands

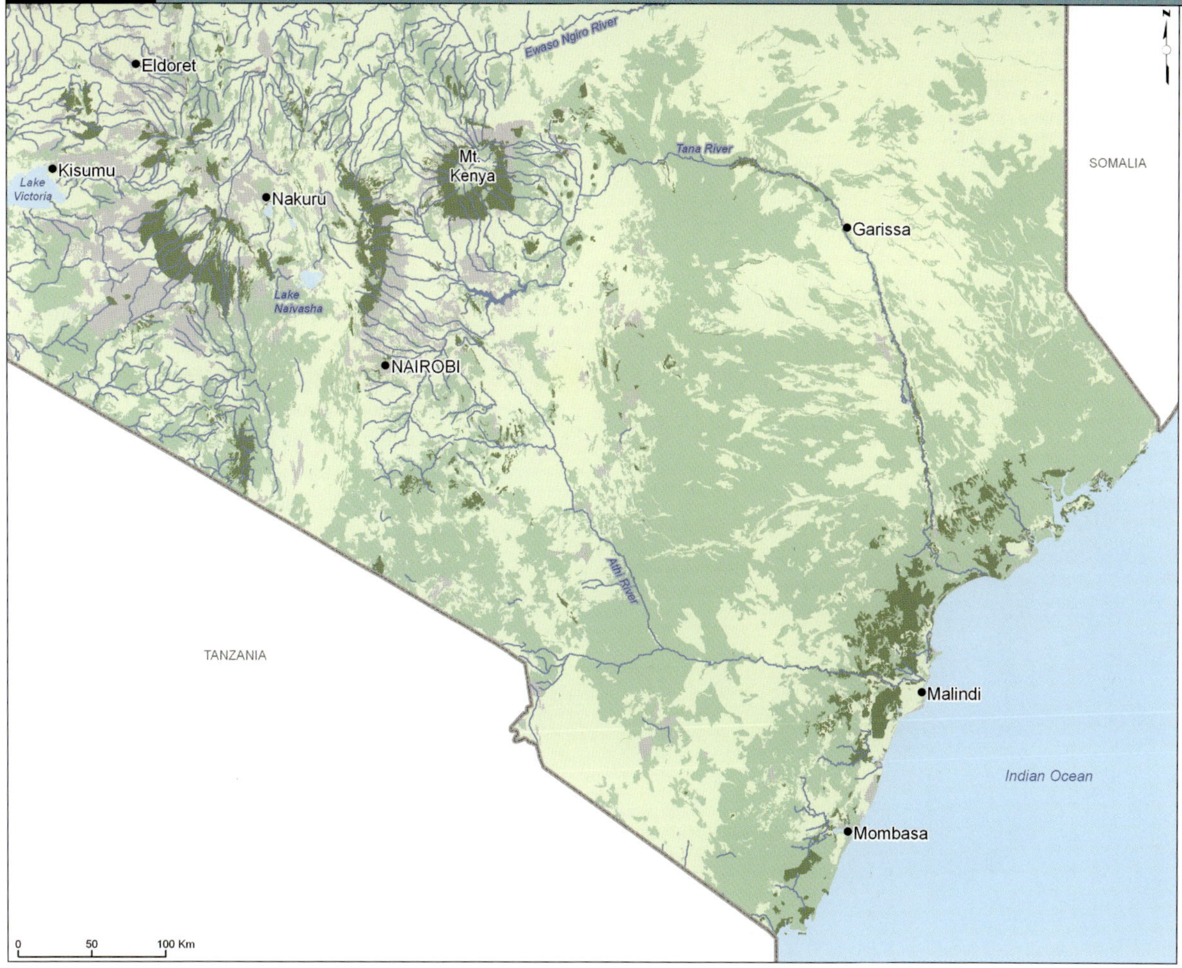

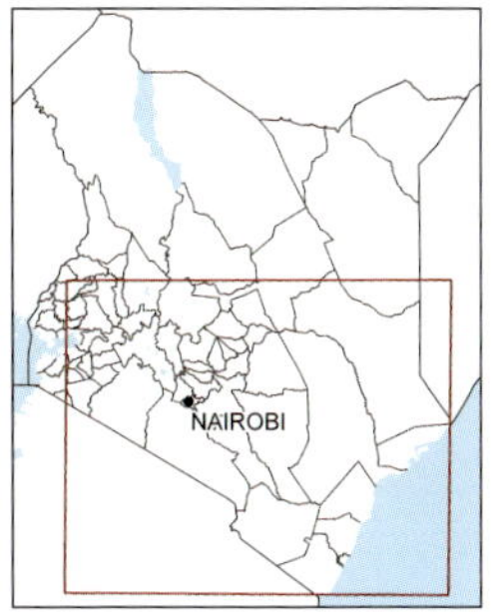

ANNUAL GROWTH OF BIOMASS OUTSIDE CROPLANDS
(cubic meters per sq. km per year)

- > 85
- 35 - 85
- <= 35
- Not applicable (intensive cropping and urban areas)

WATER BODIES AND RIVERS

- Permanent rivers
- Water bodies

Sources: Cities (SoK and ILRI 2000), water bodies (FAO 2000), permanent rivers (NIMA 1997), and annual growth of biomass for vegetation outside of croplands (WRI calculation based on MoE 2002, FAO 2000).

The map depicts, in three broad categories, a rough estimate of the amount of wood that could be sustainably harvested, that is, the annual growth of wood biomass per year from vegetation outside of croplands that could be harvested without depleting the stock. Closed forests in the mountain ranges, and forests and dense woodlands along the coast are the most productive. Areas that are generally classified as bushlands or woodlands on national maps fall in the mid-range of productivity. The areas in the lowest growth category either are a mix of cropland and natural landscapes (with little remaining natural vegetation) or have few trees, for example, grasslands.

Note: All areas classified as 'natural and semi-natural' in the *Africover* map (FAO 2000) were grouped into five broad vegetation classes (closed forest, woodland, bushland, wooded grassland, and grassland) based on their vegetation characteristics (38 different *Africover* codes). Each of the five broad vegetation classes was assigned the same average annual woody biomass growth rates as used in the Ministry of Energy (2002) study to estimate Kenya's biomass supply. *Africover* spatial units (polygons) with mixed vegetation classes (e.g., cropland interspersed with 'natural and semi-natural vegetation') were weighed by the respective area contribution. For the final map, total woody biomass growth (from standing natural biomass sources) for each *Africover* polygon was divided by its total polygon area to obtain growth of biomass in cubic meters per square kilometer per year.

Map 7.7 **Theoretically Harvestable Biomass Yield Outside of Croplands**

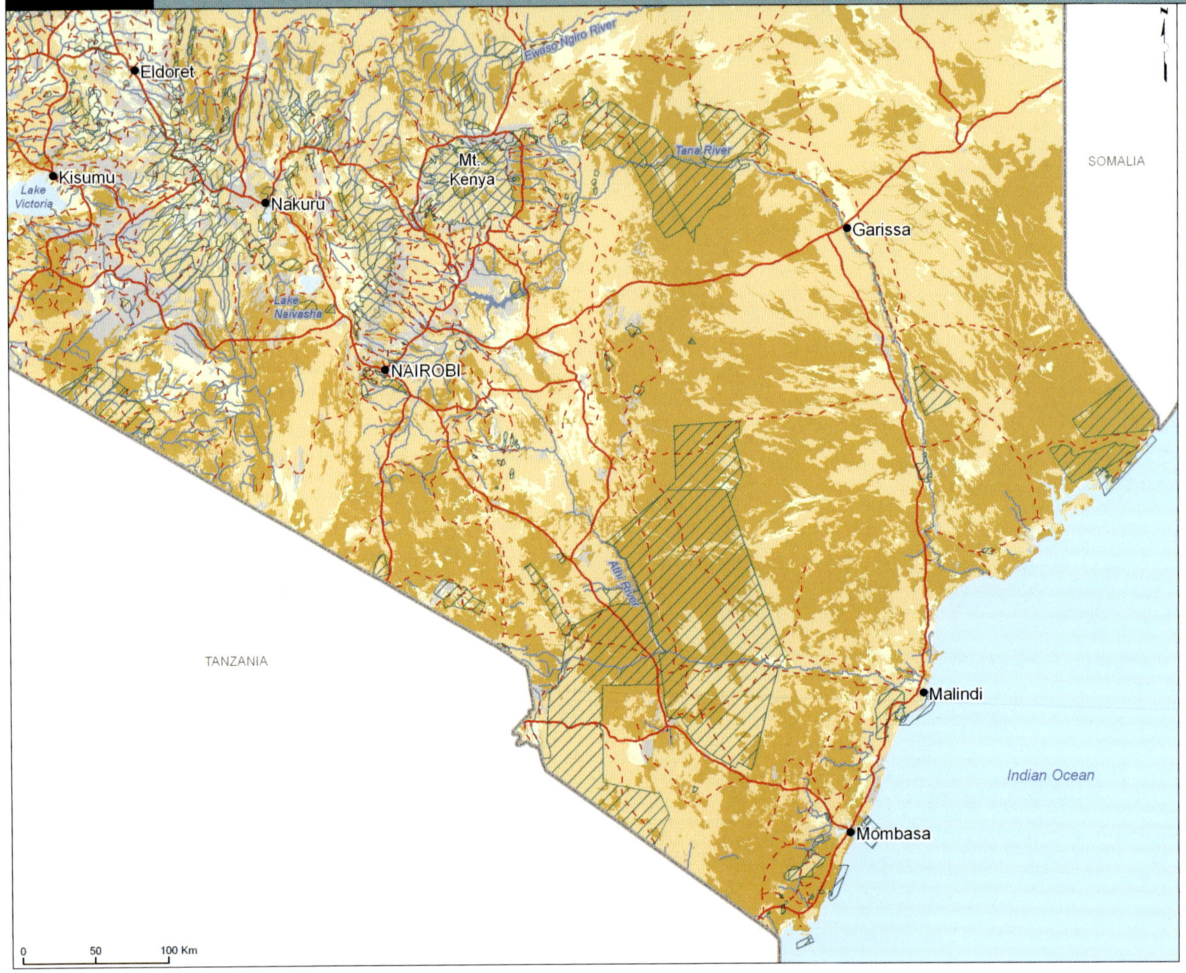

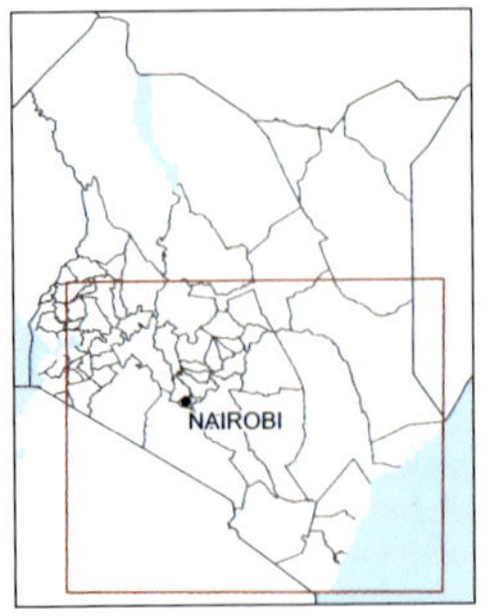

THEORETICALLY HARVESTABLE BIOMASS YIELD OUTSIDE CROPLANDS
(cubic meters per sq. km per year)

- > 10
- 5 - 10
- <= 5
- Not applicable (intensive cropping and urban areas)

OTHER FEATURES

- Primary roads
- Secondary roads
- National parks and reserves

WATER BODIES AND RIVERS

- Permanent rivers
- Water bodies

Sources: Cities (SoK and ILRI 2000), water bodies (FAO 2000), permanent rivers (NIMA 1997), parks and reserves (IUCN and UNEP/WCMC 2006), primary and secondary roads (SoK and ILRI 1997), and theoretically harvestable biomass yield outside of croplands (WRI calculation based on Map 7.6, MoE 2002, FAO 2000).

According to Map 7.7, the high-yield areas of theoretically harvestable biomass growth from natural vegetation closest to Nairobi would be the rangelands south of the city (in Narok and Kajiado Districts), but also in the southeast (in parts of Machakos District). For Mombasa, the closest areas would be the woodlands about 50 kilometers from the coast (in Kwale and Kilifi Districts), but also the tempting supplies within protected areas (Tsavo East and West National Parks). Areas close to Garissa and farther east near the Somalia border have similar high yields, but are disadvantaged by long transport distances (increasing costs), poorer roads, and a more limited supply of labor.

While Map 7.7 may provide a correct relative picture of potential woodfuel supply areas (assuming sustainable harvest levels), the map may still underrepresent the actual quantity of wood removed for energy purposes. In some areas, local energy needs may be much higher than harvestable regrowth, leading to depletion of trees and other woody vegetation. In other areas, land clearing for new farms or new cropland can result in higher, albeit short-term, supplies of wood. For example, the removal of *mathenge* (*Prosopis juliflora*, also known as mesquite) in Garissa District results in much greater local wood supplies.

However, because of legal issues (including laws governing protected areas), the distance between demand and supply centers, lack of roads, and other factors, not all the wood that theoretically grows every year is available for energy use. Map 7.7 incorporates 'accessibility' factors from the national energy study (MoE 2002). The experts behind the study assumed that only a portion of different vegetation types are available for energy use:

- 5 percent of potential growth in closed forests (a result of controlled access by the Forest Department).
- 30 percent of the growth in woodlands, bushlands, and wooded grasslands (a result of more open access, combined with smaller tree diameters).
- 10 percent of the vegetation in wooded grasslands (primarily in proximity to settlements).

Map 7.7 is a closer approximation of the theoretically harvestable biomass growth outside of croplands because it is the result of multiplying the annual growth rate (Map 7.6) with these 'accessibility' factors. By outlining the boundaries of protected areas, which prohibit but do not always manage to exclude woodfuel removal, and by indicating major roads and cities, the map can be used to delineate potential supply areas of 'sustainably harvested' charcoal, which is typically transported over long distances from rural to urban demand centers.

Box 7.1 Links to National Decision-Making: Using Maps of Ecosystem Services to Establish a More Sustainable Charcoal Industry

The most recent *National Charcoal Survey* demonstrated the economic significance of the charcoal industry in terms of employment and gross revenues. Over 2.5 million Kenyans are supported by the industry. Charcoal production or trade is carried out in almost all of Kenya's Districts. Charcoal producers capture only a small percentage of the revenues because the price at the point of production is significantly lower (Ksh 200 per bag) than the retail price (Ksh 700 per bag). Charcoal transporters have justified their high markup by citing the significant costs linked to the still illegal transport of the commodity. The study estimates that the government is foregoing Ksh 5.1 billion (US$ 72.9 million) in annual revenues by upholding the ban on production and transport of charcoal rather than taxing and regulating the industry (ESDA 2005a).

Charcoal, together with firewood, is still the dominant fuel in Kenya. About 82 percent of urban and 34 percent of rural households use charcoal as their main energy source. Demand will continue its growth in the near future (MoE 2002).

Policymakers are now acknowledging more openly that the charcoal industry is a significant contributor to rural livelihoods and that poverty and the need to generate income has been driving certain types of production patterns. They are beginning to understand that a blanket ban on production and transport has promoted inefficient production technologies, lowered producer prices, sacrificed government revenues, and led to unsustainable wood extraction in certain areas (ESDA 2005a). They also accept that, based on current economic conditions and the existing energy infrastructure, charcoal will remain an important energy source in the short to medium term. It is therefore paramount to establish a more sustainable and environmentally benign charcoal industry (ESDA 2005a).

The *National Charcoal Survey* therefore explored the potential for a more sustainable charcoal industry. It proposes specific policy and institutional changes that would put charcoal making on a more sustainable path and contribute to improved livelihoods for people involved in the industry. The Survey suggests reforming the regulatory framework, which would include specific production standards and certification processes. It recommends making the charcoal industry legal and fully integrating it into the national economy, thus making it a source of government revenue and creating better conditions for charcoal producers. The Survey also proposes establishing institutions that would oversee and audit the industry. Other important recommendations include pilot project zones of sustainable charcoal production, where different technologies and production approaches could be tested and improved, and a woodfuel fund (perhaps based on a transport levy) that could be invested in new, more sustainable production technology or support disadvantaged producers (ESDA 2005a).

Combining maps and spatial indicators of biomass energy production, energy use, other ecosystem services, and poverty can become a valuable tool for decision-makers to implement some of the main recommendations in the *National Charcoal Survey*. Below are some specific examples that link recommendations from the Survey to possible map overlays, as a first step for more detailed follow-up studies.

Support fuelwood and charcoal producers on private land. Most of Kenya's charcoal comes from private land, and a significant proportion of that supply comes from farmland owned by the charcoal producers themselves. These producers plant trees and selectively cut and prune them for charcoal (ESDA 2005a). Many of them could benefit from tree nurseries providing better-suited varieties (which, for example, yield more biomass or require less water) and knowledge of sustainable agroforestry practices. Maps can highlight where these producers are located: most of them are in higher rainfall and high potential agroecological zones. Such a production map can then be combined with maps showing which tree species are currently planted, whether households have sufficient resources to make additional investments in new species or set aside land for trees, and whether farmers have been trained in more sustainable agroforestry practices. Analyses of these relationships will provide more insight on promising locations for tree nurseries, their potential supply areas, and their demographic and poverty characteristics. Other analyses can show where to conduct agroforestry training or initiate tree-planting activities.

Improve efficiency of charcoal production. The most common technique used for charcoal production is the earth kiln. Such kilns have a very low recovery rate, requiring 100 kg of wood to produce 10–15 kg of charcoal. Changing the type of kiln, improving the stacking of wood in the kiln, and modifying the burning process, all can boost the charcoal recovery rate up to 30 kg of charcoal per 100 kg of wood (ESDA 2005a). The same set of maps listed above can assist in selecting promising sites for model community kilns and targeting training efforts.

Allocate land for sustainable charcoal production. Kenya's 2004 National Energy Policy proposes that the government dedicate 25 percent of forest land to bio-energy production (ESDA 2005a). Private landowners have also expressed interest in supplying wood for charcoal once the industry becomes fully legalized and more transparently managed (ESDA 2005a). Maps can show the locations of plantations and government lands (see Map 7.3) and can provide estimates of their current stocking levels and annual regeneration rates (as shown in Map 7.8). These maps of current supply can be combined with others of potential future supply. Such maps would delineate optimal areas for tree species well-suited for charcoal making, for example species that are endemic (and thus better adapted) and that have a very high regeneration rate (thus allowing faster rotations and recovery of investments). Combining these different layers with demographic and poverty maps, maps of energy demand and markets, and transport infrastructure, would be useful inputs to delineate the most promising areas and assist in the planning of pilot projects.

Create buffer zones for sustainable charcoal production in areas bordering protected areas. The *National Charcoal Survey* suggests creating buffer zones where charcoal production is specifically encouraged to reduce pressure on protected areas and build better livelihoods (ESDA 2005a). Maps highlighting where government land is still a major source for illegal wood, and where high local demand is outpacing local supplies, could be the first map layers used to select such buffer zones.

Establish better regulations and guidance on land use, especially when changes in land use are taking place. In the past, major land use changes—for example, from gazetted forests to croplands, or from rangelands to croplands—have led to short-term, unsustainable charcoal production because of the complete removal of all woody vegetation. Maps can highlight zones where such wood removal would have highly negative impacts on other ecosystem services (such as biodiversity or hydrology) and should not be permitted. Maps, coupled with more detailed hydrological models, can also highlight areas where massive planting of exotic tree species would greatly impact water balances, and where it would have little impact within a watershed. Such map-based analyses can become the foundation for improved land use planning and zoning in the country.

Box 7.2 Creating a Poverty and Demographic Profile for Croplands With Different Proportions of Woodlots

Map 7.8 Woodlots on Croplands in Five Provinces, 1997

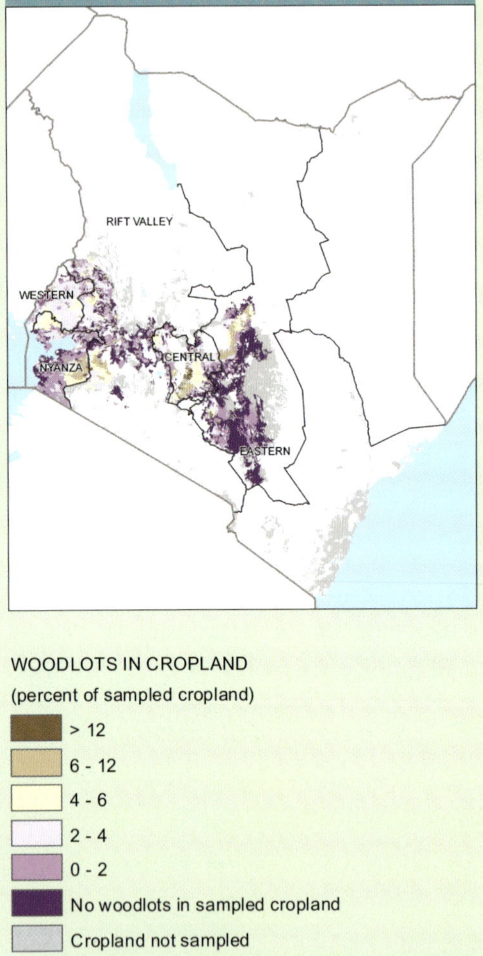

The decision to maintain woodlots on croplands illustrates the relative value that farmers assign to wood or crops. Combining poverty maps with maps that show the degree to which farmers have dedicated their cropland to woodlots can provide insights into possible relationships between wood supplies, agroforestry investments, and poverty.

Table 7.3, based on Map 7.8, classifies the land area of each Province into eight classes reflecting the degree to which croplands are covered with woodlots. Since the underlying data are in GIS format, the total population and population density for each of the eight classes can be estimated, as well as the number of poor people and the average poverty rate.

What Do the Map and Poverty Profile Show?

- The table shows that very few farmers have set aside more than 12 percent of their cropland for woodlots. In all Provinces, the class with the highest woodlot share covers the smallest area and is inhabited by fewer people compared to the other classes. Areas with shares of 2–4 percent or 4–6 percent are more extensive and include a greater number of people in most Provinces.
- The relationship between poverty rate and the share of woodlots in cropland is less straightforward. The differences in poverty rates between the five Provinces are much greater than the differences between the eight classes within each Province. At this level of aggregation, there is not a clear correlation between the percentage of cropland taken up by woodlots and the average poverty rate.
- Nonetheless, the table does reveal some noteworthy patterns. For example, areas with croplands covered by more than 12 percent woodlots are below the Provincial average poverty rate (with one exception, Western Province). In addition, some Provinces (Central and Nyanza Provinces) show a declining poverty rate from the 'no-woodlot' class to the highest woodlot class. These patterns need to be further examined at a more detailed scale. Combined with additional location-specific information (e.g., level of wood demand, presence of tree nurseries, household capital, and labor availability), this could provide insights on whether reduced poverty is the result or cause of an increased share of cropland devoted to woodlots.

Similar profiles can be constructed overlaying other woody biomass-related indicators from this chapter with poverty indicators from Chapter 2. For example, identifying all communities with high poverty rates bordering closed forest areas, and combining that information with maps on charcoal supply sources and agroforestry practices could pinpoint promising areas where tree planting and agroforestry training may reduce pressure to illegally remove wood from government reserves.

Table 7.3 People, Poverty, and Woodlots in Croplands

PROVINCE	AREAS WITHOUT CROPLAND AND SHARE OF WOODLOTS IN SAMPLED CROPLAND AREAS	AREA (SQ. KM)	NUMBER OF PEOPLE (000)	AVERAGE POPULATION DENSITY (PERSONS PER SQ. KM)	NUMBER OF POOR (000)	AVERAGE POVERTY RATE (%)	KSH NEEDED PER MONTH TO REACH POVERTY LINE[1] (MILLION)
EASTERN							
	No Cropland	118,134	670	6	359	54	93.2
	Cropland Not Sampled	13,184	650	49	391	60	112.1
	No Woodlots	14,394	1,128	78	702	62	213.8
	Woodlots 0.1–2%	4,322	563	130	357	63	109.7
	Woodlots 2–4%	1,082	294	272	166	56	44.5
	Woodlots 4–6%	2,179	277	127	154	56	39.8
	Woodlots 6–12%	2,726	540	198	290	54	72.8
	Woodlots >12%	1,474	45	31	26	57	7.1
	TOTAL 9 Districts	**157,495**	**4,166**	**26**	**2,445**	**59**	**693.1**
CENTRAL							
	No Cropland	3,675	351	96	110	31	16.3
	Cropland Not Sampled	1,819	366	201	111	30	17.2
	No Woodlots	1,564	282	180	102	36	18.0
	Woodlots 0.1–2%	1,518	366	241	123	34	17.8
	Woodlots 2–4%	1,359	439	323	145	33	19.3
	Woodlots 4–6%	1,959	854	436	263	31	32.6
	Woodlots 6–12%	942	429	455	127	29	14.5
	Woodlots >12%	388	144	371	43	30	4.5
	TOTAL 6 Districts	**13,224**	**3,231**	**244**	**1,024**	**32**	**140.3**
RIFT VALLEY							
	No Cropland	145,696	1,969	14	968	49	245.6
	Cropland Not Sampled	14,656	1,024	70	457	45	101.1
	No Woodlots	7,708	799	104	361	45	87.4
	Woodlots 0.1–2%	3,702	638	172	288	45	65.5
	Woodlots 2–4%	3,829	700	183	342	49	76.7
	Woodlots 4–6%	3,843	557	145	294	53	87.9
	Woodlots 6–12%	3,036	265	87	135	51	29.6
	Woodlots >12%	2,155	70	32	32	46	6.6
	TOTAL 6 Districts	**184,625**	**6,022**	**33**	**2,877**	**48**	**700.4**

Continued

Table 7.3 People, Poverty, and Woodlots in Croplands — continued

PROVINCE	AREAS WITHOUT CROPLAND AND SHARE OF WOODLOTS IN SAMPLED CROPLAND AREAS	AREA (SQ. KM)	NUMBER OF PEOPLE (000)	AVERAGE POPULATION DENSITY (PERSONS PER SQ. KM)	NUMBER OF POOR (000)	AVERAGE POVERTY RATE (%)	KSH NEEDED PER MONTH TO REACH POVERTY LINE[1] (MILLION)
NYANZA							
	No Cropland	806	208	258	134	65	65.9
	Cropland Not Sampled	713	182	256	120	66	43.3
	No Woodlots	1,943	385	198	256	67	88.3
	Woodlots 0.1–2%	4,849	1,189	245	755	64	253.0
	Woodlots 2–4%	1,343	417	311	259	62	79.5
	Woodlots 4–6%	1,763	869	493	542	62	161.8
	Woodlots 6–12%	1,064	603	567	392	65	124.6
	Woodlots >12%	63	12	190	7	58	2.0
	TOTAL 12 Districts	**12,544**	**3,865**	**308**	**2,466**	**64**	**818.3**
WESTERN							
	No Cropland	1,061	126	119	78	62	23.0
	Cropland Not Sampled	314	86	274	49	57	13.2
	No Woodlots	531	138	260	75	54	18.6
	Woodlots 0.1–2%	1,232	385	312	226	59	61.3
	Woodlots 2–4%	4,318	1,900	440	1,133	60	325.9
	Woodlots 4–6%	766	296	386	176	59	48.6
	Woodlots 6–12%	179	58	324	35	60	9.9
	Woodlots >12%	56	17	305	10	59	2.7
	TOTAL 6 Districts	**8,457**	**3,006**	**355**	**1,782**	**59**	**503.3**
	TOTAL 39 Districts	**376,345**	**20,290**	**54**	**10,594**	**52**	**2,855.4**[2]

Sources: Poverty and demographic estimates (1999) are WRI/ILRI calculation based on CBS 2002 and CBS 2003. Area without cropland, cropland not sampled, and area estimate for each woodlot percentage class are WRI calculation based on data for Maps 7.3 and 7.10 (ICRAF and DRSRS 2001; FAO 2000).

Note: [1] The poverty gap measures the average expenditure shortfall (gap) for the poor in a given administrative area relative to the poverty line. It is a crude estimate of the minimum amount of resources needed to eradicate poverty (see Chapter 2).
[2] The total amount to close the poverty gap for one month in the 39 Districts (Ksh 2.9 billion) equals about US$ 40.8 million (at US$ 1 = Ksh 70).

SUMMING UP

- While Kenya's forests, woodlands, bushlands, wooded grasslands, and agroecosystems supply a wide array of ecosystem services, one of their major contributions is supplying wood. Kenyans use 80–90 percent of the wood from these ecosystems for energy purposes (firewood and charcoal), and the remaining 10–20 percent for timber, posts, and poles.
- Biomass is Kenya's dominant fuel, accounting for over 80 percent of total energy consumption in 2000. Burning firewood and charcoal account for roughly equal percentages of total wood consumption.
- Estimates put Kenya's 1995 closed forest area at 984,000 hectares (1.7 percent of the land area). Other natural woody vegetation includes 2.1 million hectares of woodlands, 24.6 million hectares of bushlands, and 10.6 million hectares of wooded grasslands. Agricultural land can have a high percentage of tree cover as reflected in the high tree density in the croplands of Central Province, for example. Woodlands, bushlands, and wooded grasslands contain most of Kenya's woody biomass, albeit at much lower tree density and volume per area than the small remnants of closed forests. Closed canopy forests are only a minor contributor of woodfuel at a national level.
- The majority of wood harvested from plantations is for timber and poles. Of the 120,000 hectares designated as forest plantations, only 78,000 hectares were sufficiently stocked with trees in 1999.
- About 89 percent of rural Kenyans rely on firewood for their energy needs. More than 80 percent of households obtain firewood within a 5-kilometer radius of their home. The average length of time spent on collection is about two hours per day—a task that falls disproportionately on women and girls.
- About 8 percent of firewood supplies came from Trust Land, and another 8 percent from gazetted forests. The remaining 84 percent were supplied by agroforestry systems and on-farm sources. This consisted of firewood purchased in the market (20 percent)—most being supplied by small private farms—and other more specific agroforestry sources. The latter included vegetation along boundaries and fences (25 percent), vegetation within croplands (13 percent), woodlots (8 percent), vegetation along roadsides (5 percent), and vegetation obtained from neighbors (13 percent).
- Farmers have responded to the high demand for wood by planting woodlots in their cropland. Croplands with higher percentages of woodlots cluster more extensively in the foothills of the Aberdare Range and Mount Kenya, and in most communities of Central Kisii, Nyamira, and Buret Districts. The share of woodlots is much lower in the western parts of the country and in the more marginal cropping areas with lower rainfall.
- About 82 percent of urban households and 34 percent of rural households use charcoal regularly. Of the total national charcoal production, rural households together consume 47 percent; all urban households consume 36 percent; and cottage industries use 17 percent.
- About 82 percent of charcoal comes from private land (either farmland or rangelands) and 18 percent from public lands (including government, communal, or Trust Land). Charcoal producers in Nakuru, Nyeri, and Trans Nzoia Districts report the largest proportion of wood from government land.
- About 200,000 people produce charcoal and another 300,000 transport and vend charcoal. Gross revenues from charcoal production are estimated at Ksh 17.5–32 billion per year (about US$ 250–457 million). This puts charcoal revenues somewhere between that of horticulture exports and that of livestock. Because the charcoal industry is not fully legalized, the government is foregoing tax revenues as high as Ksh 5.1 billion (US$ 72.9 million) per year.
- Charcoal production and firewood collection is carried out in all Kenyan Districts and contributes to income in most areas. Charcoal production remains a poorly remunerated occupation with an average monthly income of Ksh 4,496 (US$ 64) for a producer. In communities of Kwale District, households obtain on average more than 20 percent of their cash from charcoal production and firewood collection—the highest in the country. The proportion of income from charcoal and firewood ranges between 10 and 20 percent in the coastal hinterlands of Malindi District and parts of Garissa District.
- The high-yield areas of theoretically harvestable biomass growth from natural vegetation closest to Nairobi would be the rangelands south of the city (in Narok and Kajiado Districts), but also in the southeast (in parts of Machakos District). For Mombasa, the closest areas would be the woodlands of Kwale and Kilifi Districts. These areas may be well suited for sustainable charcoal production once the industry becomes fully legalized and more transparently managed.

In Chapter 8

CONTENTS

LIST OF MAPS

AUTHORS AND CONTRIBUTORS

Norbert Henninger (WRI)
Dan Tunstall (WRI)
Karen Holmes (consultant)
Greg Mock (consultant)
Janet Nackoney (WRI)
Florence Landsberg (WRI)
Mohammed Said (ILRI)
Hyacinth Billings (WRI)
Stephen Adam (WRI)
Carolina de Rosas (WRI)

WHAT THIS CHAPTER SHOWS

This chapter examines maps of various ecosystem services and poverty patterns in a single region—the upper watersheds of the Tana River—to demonstrate how such maps can help to highlight the relationships among people, ecosystems, and poverty. After providing a brief overview of landforms, population distribution, and poverty patterns in the upper Tana, the spatial relationships between selected ecosystem indicators and poverty in three topic areas: water-related ecosystem services; food-related ecosystem services; and biodiversity- and wood-related ecosystem services are discussed. The chapter concludes with a detailed summary that highlights poverty for six selected ecosystem indicators and suggests possible future analyses based on the patterns observed.

8

The Upper Tana: Patterns of Ecosystem Services and Poverty

This chapter focuses on a single region of Kenya and examines a range of ecosystem services used in this region. Unlike Chapters 3–7, which paint broad national pictures of a single ecosystem service such as water or food, here we integrate data on several services to give a more holistic picture of supply and demand in a particular area. The maps show the "key supply areas" of such services as food from crops and livestock; drinking and irrigation water use; and levels of crop diversity and woodlots in agroecosystems.

This kind of analysis is important because ecosystem services are typically looked at on a sectoral basis (e.g., water, forests, agriculture), which misses the interrelationships among them. Overlapping demand for various ecosystem services may produce conflicts over resource use, requiring tradeoffs among different uses and often between different users. Alternatively, there may be opportunities for synergies among different uses of ecosystem services. Mapping and analyzing spatial patterns of the supply and demand for different ecosystem services in the same geographic area can help communities address management decisions in a more integrated manner.

Using spatial analysis to examine a range of ecosystem services in a given area also allows us to compare these with spatial patterns of poverty in the area. It can provide information on how much local communities rely on key ecosystem services, such as food, water, forest products, and wildlife. It can also offer important insights on poverty-environment relationships: It could help to identify areas where natural resource investments could boost environmental income for communities or reduce vulnerability of the poorest households from further resource degradation. Or it could help to locate better-off communities that can afford to pay for land use practices to ensure a continued supply of ecosystem services such as sufficient water for the dry season or migration corridors for wildlife.

The following three sections—water-related ecosystem services, food-related ecosystem services, and biodiversity- and wood-related ecosystem services—provide examples of how to examine these relationships between people, ecosystems, and poverty. They break new ground by showing for the first time in one publication where *key supply areas of ecosystem services coincide* and where both *poorer and better-off communities are located in relation to these supply areas*.

We acknowledge that examining poverty-ecosystem relationships by overlaying two spatial indicators can only provide limited insights: It can show *where* in the upper Tana a proposed hypothesis about the spatial relationship between selected indicators is true and where it is not. In most cases, readers will demand additional information requiring new indicators or more location-specific data. The simple map overlays portrayed here are not sophisticated enough to detect all necessary correlations or come up with conclusive answers about causal links. They represent only a first step in unraveling poverty-ecosystem relationships.

In effect, we hope to use this chapter to engage the reader in a dialogue that spurs new questions and further investigations. We see such a dialogue and analytical process as a necessary step toward managing ecosystems more wisely and identifying opportunities for poverty reduction. It will be the task of Kenya's technical institutions responsible for data collection and analytical products to take the examples in this chapter to the next level. It will also require decision-makers who are motivated to ask questions and to understand the power (and limitations) of the data and the associated tools. We hope that these examples will inspire an improved multisectoral analysis of ecosystem services and of poverty-environment relationships in the upper Tana, and will lead to more detailed cross-cutting studies in other geographic regions of the country.

LANDSCAPES, PEOPLE, AND POVERTY

Several distinctive characteristics make the upper Tana River well suited for in-depth analysis:

- Within Kenya, the upper Tana, which covers a significant proportion of Central and Eastern Province, represents an economically important region for agricultural production and experiences high demand for ecosystem services. The upper Tana region includes the Aberdare Range and Mount Kenya—two of Kenya's five major mountain ranges, and the headwaters for many of Kenya's largest rivers. These rivers are an indispensable source of water for crops, livestock, wildlife, and human use, not only within the mountain vicinity, but also farther downstream across a large expanse of arid and semi-arid lands. In fact, the Tana River is the only major river running year-round through Eastern Province.

- The upper Tana area is home to 3.1 million people (about 11.4 percent of Kenya's total population), whose livelihoods are closely intertwined with multiple ecosystem services. Most of the area is covered by smallholder agriculture. It includes important areas of cash or export crops such as tea, coffee, vegetables, and rice. The government has set aside a significant portion of the land for biodiversity and watershed protection, including Mount Kenya National Reserve, Aberdare National Park, Aberdare Forest Reserve, Meru National Park, and Mwea Reserve.
- This area also contains a broad cross-section of very poor and less poor communities. Within the upper Tana are communities with some of Kenya's lowest poverty rates; however, the area also includes several very poor communities, most of them in the drier plains below the foothills downstream of the Aberdare Range and Mount Kenya.

The yellow line in Map 8.1 and subsequent maps outlines the upper Tana area. It represents the common watershed boundaries of all the major permanent streams and rivers originating in the Aberdare Range and Mount Kenya that flow into the Tana River.

Landforms

The upper Tana encompasses some 12,500 square kilometers, with elevations ranging from 1,000 to more than 5,000 meters. Elevation and landforms strongly influence rainfall and thus vegetation and farming patterns. The 60-kilometer gradient from the top of Mount Kenya to the lower plains contains a tremendous diversity of vegetation and farming systems.

The highest peaks include glaciers and alpine habitat types surrounded at lower elevations by tropical mountain forests. Classified as mountainous, these areas make up some 20 percent of the upper Tana (see brown areas in Map 8.1).

Map 8.1 Upper Tana: Landforms and Rivers

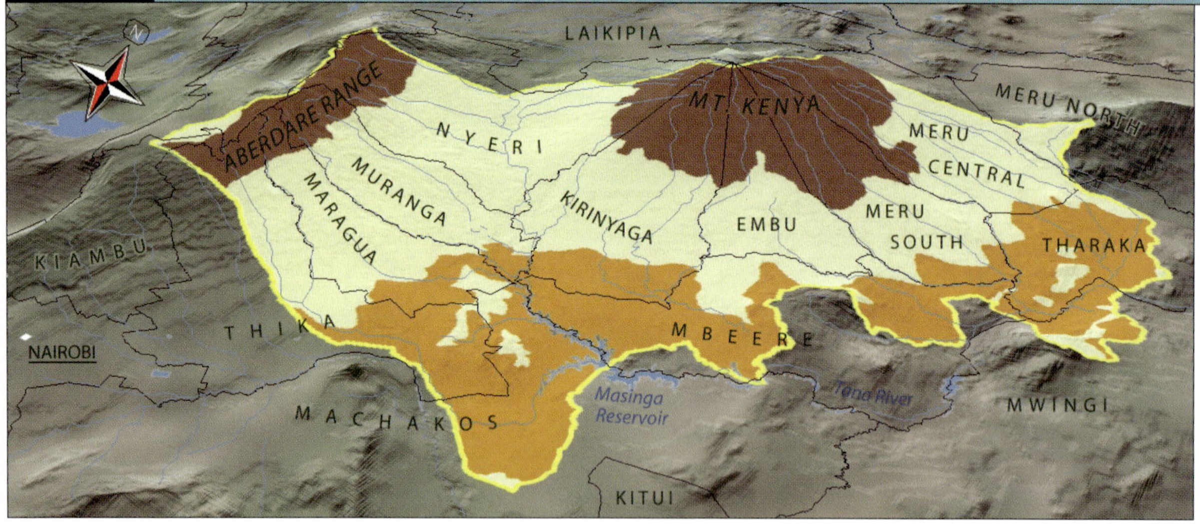

Sources: Administrative boundaries (CBS 2003), permanent rivers (NIMA 1997), 250-meter Digital Elevation Model (SoK, JICA, and ILRI 1996), subdrainage areas defining upper Tana (MoWD and JICA 1992), and landforms (FAO 2000).

LANDFORMS
- Mountains
- Footslopes, hills, and mountain footridges
- Plains

OTHER FEATURES
- Upper Tana boundary
- District boundaries

WATER BODIES AND RIVERS
- Permanent rivers
- Water bodies

Map 8.2 Upper Tana : Population Density, 1999

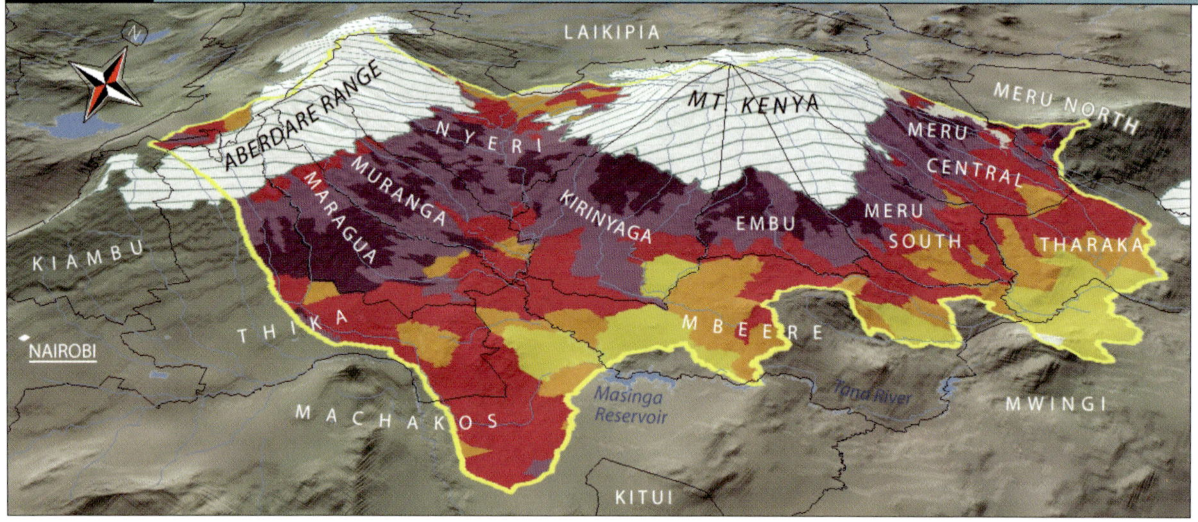

Sources: Administrative boundaries (CBS 2003), permanent rivers (NIMA 1997), 250-meter Digital Elevation Model (SoK, JICA, and ILRI 1996), subdrainage areas defining upper Tana (MoWD and JICA 1992), and 1999 population density (CBS 1999).

POPULATION DENSITY
(number of people per sq. km)
- > 600
- 300 - 600
- 100 - 300
- 50 - 100
- 20 - 50
- <= 20

OTHER FEATURES
- Upper Tana boundary
- District boundaries
- Major national parks and reserves (over 5,000 ha)

WATER BODIES AND RIVERS
- Permanent rivers
- Water bodies

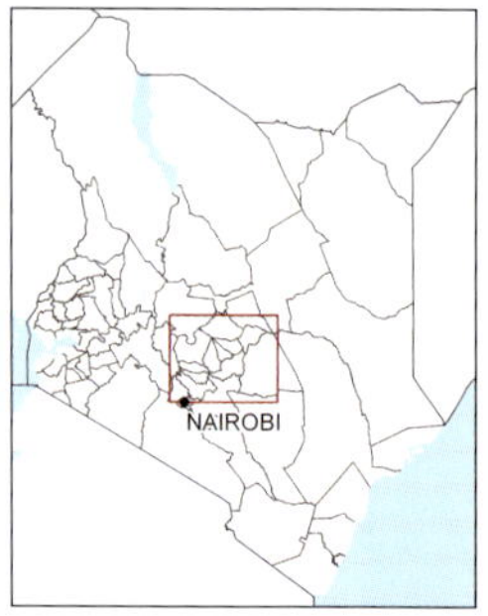

The mid-elevations are endowed with excellent soils and rainfall and are ringed by belts of tea, coffee, and other crops. This zone is less steep and often categorized as footslopes, hills, and mountain footridges (beige area in Map 8.1). It covers the largest share of the land within the Tana region. Population densities are very high, the land is intensively farmed, and average land holdings are very small.

The low-elevation sections of the Tana region are the least steep and have the lowest rainfall, segueing into the plains of Kenya's rangelands. These lands cover another 30 percent of the region (orange areas in Map 8.1). Dominant land uses are dryland agriculture (such as growing sorghum) or livestock grazing in the semi-arid rangelands.

Population, Road Network, and Administrative Units

About 860,000 households live in the upper Tana. The average population density is 250 people per square kilometer. The region includes large protected areas where settlements are not permitted and some of the most densely populated rural areas in Kenya (in Map 8.2, area in dark purple represents densities of more than 600 persons per square kilometer). Population densities in the region's lower elevation areas are generally less than 100 persons per square kilometer (Map 8.2, yellow and orange areas).

The largest towns are Thika and Nyeri. They are connected to Nairobi (45 km and 165 km from Nairobi, respectively) by a major highway. Other large towns are Embu (135 km from Nairobi) and Meru (275 km from Nairobi), connected to the Nairobi-Nyeri highway by asphalt roads. These towns host some agriculture-based industries (e.g., coffee and tea factories, flower farms, milk and cotton processing) and some small-scale timber-based industries (e.g., saw mills and furniture manufacturing). The secondary road network is denser and better developed in Thika, Maragua, Muranga, Nyeri, and Kirinyaga Districts. It is less dense in the remaining foothill Districts of Mount Kenya farther east, and is least developed in the plains.

At an administrative level, the upper Tana includes all or part of 14 Districts (as defined by 1999 census boundaries): Maragua, Muranga, and parts of Thika Districts drain the slopes of the Aberdare Range. Nyeri District includes streams from both Mount Kenya and the Aberdare Range. Kirinyaga, Embu, Meru South, and Meru Central Districts incorporate the southeast and eastern slopes of Mount Kenya. Parts of Mbeere, Tharaka, and Machakos Districts lie further downstream of Mount Kenya in the plains of the Tana River. Small slivers of Meru North District (in the far northeast corner), Nyandarua District (in the far southwest corner), and Laikipia District (just above Nyeri) also fall in the upper Tana region. Together these Districts contain 222 local administrative units (Locations) and 823 subunits (Sublocations).

Map 8.3 **Upper Tana: Poverty Rate, 1999**

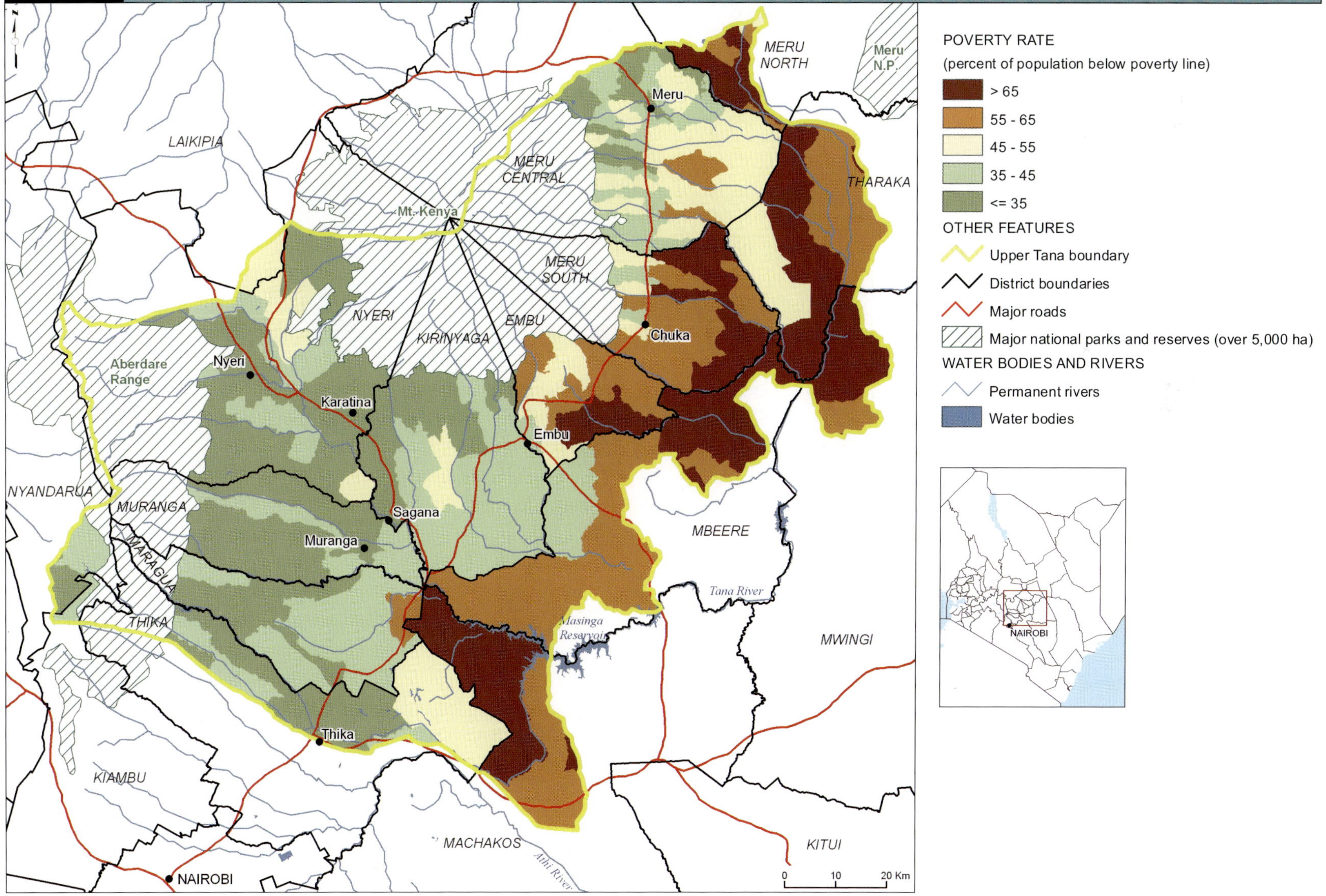

Sources: See Map 2.6.

Spatial Patterns of Poverty

Spatial patterns of poverty in the upper Tana are quite distinctive. Along the rivers that drain the Aberdare Range or Mount Kenya, Locations at higher elevations in general have lower poverty rates than the Locations further downstream.

The communities in the lower plains and the drier parts of the upper Tana have the highest poverty rates (shown in two shades of brown on Map 8.3), which are above the national rural average of 53 percent. The better off region, which contains large contiguous areas where the poverty rate is less than 35 percent (shown in dark green), is located in the foothills of Thika, Maragua, Muranga, Nyeri, and Kirinyaga Districts.

Poverty rates in the remaining three foothill Districts of Mount Kenya—Meru Central, Meru South, and Embu—reflect a more mixed picture. They are generally higher than those in foothills further west, including the slopes of the Aberdare Range. Communities in Meru Central, on average, do better than those in Meru South and Embu Districts. Meru Central includes quite a number of administrative areas with relatively low poverty rates, most of them close to the town of Meru. The spatial patterns of poverty in Embu and Meru South Districts resemble those of communities in the drier plains.

Spatial patterns of *poverty density* (Map 8.4) are quite different from those of *poverty rates*. Despite the very high poverty rates in the lower plains, the poverty density (that is, the number of poor people per square kilometer) is generally quite low in many of these dry, sparsely populated areas (see Map 8.4, areas colored in green). In contrast, some communities with the highest poverty densities (areas colored in dark brown, with more than 200 poor people per square kilometer) are located in densely populated areas with relatively low poverty rates. This reflects the situation in the nation as a whole (see Map 2.4 and Map 2.5 in Chapter 2). Map 8.4 is a reminder that analyses of spatial poverty patterns or program targeting cannot rely on poverty rates alone. That approach may overlook communities such as some spots in Maragua and Nyeri Districts that have a high number of poor, averaging more than 200 poor persons per square kilometer, but only show average poverty rates of 35–45 percent.

Map 8.4 Upper Tana: Poverty Density, 1999

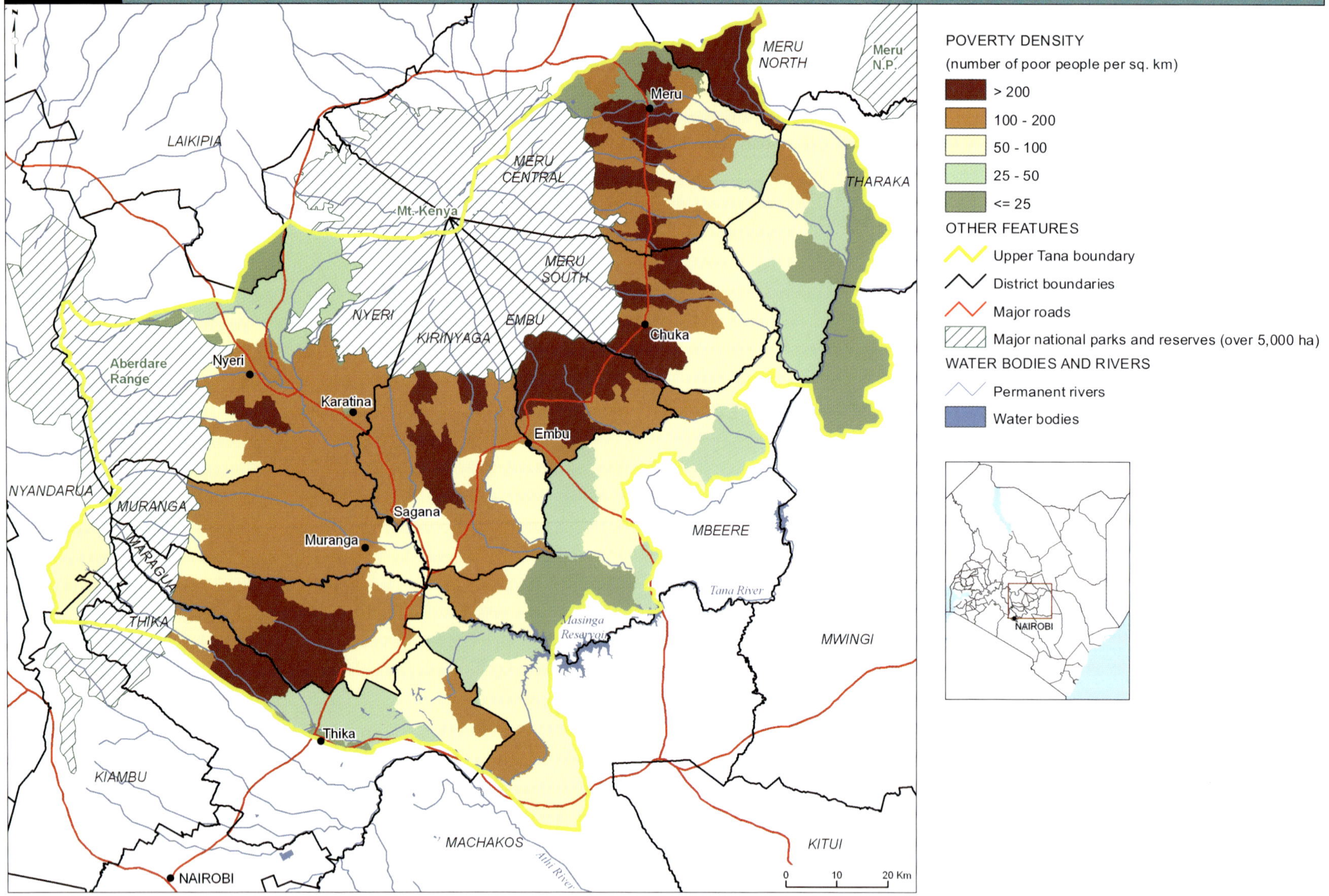

Sources: See Map 2.7.

WATER-RELATED ECOSYSTEM SERVICES

Chapter 3 examined key water issues and spatial patterns of water-related ecosystem services at a national level. Many of these issues are particularly relevant for the upper Tana. For the purposes of this section, we closely examine indicators related to drinking water and irrigation conveyed in the maps of Chapter 3. To highlight the multiple demands on freshwater systems in the upper Tana, this section also shows other uses of water such as hydropower, large inter-basin transfers to supply urban areas, and water for nature (i.e., to maintain wetlands and other natural habitats both within and outside protected areas).

While much of the country experiences marginal rainfall and conditions of general water scarcity, the Kenyan highlands, including the foothills of the upper Tana, receive ample rainfall and are relatively water-rich. The lower elevations of the upper Tana, however, receive less rainfall, making growing crops a more precarious pursuit and grazing livestock a safer bet.

Key water uses in the upper Tana include water used for agricultural production, electricity generation, household drinking supply, and maintenance of wildlife habitat. In many ways, the importance of the area's water resources takes on a national significance which transcends the value of the resources to just the Tana region itself. A large share of the nation's agricultural production occurs here, including crops for export. Hydroelectricity generated by the region's rivers is the principal electricity source for the country. And drinking water supplies from this basin are essential for Nairobi's population.

Water is also important for maintaining healthy wildlife habitats. The need to support nature-based tourism and to sustain Kenya's biodiversity thus requires a basin-wide management approach to ensure that wetlands and other habitats have enough water.

Population growth and economic development put heavy pressure on Kenya's water resources in general, especially in the upper Tana. Water demand is likely to continue to grow as urban populations rise and as the proportion of households with access to piped water increases.

Indicators Examined

The following analyses overlay maps of poverty with different water uses, making use of two water-related indicators:

- *Share of households relying on piped drinking water.* Households that benefit from piped drinking water are in theory somewhat buffered from interruptions in the quality or quantity of water (assuming well-functioning water delivery and treatment systems). Comparing poverty rates and the level of access to piped water can help identify communities that have both high poverty rates and no piped drinking water supplies. We also expect that more affluent communities are more likely to have a higher share of households relying on piped water, mostly because these communities have more resources and perhaps greater political influence to attract water infrastructure investments.
- *Presence of small-scale irrigation efforts within communities.* The presence of small-scale irrigation efforts represent investments made to generate economic benefits from increased crop productivity and to reduce vulnerability to crop failures. The purpose of overlaying poverty and small-scale irrigation efforts is to examine whether investments in small-scale irrigation have reached both poor and more affluent communities. It also highlights areas in which these investments are lacking, thus limiting livelihood options for households or making them more vulnerable to crop failure. Because of their low capital requirements, we expect small-scale irrigation efforts to be distributed throughout the upper Tana and to reach a significant share of poor communities.

The final overlay analysis in the following section examines to what degree communities with a high share of piped drinking water and communities with small-scale irrigation efforts coincide. This comparison is not so much an investigation of possible causal relationships between these two indicators. It is more to locate areas that have benefited from both types of water infrastructure investments, thus enhancing the benefits from water-related ecosystem services and buffering local livelihoods from interruptions in these services. We expect small-scale irrigation efforts to be more widely dispersed than the communities with high shares of piped drinking water sources.

Drinking Water Use and Poverty

As shown in Map 8.5, the majority of the population of the upper Tana obtains drinking water from untreated surface water, groundwater, or a combination of surface and groundwater. In areas where more than 75 percent of households depend exclusively on surface waters (shown in red), people obtain their water directly from lakes and streams or from reservoirs and ponds. In the upper Tana, such areas are mostly located in the foothills of the Aberdare Range or Mount Kenya as well as at lower elevations in the plains closer to the Tana River and its reservoirs.

Households that use surface water for drinking are particularly vulnerable to problems posed by insufficient quantity and quality of water. The quantity of surface water available at any given time depends directly on natural flows of water and the patterns of rainfall that generate these flows. Dependence on surface waters also implies direct reliance on ecosystems for their natural waste removal capacity, such as filtering by wetlands and the dilution capacity of freshwater systems.

Areas in which more than 75 percent of households depend solely on groundwater for their drinking water are shown in Map 8.5 in orange. Here people use springs, wells, and boreholes to obtain water. Such areas are located mostly in the lower plains and drier areas of the Tana headwaters. These communities are likely to be somewhat less vulnerable to water quality problems due to greater natural filtering of groundwater supplies.

Areas where more than 75 percent of households receive piped drinking water are shown in dark blue. These populations are more indirectly linked to their ecosystem and in theory could rely on modern methods of municipal water treatment to insulate them from vulnerability to drinking water contamination. They are clustered in more densely populated areas, including the towns of Nyeri, Thika, Embu, Chuka, Meru, and surrounding locations.

Map 8.6 highlights poverty rates (data are shown by Location) in communities with high access to piped water systems (more than 75 percent of households obtain their drinking water from piped water supplies). As expected, communities with a

Map 8.5 Upper Tana: Household Reliance on Ecosystems for Drinking Water

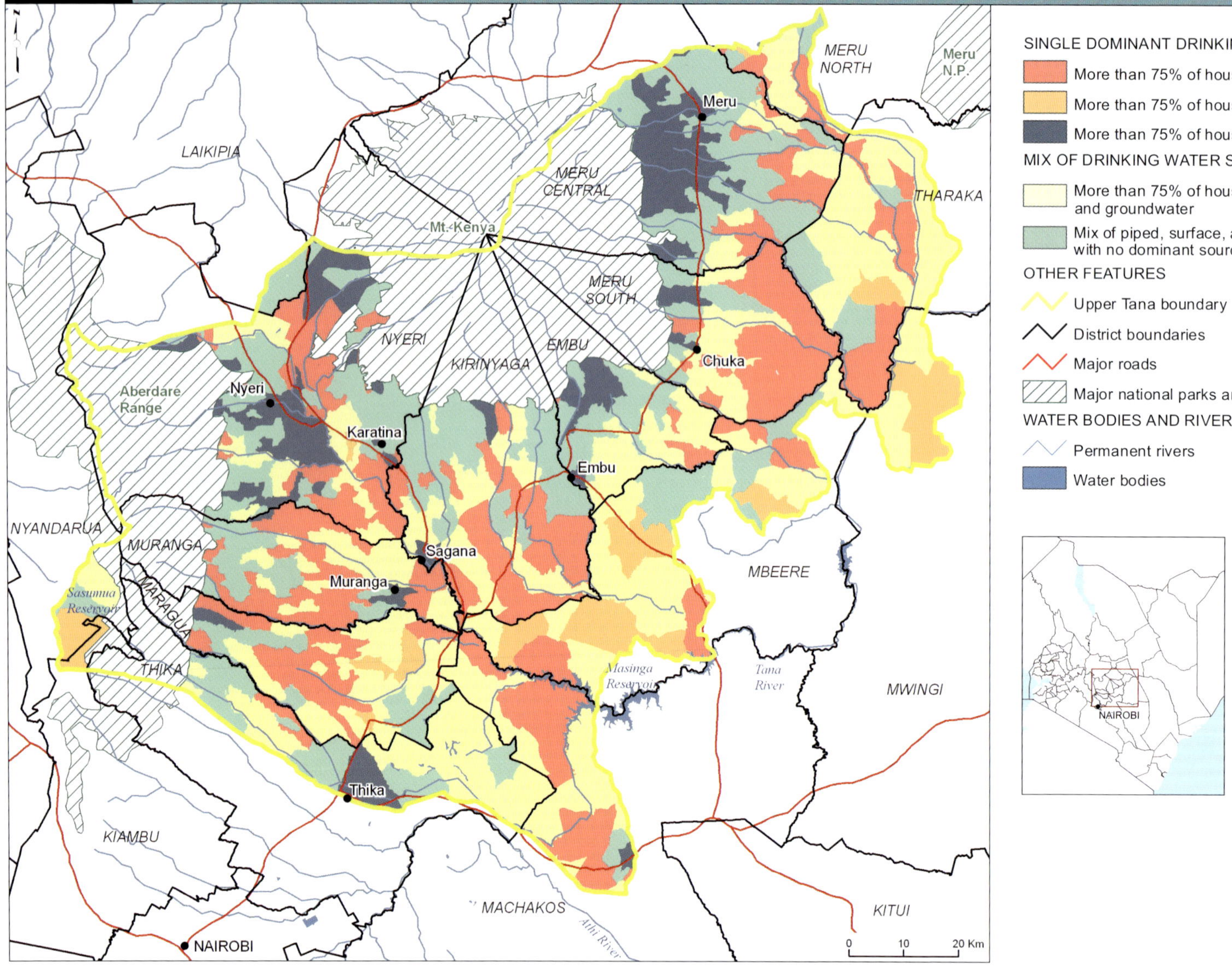

Sources: See Map 3.8.

Map 8.6 Upper Tana: High Share of Piped Drinking Water and Poverty Rate

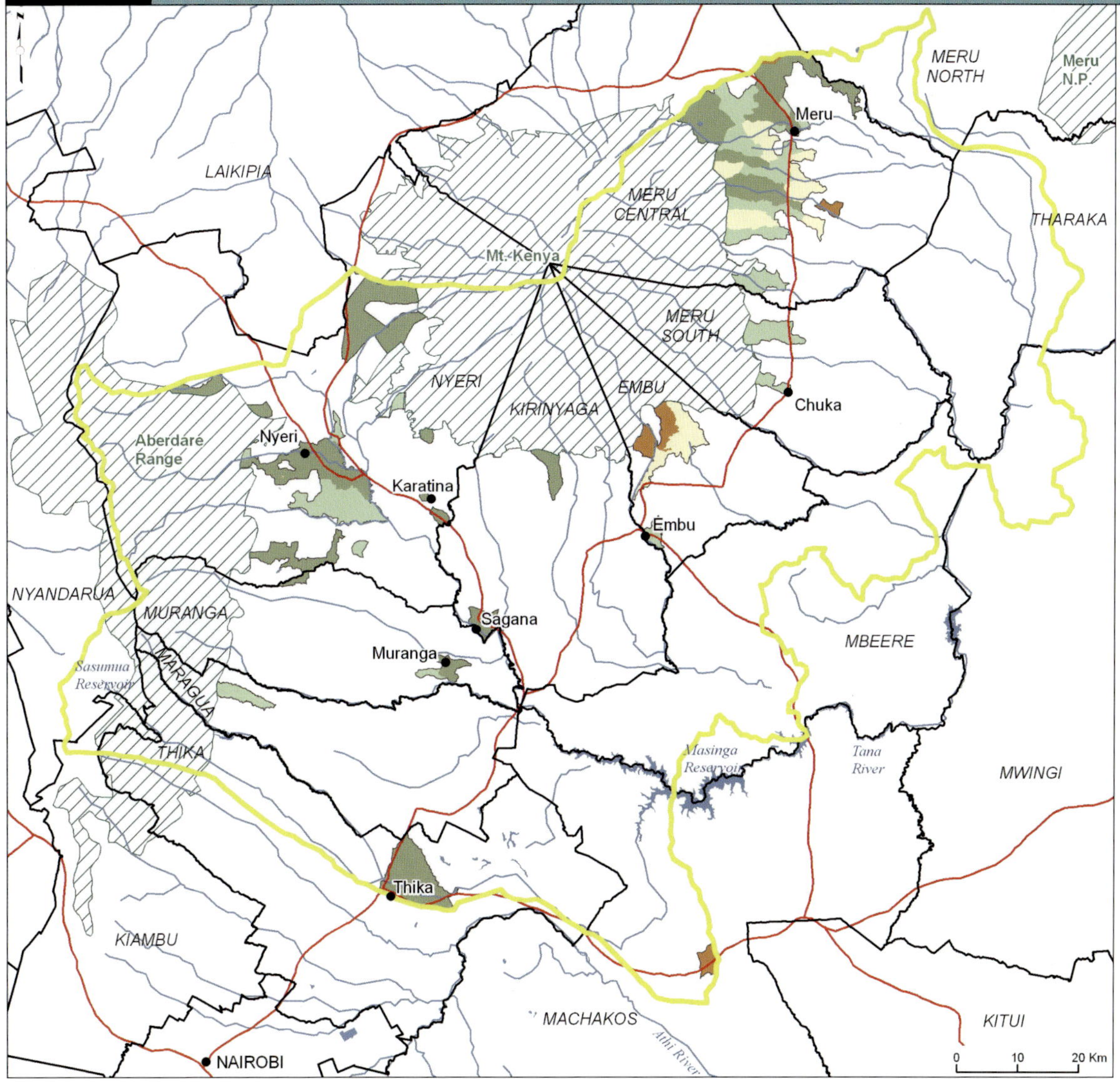

Sources: See Map 2.6 and Map 3.8.

Poverty Rate in Areas with Piped Drinking Water Access Below 75 Percent or Other Source

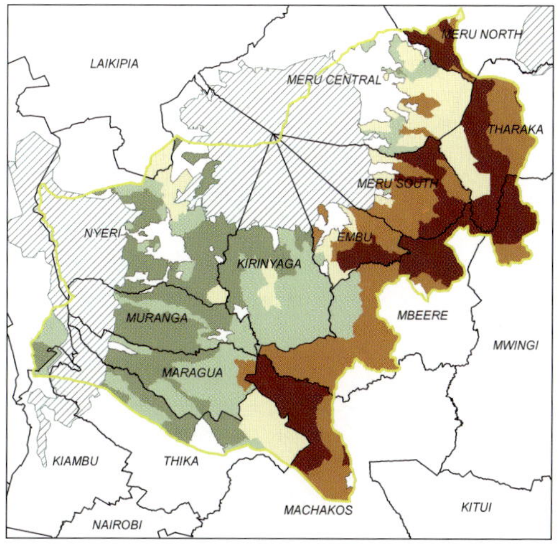

POVERTY RATE
(percent of population below poverty line)

- > 65
- 55 - 65
- 45 - 55
- 35 - 45
- <= 35

OTHER FEATURES

- Upper Tana boundary
- District boundaries
- Major roads
- Major national parks and reserves (over 5,000 ha)

WATER BODIES AND RIVERS

- Permanent rivers
- Water bodies

high share of piped drinking water are few in number, are spatially concentrated, and have low poverty rates. These communities are located in administrative areas near the towns of Nyeri, Meru, and Thika as well as in Locations in northern Nyeri, Kirinyaga, and Meru Central Districts with poverty rates below 35 percent (shaded in dark green). Locations south of the town of Nyeri, near the town of Embu, and in Meru Central and Meru South Districts have poverty rates of 35–45 percent (colored light green).

The poorest areas in the upper Tana have not benefited from investments in piped drinking water, as the lack of brown areas in Map 8.6 indicates. There are, however, a few exceptions. For instance, some Locations with a high share of piped water systems have poverty rates of 45–55 percent (shaded yellow) and even 55–65 percent (shaded light brown), mostly in Meru Central and in Embu Districts. Further analysis could investigate why piped water investments in these poorer communities were possible and whether the well-being of poor households in a community with higher piped water supplies has improved (for example, resulting in fewer cases of childhood diarrhea and more time for girls to attend school).

Although most areas with high access to piped drinking water have relatively low poverty rates, this does not imply that all Locations with low poverty have high access to improved water sources (see small inset map showing poverty rates for areas with piped water access below 75 percent, or other drinking water sources). Indeed, some Locations with quite low incidence of poverty—including extensive areas in the Aberdare foothills in Thika, Maragua, Muranga, and Nyeri Districts—have no or low access to piped drinking water (that is, fewer than 10 percent of households obtain their water from piped water systems).

Irrigation Efforts, Other Water Uses, and Poverty

Map 8.7 provides an overview of irrigation efforts in the upper Tana. Large-scale irrigation projects, shown with purple shading, include the Mwea-Tebere rice irrigation scheme, which covers some 6,100 hectares in Kirinyaga and Mbeere Districts. Small-scale irrigation, indicated by pink squares, is mostly located farther upstream on the Tana River tributaries. Many of the small-scale irrigation points serve horticultural crops, including fruit and vegetables. Most of these are concentrated at the base of Mount Kenya in the Districts of Meru Central and Meru South, as well as in Embu, Kirinyaga, and Nyeri Districts. Farmers in the foothills of the Aberdare Range rely less on small-scale irrigation, with only a few such projects, mostly located in Maragua District between the towns of Thika and Muranga.

Irrigation is only one of many water uses in the upper Tana. As shown in Map 8.7, there are multiple demands on freshwater systems in this region. Water with a low sediment content is needed for generation of energy (indicated by the shaded catchments that feed electricity-generating dams). The upper Tana region also has to handle significant water transfers to the Athi River basin and supply drinking water to Nairobi (as indicated by one of the major pipelines that connects the Sasumua reservoir to Nairobi). Water is also needed for environmental services, an often overlooked use of water resources in the region, and is represented on the map as wetland remnants and protected areas.

Due to intensive cropping patterns, very few areas of large, contiguous wetlands remain in the upper Tana. Wetlands (shown as pink-shaded areas) are located within a few kilometers of the towns of Thika and Muranga and near the boundaries of Meru National Park. In the coming years, these wetland remnants will likely face growing pressure from rising demands for land and water. Policy-makers may have to consider difficult tradeoffs—for instance, whether to allow conversion of these wetlands for irrigated crop production, or to protect them in their natural state so that they can filter runoff from intensively farmed slopes and provide habitat for wildlife.

Map 8.7 Upper Tana: Irrigation Efforts and Other Water Uses

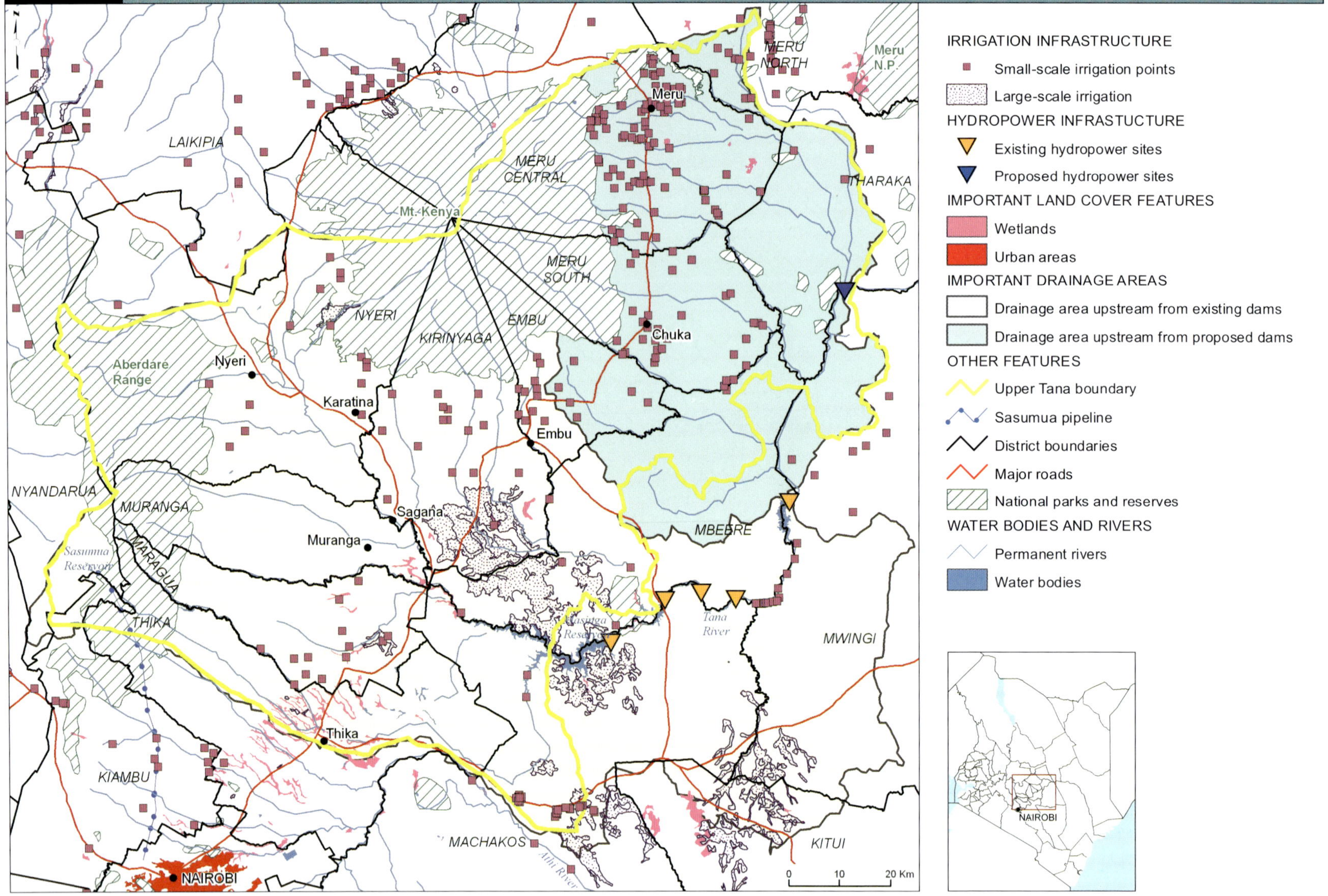

Sources: See Map 3.9, Map 3.11, Map 3.12, and Map 5.3.

Map 8.8 Upper Tana: Small-Scale Irrigation Efforts and Poverty Rate

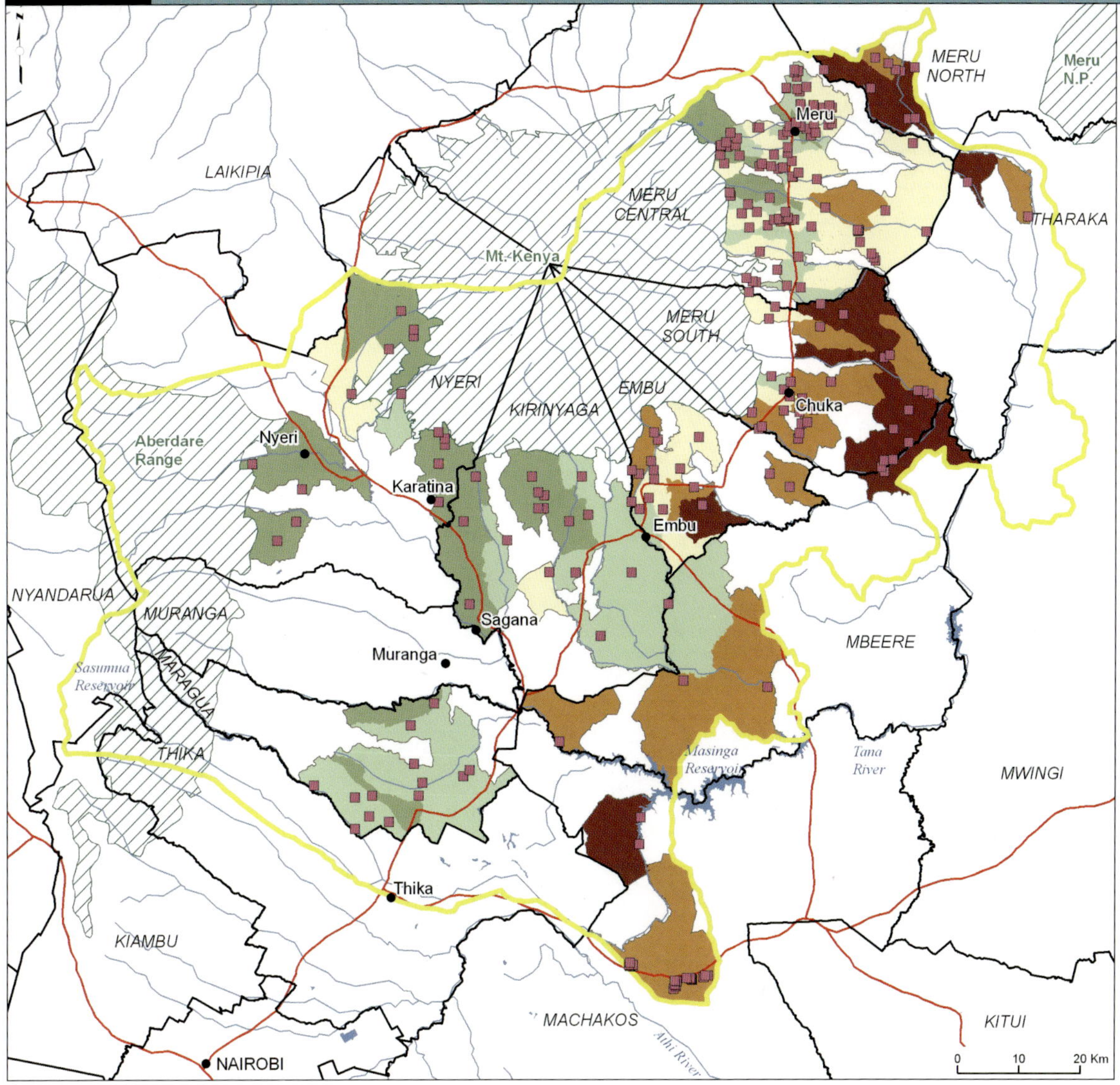

Sources: See Map 2.6 and Map 3.12.

Poverty Rate in Areas Without Small-Scale Irrigation Efforts

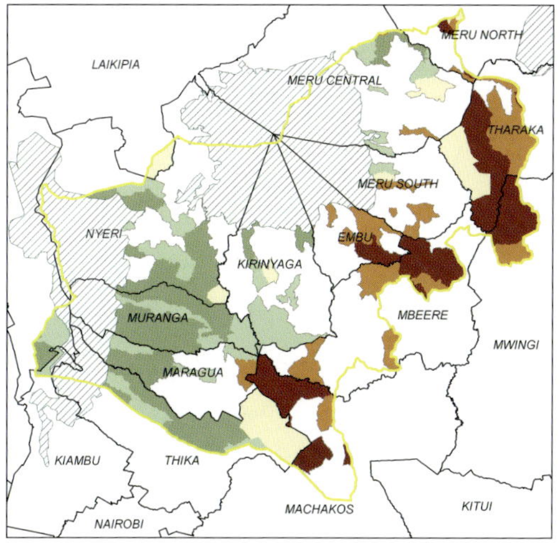

IRRIGATION INFRASTRUCTURE

- Small-scale irrigation points

POVERTY RATE

(percent of population below poverty line)

- > 65
- 55 - 65
- 45 - 55
- 35 - 45
- <= 35

OTHER FEATURES

- Upper Tana boundary
- District boundaries
- Major roads
- Major national parks and reserves (over 5,000 ha)

WATER BODIES AND RIVERS

- Permanent rivers
- Water bodies

The overlay analysis in Map 8.8 focuses on small-scale irrigation efforts because they are more widely dispersed throughout the upper Tana and require comparatively small investments, which means they can reach poorer areas and households more easily. Large-scale irrigation in the upper Tana is concentrated in a contiguous area in lower Kirinyaga and southwestern Mbeere Districts.

Map 8.8 indicates that most small-scale irrigation sites ring Mount Kenya at similar elevations and with comparable rainfall. They also reach the plains, notably in Meru South, Machakos, and Meru North Districts. In these drier areas, however, they are much lower in number and density. Compared to the map showing high shares of piped drinking water, communities with small-scale irrigation efforts are widely dispersed.

As expected, poverty rates in areas with investments in small-scale irrigation vary considerably, from Locations with quite low poverty rates (less than 35 percent) in Nyeri and Kirinyaga Districts to those with very high poverty rates (55 to 65 percent) farther east. Of all the small-scale irrigation efforts, it is those in Meru North, Machakos, and Meru South Districts that are generally in the poorer administrative areas, with poverty rates averaging 55 percent and higher. The irrigation efforts in Nyeri District are in administrative areas with much lower poverty rates, as is the case for those in the Aberdare foothills.

Map 8.8 confirms that some of the poorest communities in the upper Tana have benefited from small-scale irrigation efforts (albeit at lower numbers). Subsequent analysis focusing on these areas can pinpoint where small-scale irrigation investments have lowered poverty rates versus those areas where their contributions have not been large enough to significantly affect household income, but perhaps have increased nutritional status and food security. This could then help in targeting other poor communities in the drier lowlands, since a significant number of these communities have not benefited from small-scale irrigation yet (as can be seen in the small inset map).

High Share of Piped Drinking Water and Small-Scale Irrigation Efforts

Most communities with a high share of households relying on piped drinking water are in the more densely populated urban areas and in rural areas at higher elevations bordering the forest zone and protected areas. Meru Central District has the greatest number of rural communities with a high share of piped water access. There are still significant opportunities for investing in improved drinking water supplies throughout the upper Tana, especially where there are high rural population densities such as upper Kirinyaga and Maragua Districts (see Map 8.2 for population densities).

The foothills of Mount Kenya have the greatest number of small-scale irrigation points. Meru Central District has the greatest concentration of small-scale irrigation efforts in the upper Tana (about 40 percent of the mapped irrigation points). Meru South, Nyeri, and Machakos Districts have similar shares (around 10 percent each) of the mapped irrigation points. Only a handful of small irrigation points are located in the drier areas of Tharaka and Mbeere Districts.

The degree of spatial overlap between investments in small-scale irrigation and piped water systems varies considerably across the upper Tana. In some areas, these investments coincide, but in others they do not. For instance, in Meru Central District there is extensive coincidence of small-scale irrigation efforts and piped drinking water systems. In Meru South and Embu Districts, some overlap exists, but to a much lesser degree than that seen in Meru Central. In other Districts, however, areas with high access to piped drinking water (for example, in Nyeri, Kirinyaga, Maragua, and Thika Districts) do not show any overlap with investments in small-scale irrigation.

Examining the history of these investments and the adaptation of small-scale irrigation technology in more detail may reveal why Meru Central has benefited to a greater degree from both small-scale irrigation and piped drinking water supplies. Such an investigation could point toward possible synergies between investing in piped water systems and establishing small-scale irrigation efforts that could be instructive for neighboring Districts.

Map 8.9 Upper Tana : Small-Scale Irrigation Efforts and High Share of Piped Drinking Water

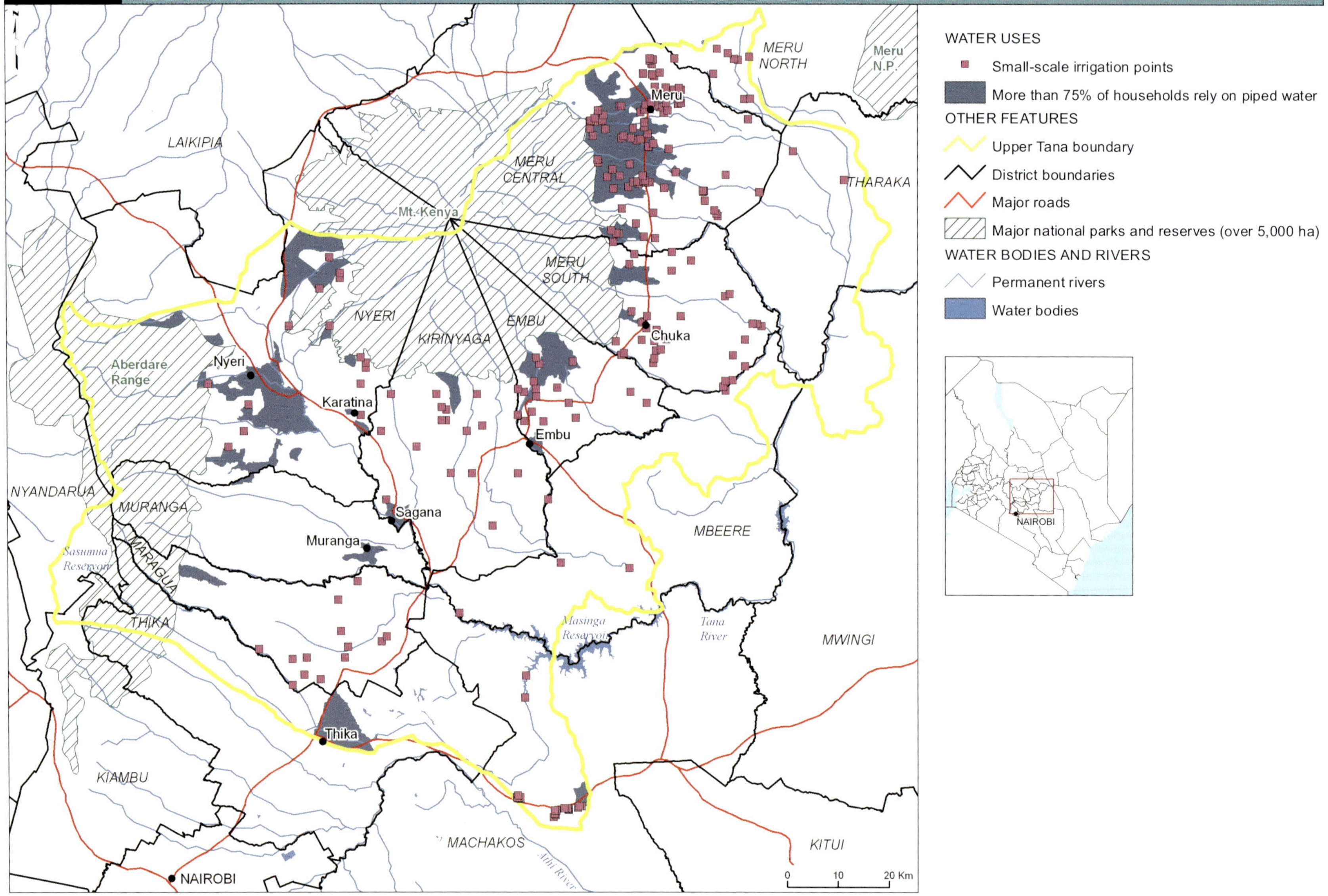

Sources: See Map 3.8 and Map 3.12.

FOOD-RELATED ECOSYSTEM SERVICES

As described in previous chapters, Kenya's croplands are concentrated in areas of reliable rainfall, including the upper Tana. Kenyan farmers grow a mixture of food and cash crops, including tea, coffee, sugarcane, tobacco, and sisal. The foothills of the Aberdare Range and Mount Kenya are an important food and cash crop supplier, with some of the longest established tea- and coffee-growing areas in the country. Over the past decade, the upper Tana has also become an important supply area of vegetables and flowers, both for export and domestic consumption.

The dominant land use for the upper Tana is smallholder agriculture. While a large number of Kenyan smallholders still grow food crops for subsistence, recent data show the growing importance of cash crops for household income. Farming families are increasingly relying on cash income and the market economy for food security (Jayne et al. 2000).

A large percentage of farming households in the foothills of the upper Tana own cross-bred dairy cattle. The animals are raised in stalls and fed cut grass, tree leaves from fodder trees, or even purchased commercial feed.

Since the soils are fertile and rainfall is more reliable in these foothills, farmers crop the available land intensively. However, because of population growth and increased subdivision of farms since Kenya's independence, average farm size has decreased, making it difficult or impossible to support a family in some areas. A longitudinal study of land use patterns since the 1950s on the eastern slopes of Mount Kenya (Embu and Mbeere Districts) found that this has prompted family members of richer households to purchase or rent land in the more marginal cropping areas at lower elevations. Other responses include investment in children's education, migration to urban areas such as Nairobi, and employment in the non-agriculture sector (Olson 2004).

Kenya's *Economic Recovery Strategy for Wealth and Employment Creation 2003–2007* (GoK 2003) seeks major reform of agricultural policies and institutions to reverse the decline in agricultural growth and productivity over the past decade. The upper Tana will be both a key region impacted by these reforms and an important pillar for future agricultural growth.

Indicators Examined

This section relies on two indicators introduced in Chapter 4 to examine the region's food crops and dairy cattle—the two major sources of food from agriculture:

- *Share of cropland under food crops.* Croplands with a relatively low share of food crops are producing a greater proportion of nonfood crops (especially coffee and tea) for cash or export. Our hypothesis is that this will correlate with lower incidence of poverty. A high share of cropland in food crops—especially when it includes the staple crop maize and very few other crops—could indicate subsistence farming, which is associated with higher poverty rates. But in some areas it corresponds with large-scale, irrigated commodity crops such as rice (upper and lower Tana), mechanized wheat farms (Narok District), high-yielding maize production (Uasin Gishu and Trans Nzoia Districts), or even more complex farming systems that produce a mix of high-value food crops including cereals, vegetables, and fruit trees.
- *Total milk production per area.* Dairy provides a source of high-quality protein and micronutrients, which are often lacking in largely cereal-based diets. Thus, we might expect areas with relatively high levels of milk production to be better off, with a greater concentration of households that can afford better nutrition. Moreover, livestock provide household savings and supplemental income for farming families. A plausible hypothesis, therefore, would be that areas with higher dairy output correlate with lower poverty rates.

For each indicator, we will first provide an overview of the major spatial patterns and then compare high production areas (high share of food cropping and high milk output) with poverty rates. Such a comparison may help formulate additional hypotheses about the relationship between food-related ecosystem services and the level of well-being in a geographic area. It can also be used to contrast areas with similar poverty levels and classify them according to their orientation toward food crops or milk production. This can then support agricultural planning, such as deciding where to target new livestock breeds or crops. In a final step, we will look at spatial overlaps between areas with high food cropping and high milk production. Such an analysis can help to delineate areas with potential conflicts or synergies between food cropping and milk production.

Food Cropping and Poverty

Map 8.10 shows how much of their cropland farmers have dedicated to food crops. The level of food cropping varies significantly across the upper Tana. In general, farmers in the higher-elevation sections of the foothills grow a lower share of food crops.

Most areas with a very high percentage of agricultural production invested in food crops (i.e., greater than 75 percent, shown as dark green) are at lower elevations. These include large tracts of irrigated rice cultivation in lower Kirinyaga and far southwest Mbeere Districts. These areas also cover non-irrigated areas in northwest Machakos and the lower parts of Muranga and Thika Districts. Small clusters of cropland with a high food share are also found farther east in Meru South and Meru Central Districts.

Areas with a greater share of cropland in non-food cash crops (orange- and yellow-shaded areas) are mostly in the foothills of the Aberdare Range or the slopes of Mount Kenya. These areas include the tea-growing zones at the highest elevations of the foothills and the coffee-growing zones on somewhat lower slopes.

Map 8.11 shows the spatial relationships between poverty (poverty rates are shown by Location) and croplands with a large (i.e., greater than 75 percent) share of production in food crops. Large areas of dark brown—signaling poverty rates of greater than 65 percent—are found in Machakos District and a few Locations in Meru South, Meru North, and Tharaka Districts. Extensive areas of light

Map 8.10 Upper Tana: Food Crops as Percentage of All Cropland

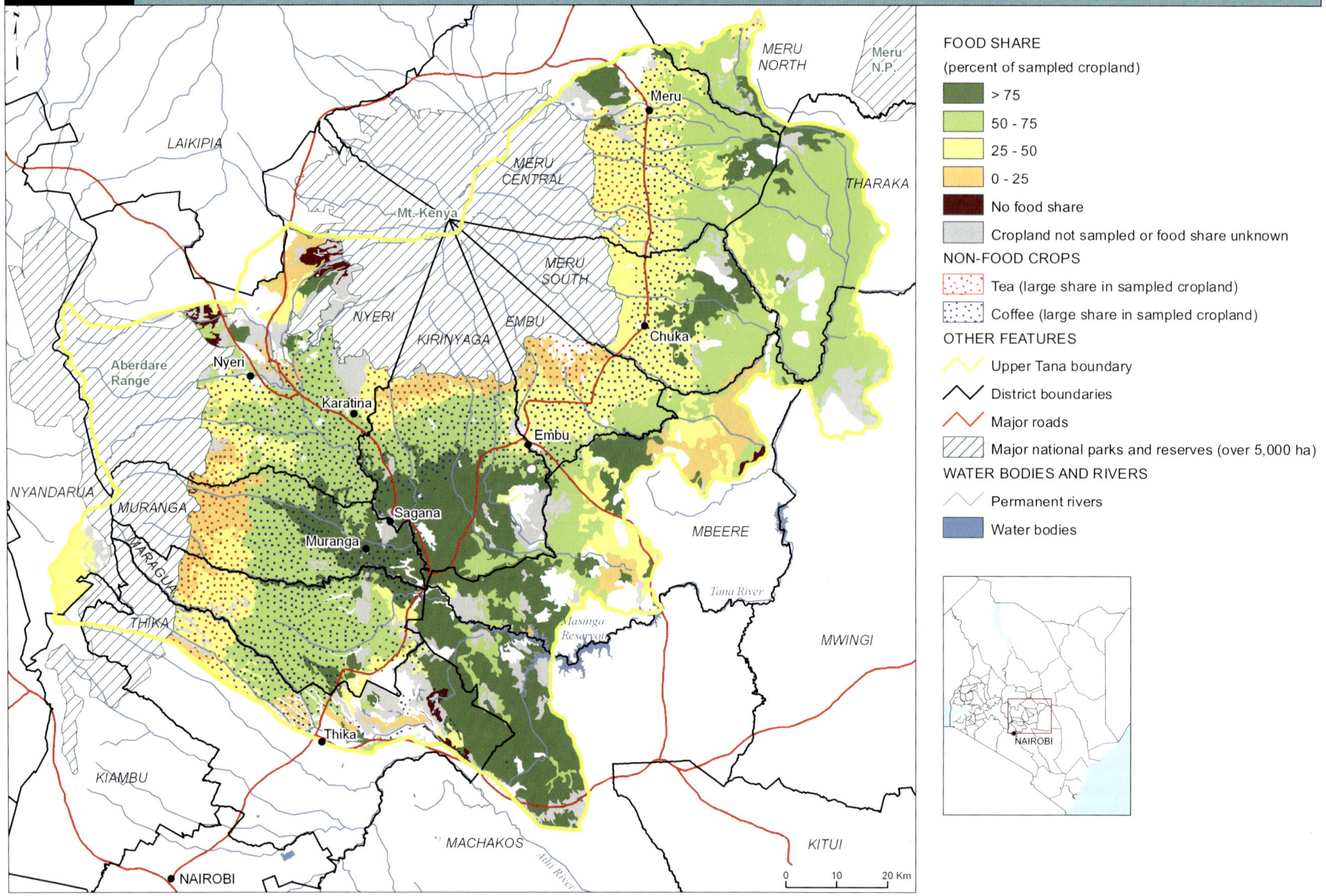

Sources: See Map 4.4.

Map 8.11 Upper Tana: High Share of Food Crops and Poverty Rate

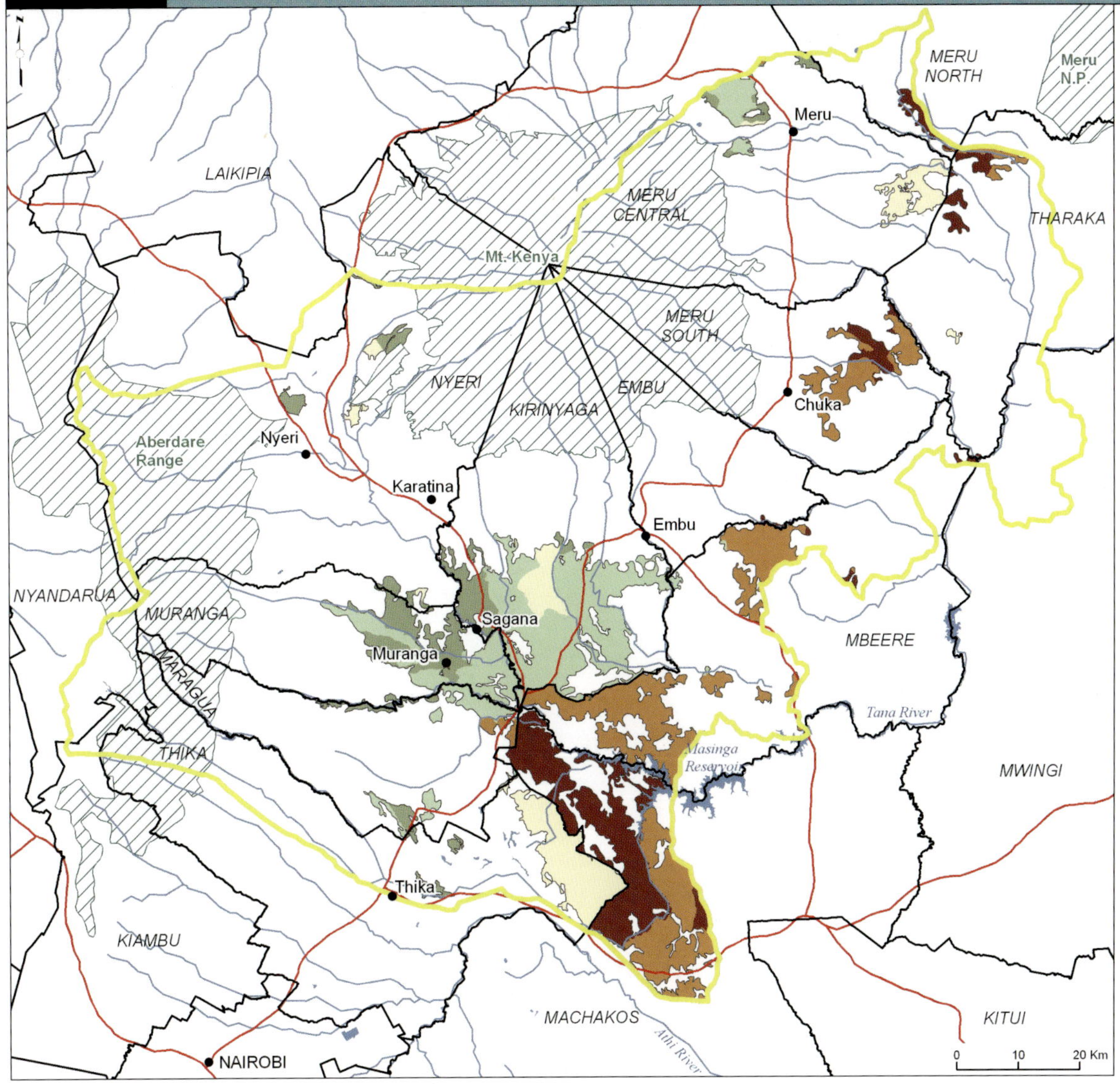

Poverty Rate in Areas with Less than 75 Percent Food Share

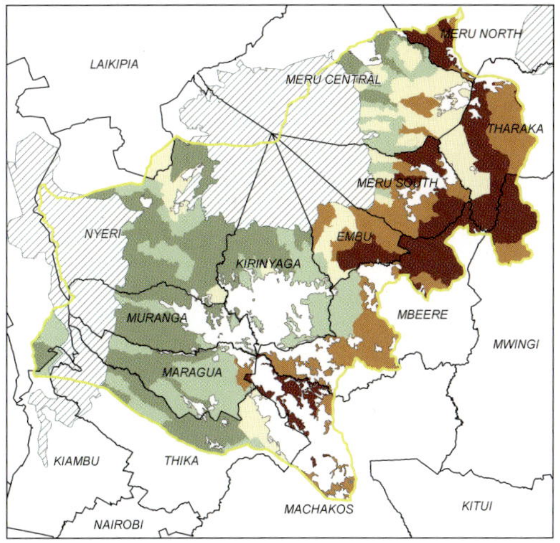

POVERTY RATE
(percent of population below poverty line)
- > 65
- 55 - 65
- 45 - 55
- 35 - 45
- <= 35

OTHER FEATURES
- Upper Tana boundary
- District boundaries
- Major roads
- Major national parks and reserves (over 5,000 ha)

WATER BODIES AND RIVERS
- Permanent rivers
- Water bodies

Sources: See Map 2.6 and Map 4.4.

brown (poverty rates of 55–65 percent) are located in Machakos, Mbeere, Meru South, and Tharaka Districts. Areas with low rates of poverty and a high share of food crops (shown in shades of green) are limited to Maragua, Muranga, and Kirinyaga Districts, along with a few Locations in Nyeri and Meru Central Districts.

Areas in the lower drier plains with a high share of food crops consistently have poverty rates above Kenya's national rural average (53 percent). While this would confirm our initial hypothesis, Locations in Kirinyaga and Muranga Districts do not support this simple, straightforward association of high poverty with a high share of food cropping. Similarly, the small inset map (showing poverty rates in areas with less than 75 percent food share) points toward a significant number of high-poverty areas with lower food shares. This suggests that additional information on the number and types of crops grown is required to illuminate the spatial patterns of food cropping and poverty in the upper Tana. For example, while areas in Maragua, Muranga, and Kirinyaga Districts have similar high food shares as areas in northwestern Machakos, Meru South, and southern Meru Central Districts, the types of crops grown and other agricultural factors may differ. In the former three Districts the food crops may include high-value vegetables and other crops destined, via good roads, to reach a large urban market such as Nairobi. In the latter three Districts the share of dryland cereal crops may be greater; the overall value of production may be lower; or the purpose for growing crops may be oriented more toward subsistence and local markets.

Milk Production and Poverty

Map 8.12 presents the spatial distribution of milk production. Areas with annual milk production greater than 100,000 liters per square kilometer (shaded purple on the map) are mostly at higher elevations in the foothills of the Aberdare Range and Mount Kenya, while areas of low milk production (colored light pink) are at lower elevations.

Map 8.13 shows the spatial coincidence of poverty (poverty rates are shown by Location) and areas with high milk production (i.e., production of more than 100,000 liters per square kilometer per year). Most of these areas are colored in shades of green, corresponding to Locations with a low incidence of poverty. Such Locations form a large expanse across the eastern foothills of the Aberdare Range and the southwestern slopes of Mount Kenya, as well as a few Locations in Meru Central District. Areas with high milk production and relatively greater incidence of poverty (greater than 55 percent) encompass comparatively few Locations. A cluster of such Locations is found in Embu District, as well as a few Locations in Meru South, Meru Central, and Meru North Districts.

The poverty pattern for most Locations with high milk production supports the hypothesis that high milk output—most likely associated with a greater number of cross-bred dairy cattle—is more prevalent in communities with lower poverty rates. This is also supported by the small inset map (showing poverty rates in areas with less than 100,000 liters per square kilometer per year), which indicates significant overlap between areas with very high poverty rates and areas with the lowest milk output

Map 8.12 **Upper Tana: Milk Production**

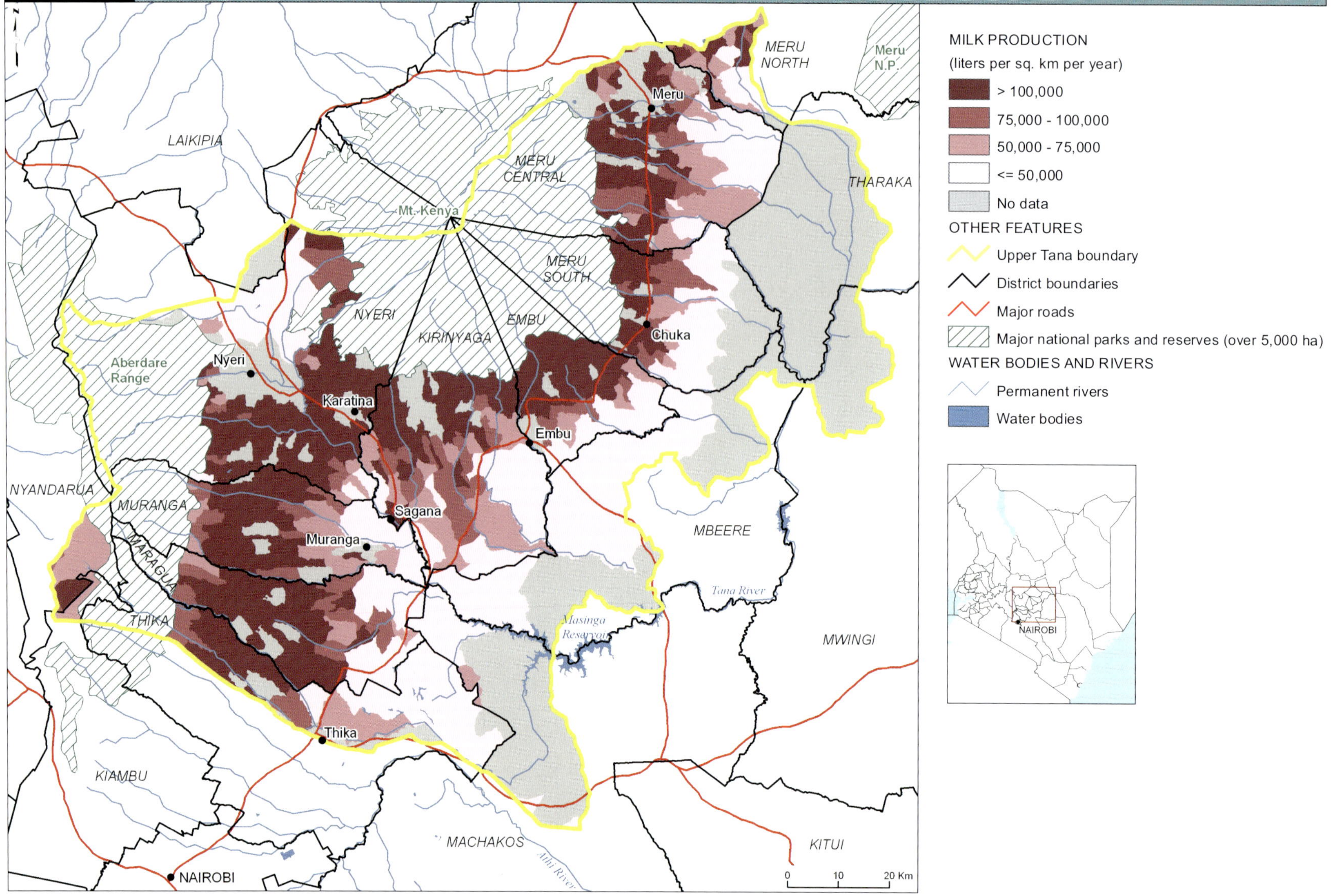

Sources: See Map 4.5.

Map 8.13 Upper Tana: High Milk Production and Poverty Rate

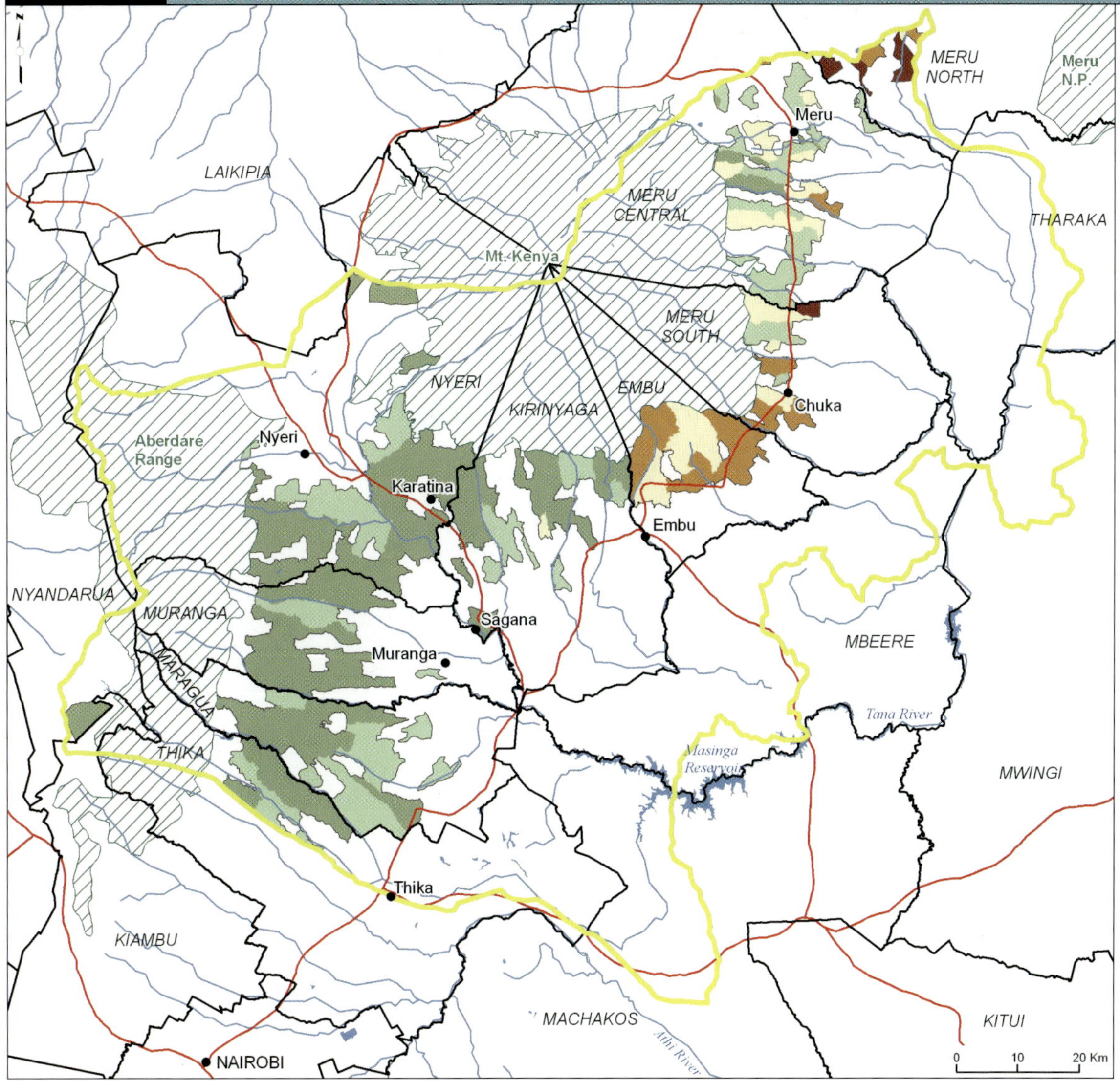

Poverty Rate in Areas with Less than 100,000 Liters per Square Kilometer per Year

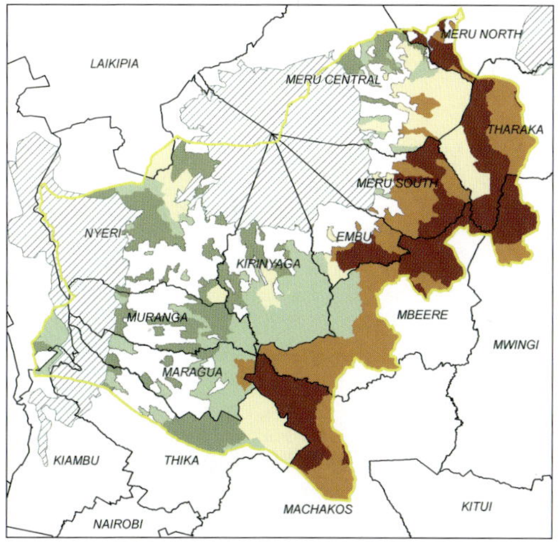

POVERTY RATE
(percent of population below poverty line)

- > 65
- 55 - 65
- 45 - 55
- 35 - 45
- <= 35

OTHER FEATURES

- Upper Tana boundary
- District boundaries
- Major roads
- Major national parks and reserves (over 5,000 ha)

WATER BODIES AND RIVERS

- Permanent rivers
- Water bodies

Sources: See Map 2.6 and Map 4.5.

(Map 8.12). Further investigation is needed to understand causal relationships and determine whether households became less poor once they became high milk producers or whether a certain amount of capital had to be in place to support a high-milk output production system.

The high-poverty and high-milk output areas in Embu District appear to contradict our initial hypothesis above. Further analysis of these areas is required to unmask the reasons why these poorer communities are such high milk producers or why higher milk output has not lowered overall poverty rates. For example, farmers may be high producers but their income may be lower because of failures in the milk market. Or farmers in the Aberdare Range may have additional and more diversified income streams than high milk producers in Embu. Such a detailed analysis could provide useful insights into the causes of high poverty rates. It could also help promote appropriate milk production technology in poorer communities in the upper Tana, for example in Meru South District.

High Food Cropping and High Milk Output

As seen in Map 8.14, there is very little spatial overlap between areas with a high share of food crops and areas with high milk production. While areas of high milk production (shaded purple on the map) form a large expanse across the eastern foothills of the Aberdare Range and the southern and eastern slopes of Mount Kenya, areas with a high share of food crops (colored green) stretch across low-elevation locations in Machakos and southern Kirinyaga Districts. Only a small number of locations (shown in orange) are intensive producers of both food crops and dairy. These areas of overlap are concentrated across the midsection of Kirinyaga District, as well as a few locations in Maragua, Muranga, Nyeri, and Meru Central Districts.

This lack of spatial overlap in Map 8.14 suggests two different livelihood strategies for farming families in the upper Tana: farmers higher up in the foothills (and to a much larger degree in the Districts east of the Aberdare Range) rely more on nonfood cash crops and high milk outputs for their income than their counterparts further downstream. Farmers at lower elevations are focused more on food crops, and the milk output per unit area in these lowlands is less. Investigating the underlying reasons for this difference—for example, less productive indigenous breeds of dairy cattle, fewer high-yielding cross breeds per area, or a less developed system for transporting and processing milk in the lowlands—could reveal where boosting milk production may improve livelihoods and well-being.

Map 8.14 Upper Tana: High Share of Food Crops and High Milk Production

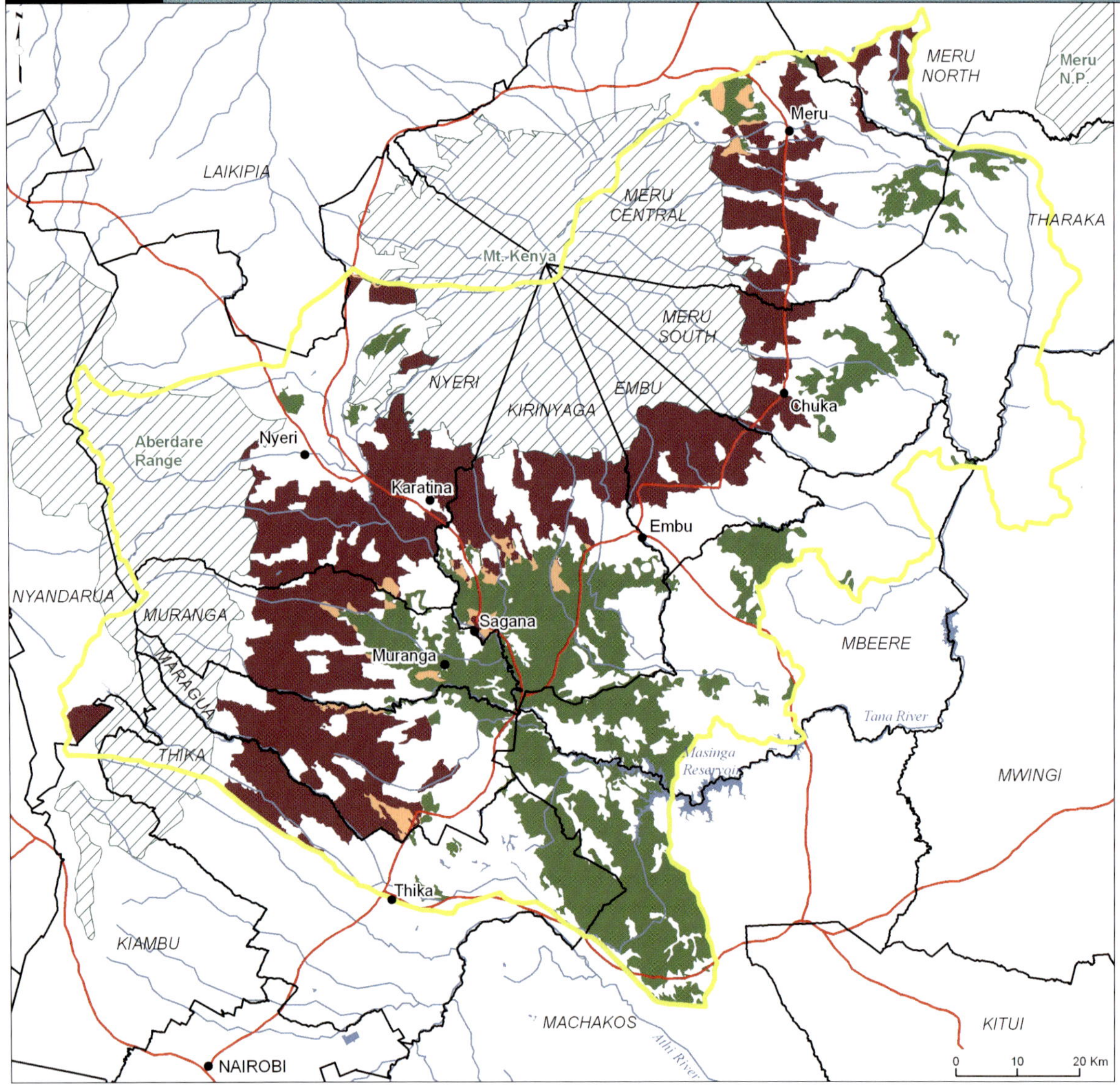

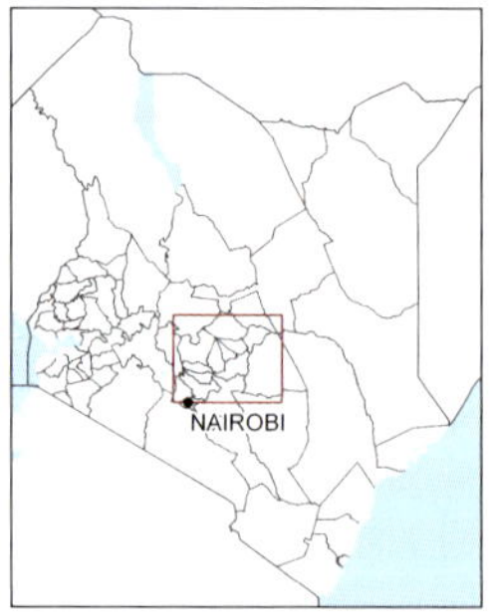

Sources: Map 4.4 and Map 4.5.

BIODIVERSITY- AND WOOD-RELATED ECOSYSTEM SERVICES

The selected upper Tana watersheds contain some of Kenya's largest tracts of indigenous forest on Mount Kenya and the Aberdare Range. Almost all of these forests are on government land—either a forest reserve or a national park set up to safeguard biodiversity or hydrological services. The area surrounding the forests of the upper Tana is densely populated and there is intensive agricultural production in the foothills of the two mountain ranges.

Over the past 200 years, much of the land in the foothills that once was forest or a mosaic of forest and other habitats has now been cleared and converted to agriculture. This has resulted in significant losses of biological diversity. For instance, most large mammals, such as large wild cats, have become rare. Elephants—which once roamed widely throughout the foothills, taking advantage of greater water availability and feed during the dry season—have retreated to protected areas or less intensively cultivated areas due to habitat loss and wildlife fences that safeguard crops and people. However, the remaining highland forests continue to provide habitat for a disproportionate share of Kenya's total biological diversity, including 50 percent of plant species, 40 percent of mammals, 35 percent of butterflies, and 30 percent of birds (KFWG 2001).

In addition to providing food and other crops, the farmlands in the foothills are an important source of wood, mostly because the remaining indigenous forests are legally protected from large-scale wood removal. Currently, at the household level, farms and woodlots in Kenya provide about two thirds of firewood for domestic use (MoE 2002).

Agricultural landscapes in the foothills also have a role to play in conserving the rich diversity of lifeforms of the Kenyan highlands. The extent to which croplands contribute to biodiversity conservation depends on how people use the land and the resulting impact on its suitability as habitat for native plants and animals. As mentioned in Chapter 5, large monocultures provide a less suitable habitat than clusters of small fields growing multiple crop species (so-called polycropping) within a patchwork of trees, shrubs, and herbaceous plants. The upper Tana is home to landscapes with some of Kenya's highest polycropping, which could contribute to conserving highlands biodiversity.

Indicators Examined

This section makes use of two indicators introduced in earlier chapters:

- *Average number of crops grown in a given farm parcel.* This indicator can be interpreted as a measure of agrobiodiversity. High incidences of polycropping would be associated with higher levels of biodiversity in agroecological landscapes. Polycropping is expected to be more prevalent in the foothills of the upper Tana than in the drier plains at lower elevations. The foothills have more reliable rainfall and a longer growing season and thus provide farmers with more options to plant a greater variety of crops. Farmers may grow different crops simultaneously because the agroclimate permits it, because there is demand for multiple products, or because they want to spread their risk from crop or market failures. We expect polycroppping to be associated with less poverty because livelihoods are based on a better agroecological endowment and more diversified risk strategy. However, not all areas where farmers grow only one or two crops are necessarily marginal farming areas with less rainfall (mostly planted with maize). They can also be highly productive areas where farmers concentrate on a single cash crop.
- *Share of woodlots within croplands.* Mapping the share of woodlots within croplands provides information about where farmlands supply wood and where farmers have made more long-term investments in agroforestry practices. Depending on the tree species and the age of the trees in the woodlot, the wood may serve as firewood, be converted to charcoal, or be used for construction purposes. Areas with less rainfall are expected to have a lower share of woodlots because it will be more difficult to grow trees. Our hypothesis is that higher shares of woodlots in cropland are associated with lower poverty rates—not necessarily because farmers realize higher returns from wood, but because farmers and communities that are better off have a greater financial ability to dedicate some of their land to wood production.

These two indicators, when combined with indicators of the average size of farmers' fields and the extent of tree cover in croplands (as shown in Chapter 5), can provide an overall measure of agrobiodiversity in agricultural landscapes. These measures shed light on the extent to which agricultural land uses and configurations could help relieve pressure on remaining natural forest areas and forest-related biodiversity.

Number of Agricultural Crops and Poverty

Map 8.15 shows the spatial pattern of crop diversity in the farmlands of the upper Tana. Areas where an average of more than four different crop types are being grown simultaneously (shaded green) extend across most of upper Kirinyaga District on the southern slopes of Mount Kenya, as well as in Meru South and Meru Central Districts on the eastern footslopes. A few areas in the lower Aberdare foothills in Thika, Maragua, and Muranga Districts also have relatively high crop diversity levels. Across much of the remaining cropland, especially at higher elevations, farmers grow, on average, two to four crop types in a growing season (yellow areas). Farms located at lower elevations, including rice-growing areas under large-scale irrigation, tend to produce on average one or two crops simultaneously (light brown areas).

Map 8.15 highlights the extremely diverse cropping patterns in the upper Tana. Landscapes are a patchwork of multiple crops—the majority of them in very small fields. Overall, the farmers on the footslopes of Mount Kenya favor a greater number of crops compared to farmers at similar elevations in the Aberdare foothills (except for a cluster of locations in Thika, Maragua, and Muranga Districts). A closer examination of the types of crops grown, their relative prices, their contribution to safeguarding against possible market risks (price declines) or weather risks (drought or flooding), and institutional and land use policy issues could shed more light on the reasons for this specific spatial pattern.

Map 8.15 Upper Tana: Average Number of Crops Grown in Cropland

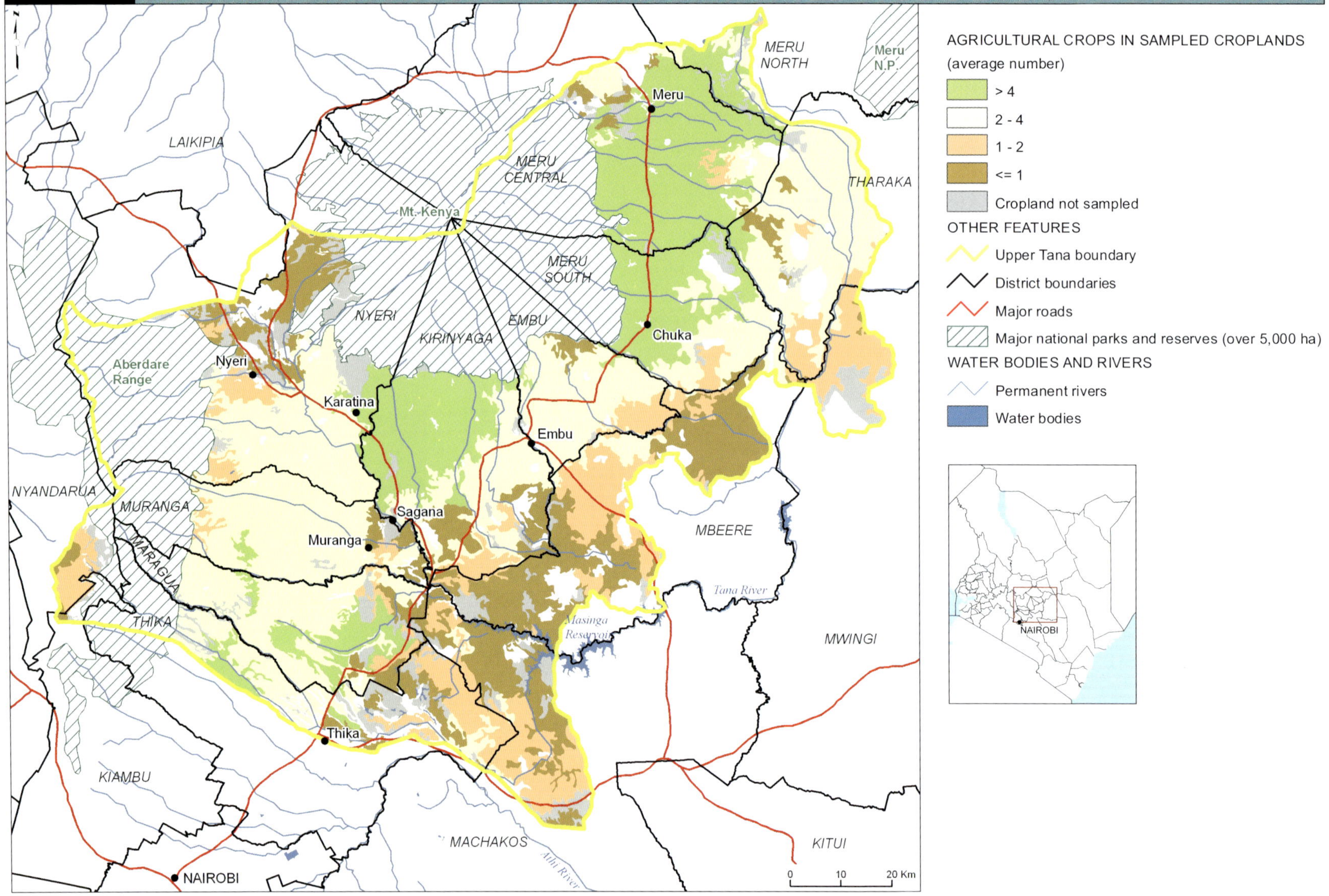

Sources: See Map 5.5.

Map 8.16 Upper Tana: High Average Number of Crops Grown in Cropland and Poverty Rate

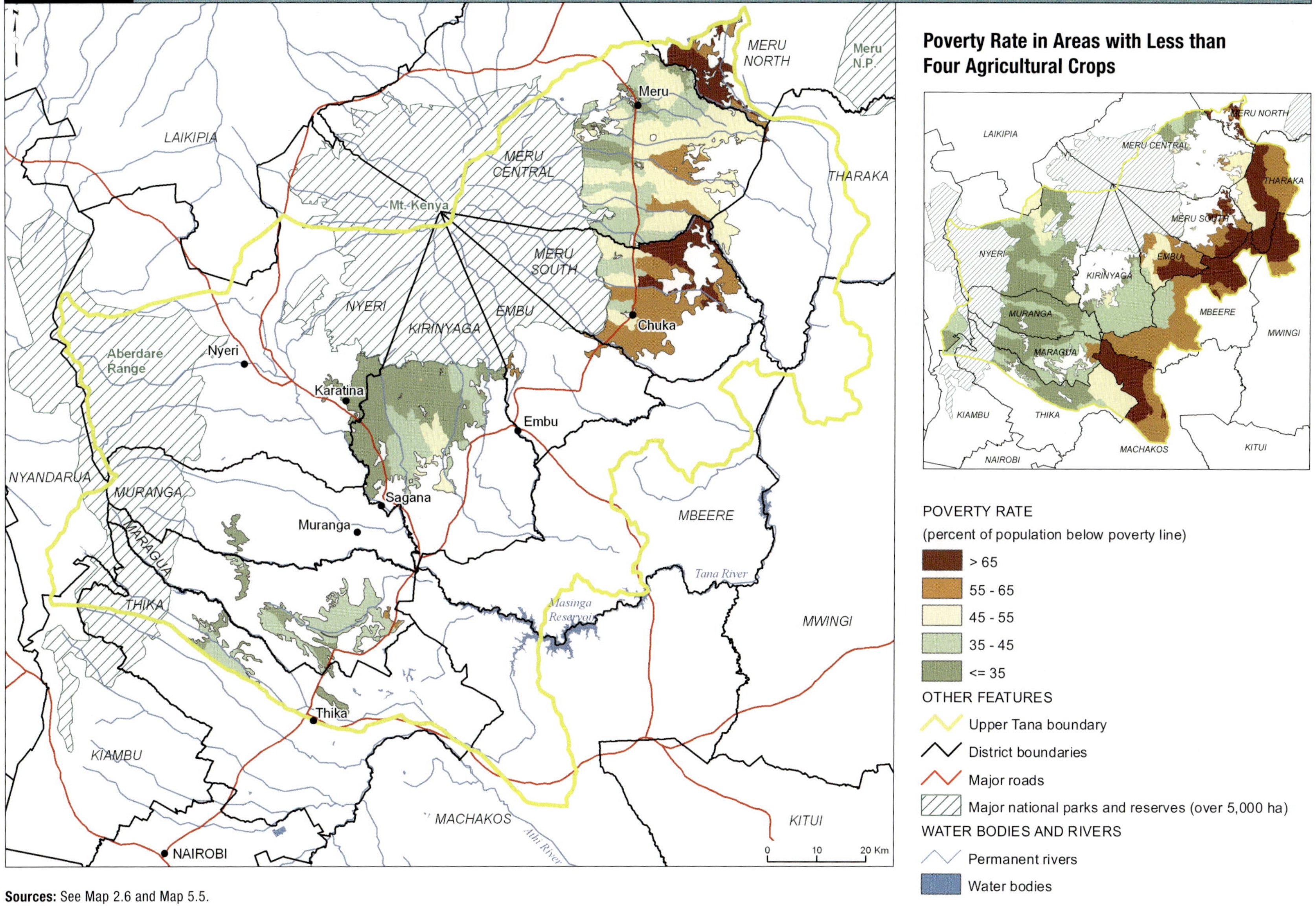

Sources: See Map 2.6 and Map 5.5.

Map 8.16 compares the spatial distribution of areas with high rates of crop diversity (average number of crops is greater than four) with spatial patterns of poverty (poverty rates are shown by Location). Many such high-diversity areas have low poverty rates, including a large expanse on the southern slopes of Mount Kenya in Kirinyaga District, as well as clusters of low-poverty Locations in the foothills of the Aberdare Range, and a few Locations near the town of Meru. However, some areas with high rates of polycropping are found in zones with moderate poverty rates (especially in Meru Central District) as well as in high-poverty areas (in Meru South District).

Further comparison of poverty rates in areas with lower crop diversity (see small inset map showing poverty rates in areas with less than four agricultural crops) indicate that in the Aberdare foothills, very low crop diversity (tea growing areas in Map 8.10) corresponds with very low poverty rates. Inversely, low crop diversity (see Map 8.15) in the drier lowlands (more marginal cropping of maize) corresponds with high poverty rates. This confirms that analysts need to distinguish between marginal and high potential croplands when comparing levels of polycropping and poverty.

High levels of polycropping are therefore not automatically associated with certain poverty rates in the upper Tana. Explaining these spatial patterns of poverty and the number of crops grown will require gathering information on the specific crops being grown and the reasons for selecting them, which could be driven by market demand or household needs for food security.

Woodlots in Cropland and Poverty

Map 8.17 shows the share of woodlots in cropped areas of the upper Tana. Areas in which more than 12 percent of the cropland is allocated to woodlots (shown in dark brown), are clustered at high elevations in the Aberdare foothills in Thika, Maragua, and Muranga Districts. A large band of lighter brown, indicating areas in which 6 to 12 percent of cropland is devoted to woodlots, stretches across the foothills of Mount Kenya in Embu, Meru South, and Meru Central Districts. Croplands that contain no woodlots at all (dark purple areas) occur at lower elevations in the drier plains.

Throughout the foothills of the upper Tana, most farmers include wood as one of their crops. A complex set of factors, such as the size of local or urban market demand for wood, availability of labor to grow other more labor-intensive crops, returns on investment of tree crops versus other crops, and even efforts to promote tree planting (e.g., women of the Green Belt Movement), all have to be taken into consideration when analyzing why certain locations in the Aberdare foothills and along the Embu-Meru road have become more significant supply areas.

Map 8.18 depicts spatial patterns in the relationship between poverty (poverty rates are shown by Location) and the share of farmland devoted to woodlots. Areas where farmers set aside a relatively large share of cropland (6 percent or more) as woodlots are found across diverse areas

Map 8.17 Upper Tana: Share of Woodlots in Cropland

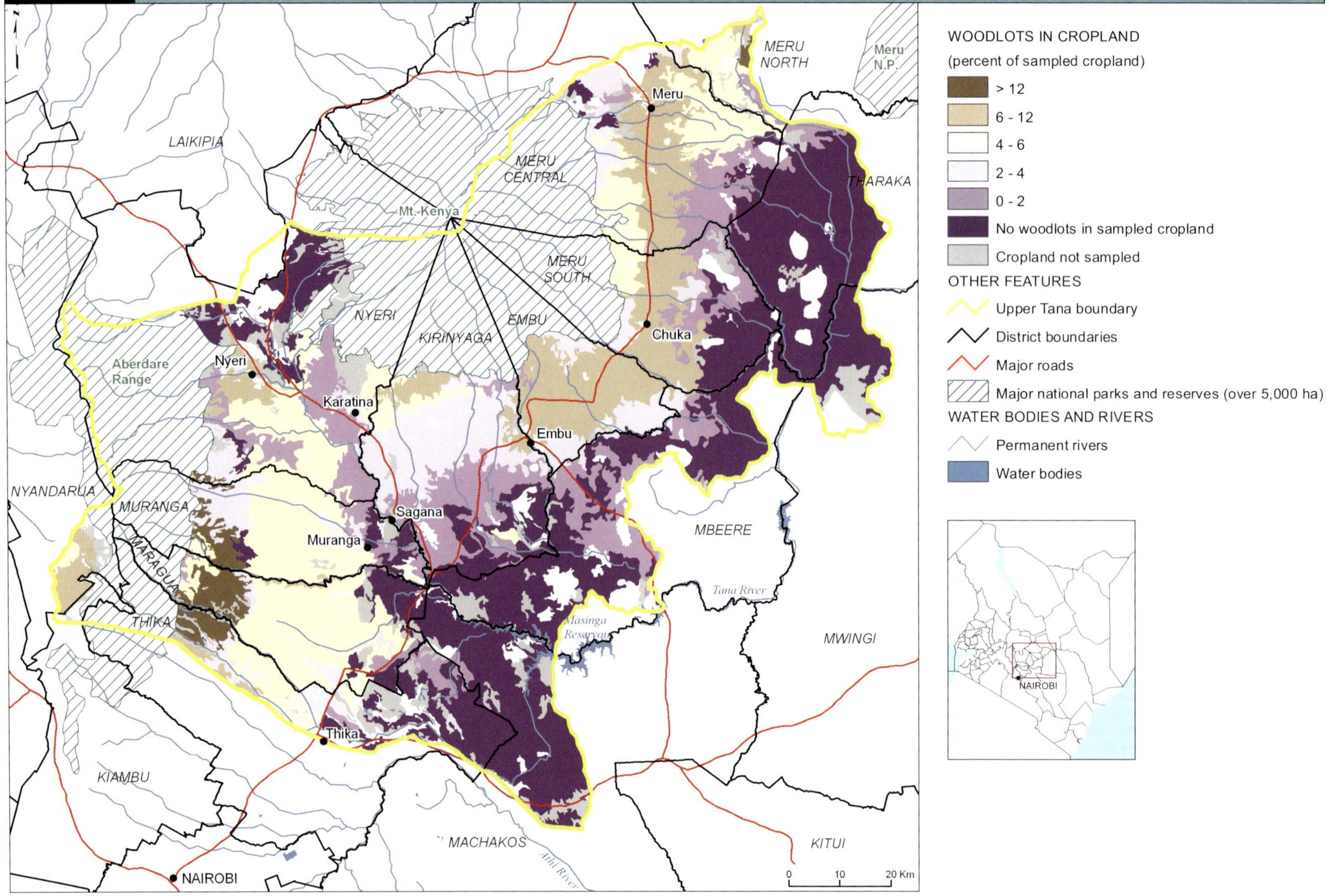

Sources: See Map 7.3.

Map 8.18 Upper Tana: High Share of Woodlots in Cropland and Poverty Rate

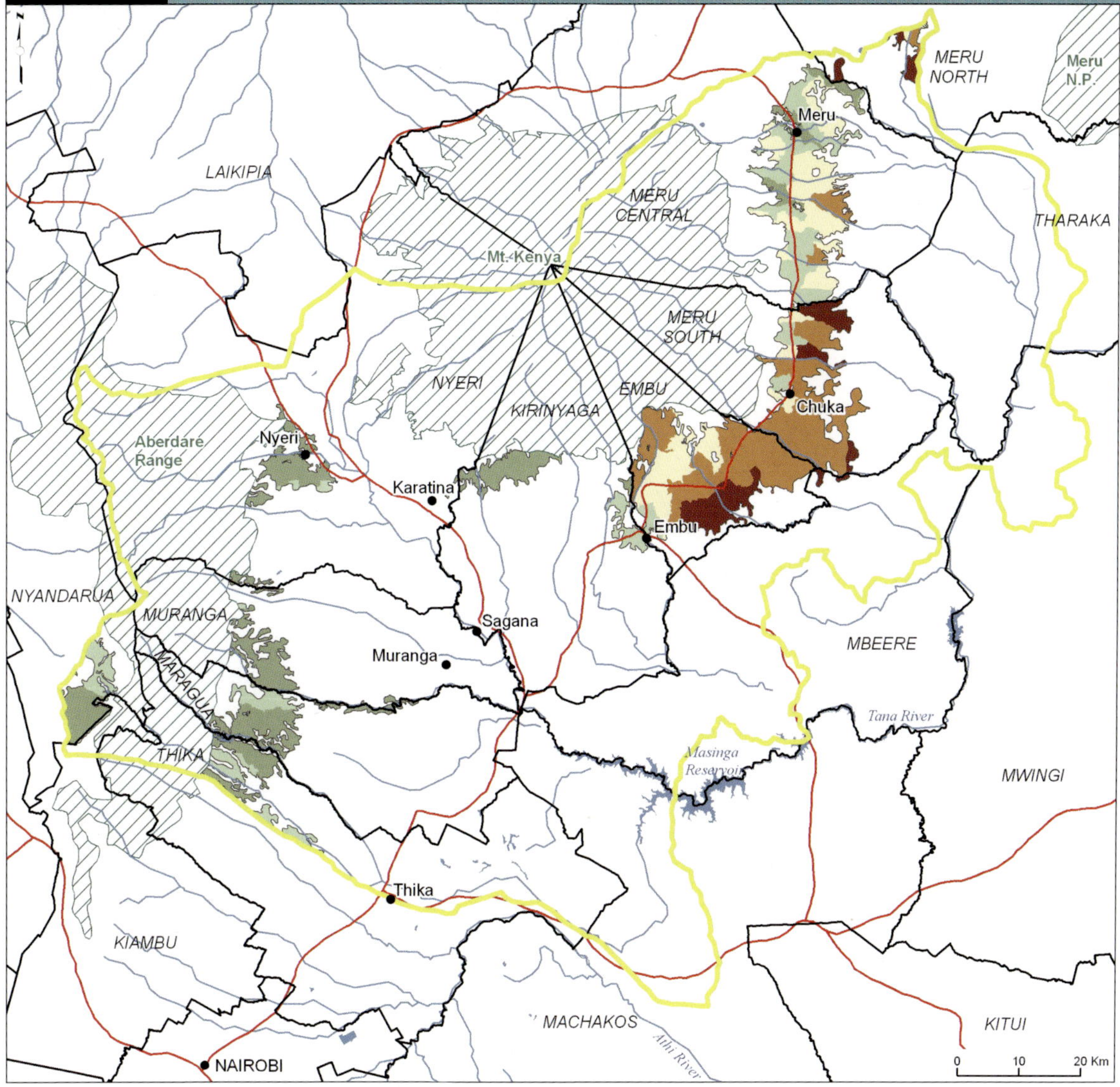

Poverty Rate in Areas with Less than 6 Percent Woodlots in Cropland

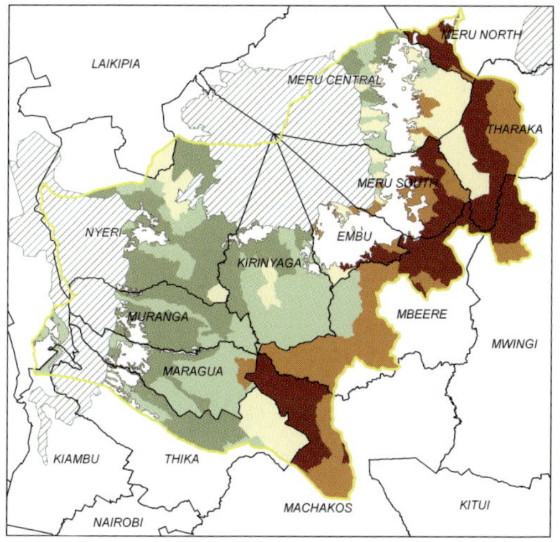

POVERTY RATE
(percent of population below poverty line)
- > 65
- 55 - 65
- 45 - 55
- 35 - 45
- <= 35

OTHER FEATURES
- Upper Tana boundary
- District boundaries
- Major roads
- Major national parks and reserves (over 5,000 ha)

WATER BODIES AND RIVERS
- Permanent rivers
- Water bodies

Sources: See Map 2.6 and Map 7.3.

of the upper Tana and coincide with low, moderate, and high rates of poverty. In the foothills of the Aberdare Range, areas where a large share of farmland is allocated to woodlots tend to be found in Locations with the lowest poverty rates (dark green-shaded map areas, with poverty rates of less than 35 percent). Locations farther downstream in the Aberdare foothills with poverty rates of 35 to 45 percent (light green areas in the small inset map showing poverty rates for areas with less than 6 percent woodlots) appear less likely to contain cropland with a large share of woodlots.

On the southeastern and eastern slopes of Mount Kenya, areas where a large share of cropland is set aside as woodlots are found in Locations with poverty rates ranging from the very low to the very high. These Locations occur in a large band stretching from the town of Embu to the town of Meru. There is very little apparent difference in the incidence of poverty within this band relative to surrounding areas (see small inset map) where a smaller proportion of farmland is devoted to woodlots.

Thus, the pattern of poverty rates in Map 8.18 indicates a more ambiguous relationship between the share of woodlots in croplands and levels of poverty. It is not clear from the maps alone what factors might account for the differences in poverty rates. For example, the purpose of these woodlots—producing wood for household use, for sale in local markets, or for sale in nearby urban markets—could result in different household incomes and affect overall poverty rates. Such information, coupled with additional analysis, might help identify opportunities for increased wood production on farmlands in poorer communities, perhaps in the lower and drier regions.

High Number of Agricultural Crops and High Share of Woodlots in Cropland

Unlike the map comparing high food cropping and high milk output (Map 8.14), Map 8.19 shows greater spatial overlap between areas with high polycropping and areas with a high share of woodlots in cropland. Nevertheless, the majority of high-supply areas for both indicators do not coincide: high-supply areas of woodlots (shaded in brown) dominate in the Aberdare foothills and in Embu District; areas with a high number of agricultural crops (shaded in green) are predominantly found on the slopes of Mount Kenya in Kirinyaga, Meru South, and Meru Central Districts.

Investigating the different local factors influencing farmers' choices in Maragua, Murunga, and Nyeri—all Districts with high shares of woodlots in cropland—could help to identify opportunities for boosting wood production. For example, wood demand for tea processing (tea-growing areas are in close proximity) or urban energy needs (in nearby Nyeri Town) could be behind these production patterns. Similar factors may explain why farmers in Embu chose to grow a higher share of woodlots than the neighboring communities in Kirinyaga and Embu Districts (with almost identical agroecological conditions). Investigating the underlying reasons for these differences—for example availability of seedlings, training, or perhaps lack of capital—could reveal where introducing new crops or agroforestry practices may improve livelihoods.

The areas of overlap between a high share of woodlots in cropland and high polycropping (shaded in red) stretch along the Chuka-Meru road in Meru South and Meru Central Districts, as well as some more isolated locations in Kirinyaga District. These could become priority areas to increase biodiversity in agroecological landscapes of the upper Tana.

None of the croplands in the drier plains at lower elevations appear as high-supply areas. This may indicate opportunities for future interventions, which may require new crop varieties or tree species that are better adapted and more suitable to the drier conditions.

Map 8.19 Upper Tana: High Average Number of Crops Grown and High Share of Woodlots in Cropland

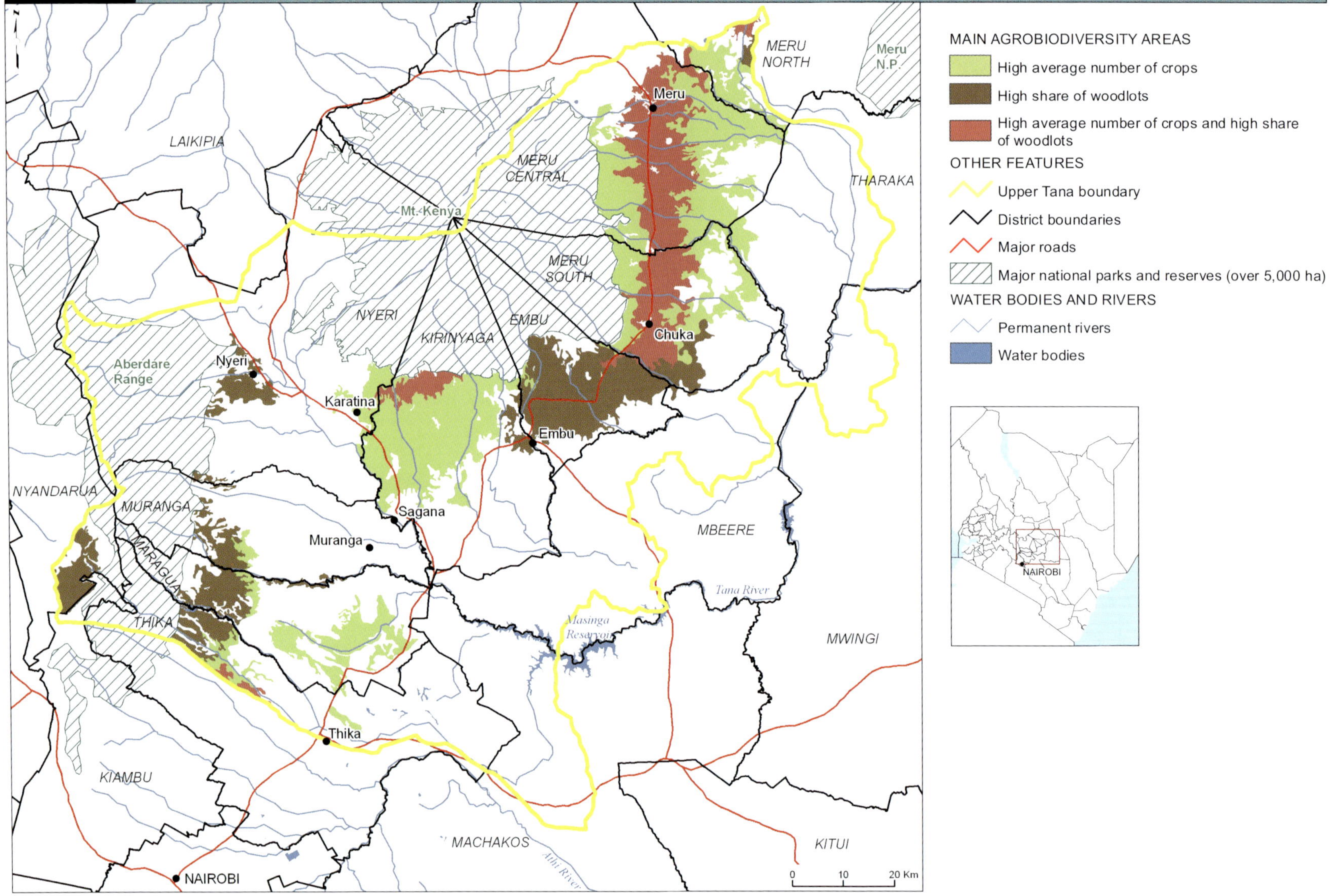

Sources: See Map 5.5 and Map 7.3.

SUMMING UP

Overview

- Within Kenya, the upper Tana represents an important supplier and consumer of ecosystem services. The selected watersheds for the upper Tana fall roughly into three major physiographic regions—mountains, foothills, and plains.
- About 3.1 million people live in the upper Tana area, representing 11.4 percent of Kenya's total population. Smallholder agriculture is the dominant land use and is concentrated in the foothills of the Aberdare Range and Mount Kenya. The government has set aside a significant portion of the land for biodiversity and watershed protection, most of it in the mountainous areas.
- About 1.3 million poor people live in the upper Tana, and the average poverty rate for the region is 43 percent (that is 10 percentage points better than Kenya's rural national average). The area contains a broad cross-section of very poor and less poor communities that have some of Kenya's lowest poverty rates. Most of the poorest communities are located in drier plains downstream of the foothills of the Aberdare Range and Mount Kenya.

Table 8.1 Upper Tana: Demographic and Poverty Characteristics for Areas Outlined by Selected Ecosystem Indicators

INDICATOR	NUMBER OF LOCATIONS OVERLAPPING WITH SELECTED AREAS	SHARE IN THE TOTAL NUMBER OF UPPER TANA LOCATIONS (PERCENT)	NUMBER OF PEOPLE LIVING IN SELECTED AREAS (MILLION)	SHARE IN THE TOTAL NUMBER OF PEOPLE IN UPPER TANA (PERCENT)	NUMBER OF POOR LIVING IN SELECTED AREAS (MILLION)	SHARE IN THE TOTAL NUMBER OF POOR IN UPPER TANA (PERCENT)	AVERAGE POVERTY RATE IN SELECTED AREAS (PERCENT)	LOWEST POVERTY RATE IN SELECTED AREAS (PERCENT)	HIGHEST POVERTY RATE IN SELECTED AREAS (PERCENT)
Water									
High Share of Piped Drinking Water (> 75 percent)	56	25	0.8	25	0.3	22	38	18	64
Small-Scale Irrigation Efforts	107	48	1.5	47	0.7	49	45	21	76
Food									
High Share of Food Crops (> 75 percent)	91	41	1.3	41	0.6	60	46	21	76
High Dairy Output (> 100,000 liters per sq. km per year)	130	59	2.2	69	0.8	52	37	18	71
Wood and Biodiversity									
High Number of Agricultural Crops (> 4)	116	52	1.7	54	0.7	51	41	18	76
High Share of Woodlots in Cropland (> 6 percent)	111	50	1.6	51	0.7	51	42	18	80
Total Upper Tana	**222**	**100**	**3.1**	**100**	**1.3**	**100**	**43**	**18**	**80**

Source: WRI calculation based on Map 8.2, Map 8.3, Map 8.6, Map 8.8, Map 8.11, Map 8.13, Map 8.16, and Map 8.18.

Note: All estimates (rounded to the nearest 100,000; percentages are based on unrounded numbers) of the number of people and the number of poor represent averages for administrative units (Locations) that overlap with the areas delineated by the six indicators. These averages may conceal important poverty linkages at the household level. For example, about 800,000 people (of which 300,000 are poor) live in Locations in which more than 75 percent of the households rely on piped drinking water. This does not automatically mean that 300,000 poor individuals have access to piped drinking water. In fact, it is more likely that the share of poor among the 25 percent of households without piped drinking water is greater than among the 75 percent benefiting from it.

Water, Food, Crop Diversity, and Woodlots

- A large number of communities in the upper Tana rely directly (and exclusively) on ecosystems to filter their drinking water and provide it in sufficient quantity. This is indicated by the great number of communities in which more than 75 percent of households rely on surface water as their primary drinking water source.
- Communities with a high share of piped water (greater than 75 percent of all households) are few in number and are spatially concentrated (including larger towns such as Thika, Nyeri, and Meru).
- There are multiple demands for water in the upper Tana. Most agriculture is rainfed. Water is needed for irrigation, hydropower, drinking water, inter-basin water transfers to Nairobi, and for sustenance of nature (i.e., wetlands and wildlife).
- Large-scale irrigation efforts are concentrated in the plains of two adjacent Districts (lower Kirinyaga and Mbeere) and include Kenya's largest rice irrigation scheme.
- Most small-scale irrigation efforts exist in a ring-like pattern around the base of Mount Kenya, with the largest numbers concentrated in Meru Central Districts. There are fewer small-scale irrigation sites in the Aberdare foothills.
- Most areas with a very high percentage of cropland (more than 75 percent) in food crops are located at lower elevations, including the plains.
- Higher elevations in the foothills—representing the tea- and coffee-growing zone—have generally lower shares of food crops.
- Areas with high milk production are located at higher elevations in the foothills of the Aberdare Range and Mount Kenya.
- Milk production in the drier plains is low.
- Farmers in the foothills of Mount Kenya favor growing a greater number of crops compared to farmers at similar elevations in the Aberdare foothills.
- Areas of high polycropping (where the average number of crops grown is greater than four) extend across most of upper Kirinyaga, Meru South, and Meru Central Districts.
- Most farmers throughout the foothills include wood as one of their crops, as indicated by the share of cropland set aside for woodlots.
- Few croplands at lower elevations in the drier plains contain woodlots.
- The highest share of woodlots in cropped areas are clustered in upper Thika, Maragua, and Muranga Districts. Embu, Meru South, and Meru Central Districts contain croplands with significant woodlot shares as well.

Relationships between Selected Ecosystem Indicators

- For large parts of the upper Tana, communities with a high share of piped drinking water and small-scale irrigation efforts do not overlap, except for a relatively large number of communities in Meru South District.
- There is practically no overlap between areas with a high share of food cropping and areas with high milk production. Farmers higher up in the foothills (and to a much larger degree in the Districts east of the Aberdare Range) rely more on nonfood cash crops and high milk outputs for their income than their counterparts further downstream (where production is focused more on food crops and where milk output per unit area is lower).
- Along the Chuka-Meru road in Meru South and Meru Central Districts there is significant overlap between areas with a high average number of agricultural crops and areas with a high share of woodlots in cropland. These areas may thus hold the potential to boost agrobiodiversity.

Map 8.20 Upper Tana: Summary of Poverty Rates in Areas Outlined by Selected Ecosystem Indicators

High Share of Piped Drinking Water and Poverty Rate

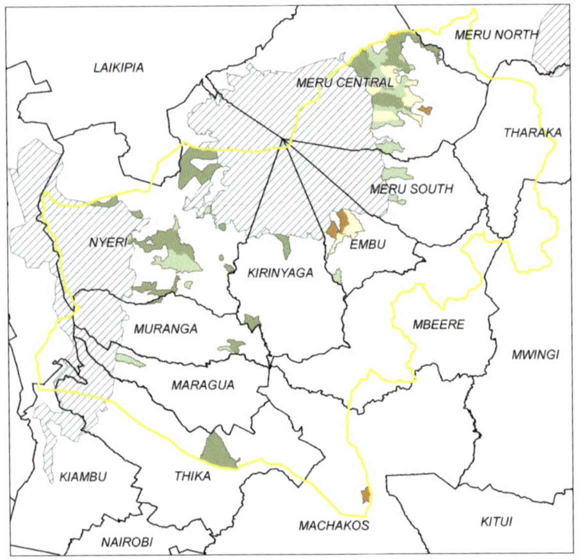

High Share of Food Crops and Poverty Rate

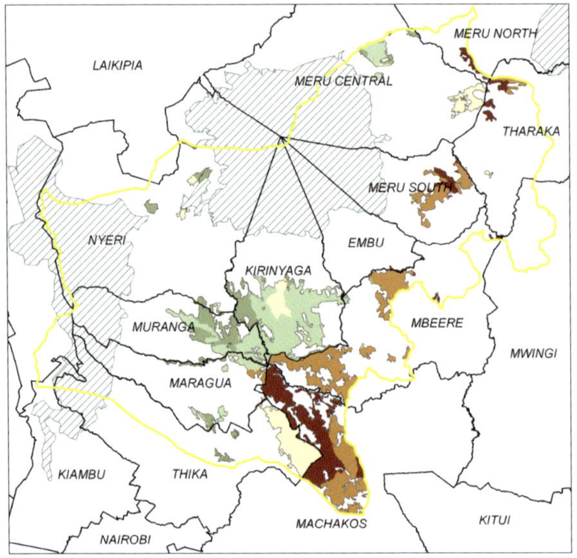

High Average Number of Crops and Poverty Rate

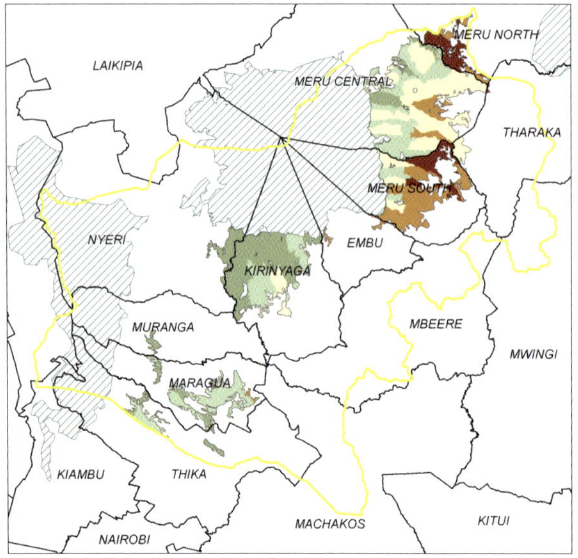

POVERTY RATE

(percent of population below poverty line)

- > 65
- 55 - 65
- 45 - 55
- 35 - 45
- <= 35

OTHER FEATURES

- Upper Tana boundary
- District boundaries
- Major national parks and reserves (over 5,000 ha)

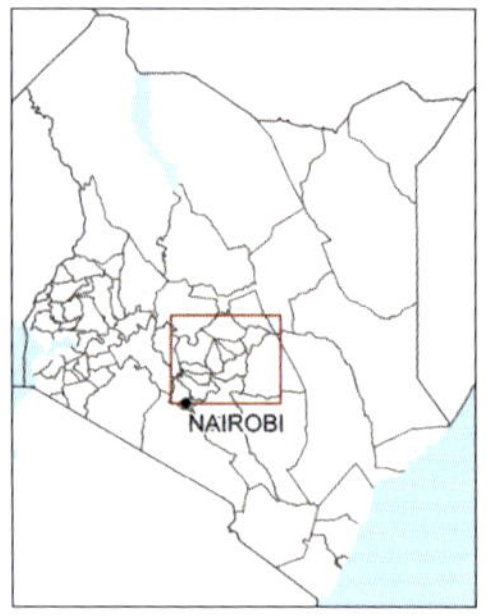

Small-Scale Irrigation Efforts and Poverty Rate

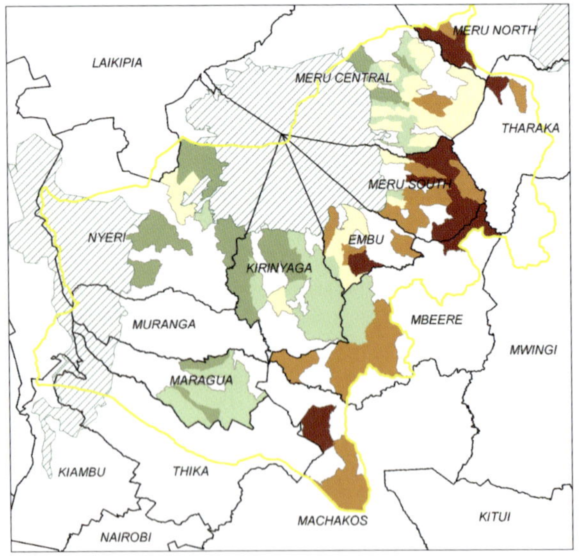

High Milk Production and Poverty Rate

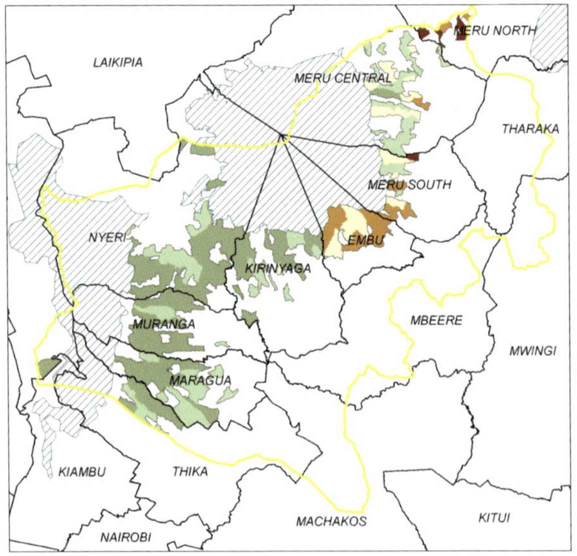

High Share of Woodlots in Cropland and Poverty Rate

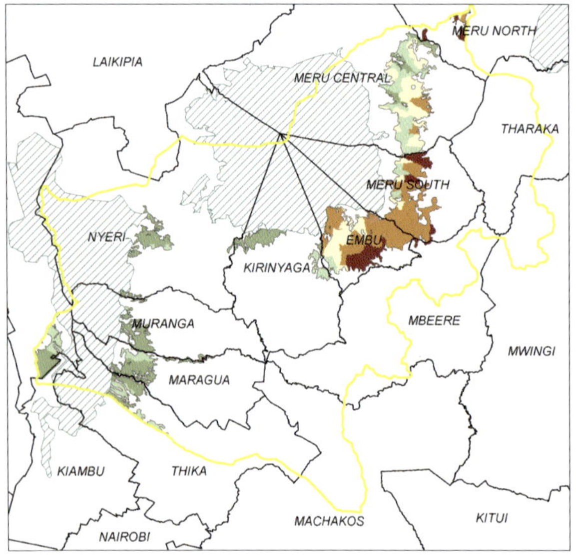

Sources: See Map 8.6, Map 8.8, Map 8.11, Map 8.13, Map 8.16, and Map 8.18.

Side-by-Side Comparison of Poverty-Ecosystem Relationships for Selected Indicators

For the purpose of this summary, Map 8.20 brings together the six indicators from the previous maps in this chapter: high share of piped drinking water, presence of small-scale irrigation efforts, high share of food crops in cropland, high milk production, high number of crops grown, and high share of woodlots in cropland. They reflect either investment areas in water infrastructure (to enhance water-related ecosystem services) or represent important supply areas of food-, wood-, and biodiversity-related ecosystem services. Such a side-by-side comparison is useful for describing poverty-ecosystem relationships and identifying locations where key supply areas and poverty patterns coincide. The following bullets show that, for some of the selected indicators, distinct spatial patterns emerge. They also show that for many of the selected indicators, the key supply areas are not automatically associated with lower or higher poverty rates, suggesting determinants that are outside of the selected variables and not necessarily related to geography.

- Communities with a high share of piped drinking water (greater than 75 percent) are concentrated in more affluent areas (Locations with poverty rates below 35 percent). The average poverty rate of administrative areas intersecting with communities that have a high share of piped drinking water is 38 percent, significantly below the average 43 percent for the upper Tana (Table 8.1).
- The poorest areas in the upper Tana have not yet benefited in a major way from piped drinking water.
- Low poverty rates are not automatically associated with higher shares of piped drinking water supplies. This is indicated by communities in the Aberdare foothills that have poverty rate of less than 45 percent but still rely on surface water or have very low shares of piped drinking water supplies in their administrative area (less than 10 percent of the households in the respective areas).
- Small-scale irrigation efforts have reached both poor and more affluent communities as indicated by the great variation of poverty rates for Locations with small-scale irrigation efforts.
- Small-scale irrigation efforts have reached some of the poorest communities, but the number and density in poorer communities is lower than in better-off areas (this does not necessarily mean that they also reached the poorest households in these communities with high average poverty rates).
- A large number of very poor areas in the lower, drier plains have not benefited from small-scale irrigation efforts.
- Areas in the lower, drier plains with a high share of food crops consistently have poverty rates below Kenya's rural national average of 53 percent.
- Locations in Kirinyaga and Muranga Districts do not confirm the simple association between high poverty and high food share—they have a high food share and low poverty rates.
- High milk production in general is more prevalent in communities with lower poverty rates. The average poverty rate for the administrative areas intersecting with high milk production areas is 37 percent (Table 8.1).
- Three areas in the Districts of upper Embu, parts of upper Meru South, and parts of Meru North diverge from this association between high milk output and lower poverty rates—here the poverty rates range between 45 and 65 percent.
- Many areas with high polycropping have low poverty rates and include Locations in Kirinyaga and Meru Central Districts, as well as a few Locations in the Aberdare foothills. However, some areas with high polycropping and moderate and high poverty are found in Meru Central and Meru South District. Therefore, high levels of polycropping are not automatically associated with certain poverty rates.
- The relationship between high share of woodlots in cropland and poverty is ambiguous. In the Aberdare foothills, the highest share of woodlots tend to be in Locations with the lowest poverty rates, and poverty rates are slightly higher in areas with lower woodlot shares. In the Mount Kenya foothills, poverty rates range from very low to very high in areas where a large share of cropland is dedicated to woodlots.

Further Analysis that Would Enhance Understanding of Poverty-Ecosystem Relationships Suggested by the Maps in this Chapter

- Investigate why some communities in Embu and Meru Central Districts with poverty rates between 45 and 65 percent have a high share of piped drinking water.
- In communities that have both small-scale irrigation efforts and high to medium-high poverty rates, find out whether these investments have had a noticeable impact on income, poverty levels, or food security (at more local scale or household level).
- Examine why high-poverty communities in the drier plains have not benefited from small-scale irrigation investments and whether future investments are technically and socially feasible.
- Analyze further the relationship between high share of food crops and poverty in certain areas. Include specific information on the number and type of food crops grown in the analysis and differentiate between high potential and more marginal croplands. Examine whether farmers in one or the other prefer higher-value food crops (e.g., vegetables and fruit) to maize or dryland cereal crops.
- Find the reasons behind the association of higher poverty rates and high milk output in Embu District.
- Determine the obstacles to higher milk output in poorer communities. Examine whether higher milk production is feasible in the poorer communities where obstacles such as availability of fodder and water, milk demand, availability of capital, etc. are present.
- Further examine the relationship between levels of polycropping and poverty. Distinguish between marginal and high-potential croplands and incorporate information on specific crops and reasons for selecting them.
- Search for additional factors that may explain the high share of woodlots in parts of the Mount Kenya foothills (e.g., purpose of wood, labor availability, and returns on investment).
- Examine why farmers in upper Maragua, Murunga, and Nyeri Districts are dedicating such a high share of their cropland to woodlots, and compare it to neighboring communities with similar agronomic conditions.
- Determine the reasons behind the low share of woodlots in poorer, drier lowlands and whether they are linked to agronomic, environmental, economic, and social factors.
- Investigate why a large proportion of communities in Meru South District have benefited from both piped drinking water supplies and small-scale irrigation efforts.
- Find out why farmers in Meru South and Meru Central grow a high number of agricultural crops and dedicate a high share of cropland to woodlots; compare this to neighboring Districts such as Embu.

Lessons Learned and Next Steps

As the previous chapters demonstrate, it is possible to compile maps of Kenya that show patterns of ecosystem service availability and use, and explore the relationships that these services have with human well-being and poverty. Kenya has made substantial investments to map many of its most important natural resources using wildlife and resource survey data. Much of this information is available to the public for use in monitoring, assessing, and managing the country's ecosystems.

At the same time, Kenya's technical institutions have established a good track record of providing maps that show the extent of poverty across the country and at various scales. The establishment of the Poverty Analysis and Research Unit at the Central Bureau of Statistics in the Ministry of Planning and National Development and its steady release of maps showing the geographic dimensions of well-being is evidence of the country's commitment to timely and accurate poverty mapping. Kenya thus has the capacity and information to map poverty and other dimensions of well-being across the country and at a scale that allows meaningful examination of its location, the ecosystem services that are nearby, and some indication of how those services influence life in Kenya.

As a result, the country has established a good foundation for analysts to use to examine the spatial relationships between poverty and selected ecosystem services, and for decision-makers to increase their understanding of poverty-environment linkages in specific locations.

LESSONS LEARNED

The following conclusions constitute general findings on the use of the maps presented in this atlas for sociogeographic analysis. More specific observations about selected ecosystem services and poverty can be found in the 'Summing Up' section at the end of the previous chapters.

1. **By combining existing maps and data on ecosystem services and human well-being, analysts can create new ecosystem-development indicators.**
   For example, Chapters 3 through 7 present poverty and demographic profiles for the upper watersheds of Kenya's 'water towers'; the communities within 25 kilometers of the most visited national parks; and croplands with high shares of food crops or woodlots in five Provinces. Each of these indicators captures a certain relationship between resources and residents that can shed light on development in these regions. This approach can now be used to analyze many other ecosystem-development relationships such as: communities within a certain distance of rivers, lakes, and reservoirs; high poverty areas and access to intensively managed cropland; or physical infrastructure, poverty, and major ecosystem services.

2. **Decision-makers can examine the spatial relationships among different ecosystem services to shed light on possible competition (i.e., tradeoffs) and synergies among various ecosystem services.**
   The maps in Chapter 8 overlay different indicators of ecosystem services such as surface water as a dominant source for drinking water, water used for small-scale irrigation, food crop production, milk output, crop diversity, and woodlot densities. These overlays suggest how analysts and policymakers can compare the spatial patterns of various ecosystem-related indicators. This is the first step to more closely examine potential synergies and tradeoffs among different ecosystem services.

3. **Decision-makers can examine the spatial relationships between poverty and combinations of ecosystem services.**
   The overlay of poverty and selected ecosystem services, shown in Chapter 8, highlights whether spatial patterns of selected ecosystem services parallel those of poverty. Decision-makers and analysts can begin to ask questions, such as: Do areas with high poverty rates coincide with areas of low food cropping? Where are the exceptions? For example, in which parts of the upper Tana River watershed is there high milk output but still relatively poor communities?

4. **In spite of the usefulness of overlaying maps of ecosystem services and poverty, there are limitations to this approach.**
   These include:
   - *Lack of data to map a comprehensive set of ecosystem services for all of Kenya.*
     Data collection systems for natural resources generally focus on sectors and commodities with high economic value or important political constituencies. They typically concentrate on the provisioning aspect of ecosystems such as the supply of food and non-food crops, timber, and fish. Data that capture non-timber forest products or reflect the local use of wetlands or mangrove-coral ecosystems, for example, could correct for some of the bias in the available data. Information on regulating services would also be useful, such as spatial data delineating groundwater recharge zones or areas where rapid changes in vegetation would greatly affect hydrological flows.
   - *Inherent limitations of spatial analyses (i.e. map overlays).*
     Analysts often lack scientifically valid models with which to link human behavior, ecosystem services, and human welfare. This means that even though they may be able to identify spatial correlations, they may not always be able to pinpoint the cause of poverty or the threats to ecosystem sustainability.

- *Limitations in the fundamental knowledge of ecosystems and their value.*
  Some of the shortcomings in mapping ecosystem services are a result of important gaps in basic ecological science and economics. The current understanding of how various ecosystem processes interact with human interventions is still limited, as is a comprehensive estimation of the economic value of ecosystem services in Kenya.
- *Complexity of measuring and monitoring poverty and livelihoods.*
  Kenya's poverty maps, based on combining household expenditure information with census data, can only capture certain aspects of human well-being and a limited set of poverty dimensions. Likewise, even though this atlas maps—for the first time—important livelihood components such as hunting, wood gathering, and charcoal production, it cannot adequately represent the variability and complexity of the livelihoods of poor families.

5. **There are important institutional barriers to measuring and mapping poverty-ecosystem relationships and using this information to inform national policies and decision-making.** These barriers include:
   - *Lack of awareness about ecosystems and ecosystem processes.*
     The findings of the Millennium Ecosystem Assessment, a global effort to assess ecosystem conditions and the links to human well-being, were released in 2005. The southern African component of this assessment demonstrated that ecosystems can be examined at various scales (including multiple countries, a large river basin, the area surrounding a protected area, and local communities), and that the resulting information can be linked to national development goals (Scholes and Biggs 2004; Biggs et al. 2005). In spite of this success, most countries have not fully adopted the ecosystem-oriented approach whose usefulness the Millennium Ecosystem Assessment proved. This is true in Kenya as well, where ecosystem thinking is still vying with a traditional sectoral focus.
   - *A sectoral mandate among government institutions that works against cross-cutting analysis involving multiple ecosystem services and poverty.*
     Mapping a set of ecosystem services and examining the links between these services and poverty requires data and expertise from a number of institutions within and outside government. However, the mandate of many government institutions focuses narrowly on sectors in the economy such as agriculture, fisheries, urban affairs, transportation, water, forests, etc. Central government budgets are designed to support these mandates, generally leaving a relatively small amount of funds and staff support for more integrated cross-sectoral work, such as environmental reporting and ecosystem mapping.
   - *Insufficient promotion of interdisciplinary analysis.*
     Mapping poverty and ecosystem services and analyzing the linkages between them requires an interdisciplinary approach, since no single individual generally has the wide range of expertise needed. Currently, the commitment to such an approach—in training and resources—is often lacking.

NEXT STEPS

Using the data and concepts demonstrated in this atlas, analysts and decision-makers in Kenyan institutions can initiate a comprehensive accounting of ecosystem services for the country. They can continue to develop new approaches to better integrate poverty-ecosystem relationships in national policies and decision-making. They can foster a better understanding among legislators of these poverty-ecosystem links. And they can apply ecosystem principles and the approach taken by the Millennium Ecosystem Assessment to national and local environmental reporting.

Accomplishing this would result in programs for poverty reduction that take into account where the poor live and what ecosystem services they depend upon, how these are changing and what opportunities exist to invest in enhancing ecosystem services to support sustainable rural livelihoods. It would improve the targeting of social expenditures and ecosystem interventions so that they reach the areas of greatest need. And it would make available to decision-makers—both in the public and private sectors—an array of spatial information that could inform their decisions on a range of resource and social issues.

Achieving such outcomes will require leadership by the Ministry of Planning and National Development and the Ministry of Environment and Natural Resources, as well as creative contributions from actors outside of government. It will require actions in four areas:

1. **Use and communicate the atlas.**
   Many organizations can use this atlas and its underlying data. The following activities would help to create a network of users:
   - *Make the underlying spatial data in this atlas publicly available.*
     Making these data available at no cost can create opportunities for developing new products, conducting new analyses, and exploring other opportunities for integrating poverty and ecosystem data. The collaborating institutions have agreed to make the core spatial data sets available on the Internet once the atlas has been published.
   - *Encourage development and dissemination of additional products.*
     Presentation slides of key maps can increase their use by senior decision-makers. Incorporating maps and articles into newspapers, magazines, and television and radio programs will enhance the communication of key messages to selected target audiences and the public. The collaborating institutions have agreed to seek opportunities to widen the use of the atlas.
   - *Incorporate maps and information on ecosystem services in Kenya's next state of the environment report and other environmental reporting efforts.*
     Periodic reports on the state of the environment can benefit from the use and application of the spatial information contained herein. Furthermore, environmental profiles of Districts and other subnational administrative assessments can adapt poverty and ecosystem maps using the GIS files from this project.
   - *Introduce poverty and ecosystem services maps into sectoral reporting.*
     Sector assessments on agriculture, water resources, biodiversity, wildlife, forestry, and others can take advantage of the data and analyses to highlight poverty-ecosystem relationships in considerably more detail.
   - *Inject maps and information on ecosystem services into future poverty analyses.*
     The second volume on the geographic dimensions of well-being in Kenya (CBS 2005) examined relationships between education levels and poverty, and between gender-specific variables and levels of poverty. The Poverty Analysis and Research Unit could take the lead and work with other government agencies to

better integrate maps and information on ecosystem services in their future work. Research organizations such as the Kenya Institute for Public Policy Research and Analysis, World Agroforestry Centre, and International Livestock Research Institute could draw on some of the underlying environmental data and use them to investigate to what degree geographic factors (e.g., remoteness and agroecological endowment) determine poverty patterns in Kenya (i.e., studies on the spatial determinants of poverty).

- *Integrate maps and information on ecosystem services into coursework.*
  Professors and lecturers can use the data and materials from this atlas in courses on environment, development, and planning. These and other public data can help students to improve the relevance of their research projects to various sectoral areas.
- *Prepare guidance and training materials to enable other countries to develop their own maps.*
  Encourage development cooperation partners to coordinate funding for such materials and mapping efforts.

2. **Build the knowledge base for mapping ecosystem services and for examining the relationships between poverty and ecosystem services.**
   There are numerous ways to improve upon this atlas and expand into new areas of research and analysis. Some of the efforts proposed below are directly applicable to ongoing government planning and decision-making. Others are more fundamental and long term, requiring leadership from universities and national and international research centers. They include the following activities:

- *Expand mapping and spatial analyses to include more ecosystem services.*
  Mapping an expanded range of ecosystem services (e.g., areas important for water regulation, water purification, or climate mitigation; important supply areas of wild plants for food security) could directly contribute to several government programs now under way. For example, a few of the six Regional Development Authorities under the Ministry of Regional Development Authorities have begun implementing catchment conservation programs or have mapped resource availability and use for long-range, integrated regional development master plans (RoK 2006). Additional information on which areas are important for hydrological services or other important regulating services could greatly enhance these plans. Similarly, the National Environment Management Authority could commission studies to map some of these services and present a more comprehensive picture of ecosystem accounting in Kenya's next state of the environment report.
- *Integrate ecological processes into future mapping of ecosystem services and use more sophisticated tools to analyze patterns and spatial relationships.*
  It is clear that maps reflecting a deeper understanding of ecological processes such as soil erosion, nutrient flows, and hydrological processes can provide an enhanced picture of whether ecosystems can continue to produce food, fiber, and other services. Similarly, analysts can adopt tools such as spatial econometrics to understand the complex interactions between resource use and well-being. Such efforts go beyond the mandate, resources, and skills of most government agencies, but several international research organizations, such as the World Agroforestry Centre and the International Livestock Research Institute in collaboration with national partners, are already carrying out work in these areas. These research organizations could continue to refine and extend the mapping and spatial analysis undertaken here in order to clarify the role of environmental resources in reducing poverty and creating economic opportunities.

3. **Use geospatial information to inform policy, planning, and implementation.**
   The maps in this atlas provide insights into national development patterns and can be used to plan and implement policies and programs aimed at locations that have high poverty rates. The text boxes titled 'Linking the Maps to Decision-Making' in Chapters 3–7 include suggestions on how maps and spatial analyses could be used to address broad national strategies and plans (see the chapters on water, food, and tourism) or to address issues such as wildlife management, preservation of biodiversity, or the charcoal industry (see the chapters on biodiversity and wood).
   While there are numerous opportunities to adapt the underlying spatial data and ideas to specific policy and planning processes, efforts in three general areas would particularly benefit from the approach used in this atlas:

- *Shaping national strategies and plans such as the Economic Recovery Strategy and the Millennium Development Goals (MDGs).*
  A follow-up to Kenya's *Economic Recovery Strategy* (GoK 2003) will need to be developed in 2007. The report on *Millennium Development Goals in Kenya, Needs & Costs* has already pointed out the investments required to close the country's information gap regarding ecosystem services (MoPND et al. 2005). Plans to implement the MDGs could benefit from a more systematic examination of the linkages between different MDG targets. For example, are the planned investments to promote higher food production, increased water use, and income generation through growth in the agriculture and tourism sectors in line with the capacity of ecosystems to provide these services? The Millennium Ecosystem Assessment carried out such an examination, finding that at least four of the eight MDGs (i.e., reducing hunger, lowering child mortality, combating diseases, and ensuring environmental sustainability) could not be met unless action was taken to stabilize the supply of ecosystem services (MA 2005). It is recommended that ecosystem services mapping take on a greater role in the process of determining what actions might be effective in stabilizing ecosystem services and balancing needed growth in agriculture, energy production, and tourism.
- *Formulating cross-sectoral policies.*
  Developing and implementing food security policies and formulating a new wildlife policy are examples of cross-sectoral policymaking. Such cross-cutting decisions require consideration of a range of resource and social issues. For example, to formulate a new wildlife policy, issues of land tenure, land use and zoning, forest management, water use and water quality, poverty reduction, and pastoralism have to be taken into account. Such policies also require integration with other related ones, such as the Forest Bill of 2005, the draft Livestock Policy, the Arid and Semi-Arid Lands Policy, the Environmental Policy, and the Land Policy. In addition, they must be aligned with national strategies like the *Economic Recovery Strategy* and plans outlined in the National Session Papers. To support such cross-cutting work, it is recommended that the policymakers and technical agencies involved take advantage of already existing spatial information on ecosystem services and poverty. With the help of additional analysis and information products that could be derived from these maps, these actors will be able to move to more fact- and evidence-based policy processes.
- *Improving local land use planning, zoning, and management plans.*
  The idea of mapping key supply areas for ecosystem services and the use of spatial overlays to link poverty and environmental issues can

be adapted to the local level, although many local planning activities will require more detailed data in addition to what is provided in this atlas. It is recommended that local actors responsible for these planning efforts look carefully at some of the ideas and examples in this publication.

4. **Strengthen institutions to research and study poverty-ecosystem relationships.**
Enhancing the research and analytical skills needed to examine poverty-ecosystem relationships will require the following efforts:

▶ *Continue to develop technical and analytical skills for spatial analysis within Kenyan institutions.*
Building technical capacity to collect data, compile maps, and carry out further analyses of poverty-environment linkages will be valuable for sectoral planning and reporting. Strengthening institutions such as Kenya Wildlife Services, Department of Resource Surveys and Remote Sensing, National Environment Management Authority, Forest Department, Kenya Agriculture Research Institute, and other national research centers will advance the analyses and understanding of poverty-ecosystem relationships.
It is equally important to expand the use of the ecosystem service approach in ministries mandated to promote industrial, transport, housing, and urban development. It is these agencies (and the private sector) that will have the greatest impact on the extent and condition of ecosystems. This will not only help in formulating sector-specific policies, but will also assist with better implementation, and will be useful for cross-sectoral work.
It is recommended that the chief executives of the above-mentioned institutions continue to invest in developing GIS data and spatial analytical skills to support more effective and efficient natural resource use and better integration of poverty-environment issues. These individual sectoral investments need to be well coordinated to avoid duplication in GIS data collection and to fit within Kenya's overall effort to build its national spatial data infrastructure.

▶ *Establish a technical working group to promote integrated spatial analyses for implementing the MDG needs assessment and the Economic Recovery Strategy (and its successor strategy).*
Such a technical working group would include key data providers and research centers. The technical staff and the chief executives of the institutions contributing to this atlas could form the nucleus of such a team. This group could foster data exchange and promote integrated analysis to better understand the relationships between poverty and ecosystem services. They could also be a catalyst for enabling easier and more direct data sharing and for formulating a national data and information policy supporting this objective.

▶ *Establish a new technical unit that could spearhead more integrated and cross-cutting work involving multiple ecosystem services and poverty.*
Experience shows that investments in collecting census and household survey data, building technical skills to produce poverty maps, and funding and staffing a poverty analysis unit within the Ministry of Planning and National Development can produce information that is useful far beyond the financial or macroeconomic sector. These investments have led to a much better understanding of the prevalence and severity of poverty in the country. And they have led to improved national planning for resource allocation to the poor, for example, by putting forward 'objective' criteria to allocate funds under the Constituency Development Funds. These criteria can now be debated and modified, thus making the process more transparent and more effective.
Kenya's successful development and use of poverty maps should serve as an incentive to create maps of ecosystem services and poverty-environment overlays. However, this will require institutional changes and resources that foster cross-sectoral collaboration.
It is recommended that high-level decision-makers actively search for opportunities to establish a cross-cutting unit or expand and better coordinate the mandates of existing units. The latter include: the Poverty Analysis and Research Unit at the Central Bureau of Statistics in the Ministry of Planning and National Development; the Geo-Information Unit of the Department of Resource Surveys and Remote Sensing in the Ministry of Environment and Natural Resources; the Agricultural Sector Coordination Unit; the Agricultural Information Resource Center; and the Arid Lands Resource Management Project.

▶ *Seek better integration of spatial information in monitoring and evaluation efforts.*
Various institutions responsible for activities in the agriculture and rural development sector have indicated that they are having difficulty establishing effective monitoring and evaluation systems for their programs (RoK 2006). These institutions could examine how investing in more compatible monitoring efforts and additional data collection can help to address some of these constraints.
In the same way, national monitoring and evaluation efforts can become the driver for better-integrated spatial information that would enhance analysis of poverty-environment relationships. Selected monitoring and evaluation activities led by the Ministry of Planning and National Development are covering a broad set of ecosystem and human well-being indicators. For example, a new Monitoring and Evaluation Department has been established to assess progress toward the MDGs (MoPND 2005). Similarly, the Central Bureau of Statistics collects data for MDG-related indicators, provides statistical support to measure progress on the *Economic Recovery Strategy*, and produces regular statistics on the spatial patterns of poverty and well-being in Kenya.
It is recommended that policymakers and technical agencies responsible for establishing national monitoring and evaluation systems reassess the role of spatial information in these efforts and identify opportunities where better integration of spatial information would strengthen these systems.

Acknowledgments

Nature's Benefits in Kenya: An Atlas of Ecosystems and Human Well-Being was possible because of financial support from the Swedish International Development Cooperation Agency; the Netherlands Ministry of Foreign Affairs; the United States Agency for International Development; the International Livestock Research Institute; Irish Aid at the Department of Foreign Affairs; and the Danish International Development Agency at the Ministry of Foreign Affairs. We deeply appreciate their support. A special thank you goes to Mats Segnestam, Michael Colby, and Carrie Stokes for their early interest and their consistent support for this project.

We wish to express our gratitude to the following institutions which contributed generously with data, maps, staff, or expert advice: the Department of Resource Surveys and Remote Sensing at the Ministry of Environment and Natural Resources (Kenya), the Central Bureau of Statistics at the Ministry of Planning and National Development (Kenya), the Food and Agriculture Organization of the United Nations and its *Africover* staff in Kenya, the International Livestock Research Institute, the Arid Lands Resource Management Project (Kenya), the United Nations World Food Programme and its Vulnerability Analysis and Mapping branch in Kenya, the Famine Early Warning Systems Network in Kenya, the Ministry of Agriculture (Kenya), Energy for Sustainable Development Africa, and the World Agroforestry Centre.

This atlas is the result of many efforts—large and small—of a larger team. We would like to express our appreciation to: Patti Kristjanson for her generosity, team-building skills, and ability to convince busy colleagues to join our efforts; Mohammed Said for his detailed knowledge of data, science, and policy processes, and his consistent support ranging from the highly technical to the diplomatic; Robin Reid for providing funding, ideas, and institutional support that came always at the right time; Paul Okwi for his ideas, patience and persistence; Jaspat L. Agatsiva for being an early champion, advisor, and enthusiastic supporter; Anthony K.M. Kilele for his guidance and institutional support; Godfrey Ndeng'e for his consistent input and support despite many other commitments; Antonio Di Gregorio and Craig von Hagen for their advice and support enabling early access to the *Africover* database; Nancy Mutunga and James Oduor for sharing livelihood zones data; Bancy Mati and David Mungai for their water expertise; Stephen Mutimba, Evans Kituyi, and Rob Bailey for their insights on firewood and charcoal use; Moses Ikiara, Fred Kaigua, and Sandra van Dijk for their tourism expertise; Brent Swallow and Frank Place for their advice and support when this project was in its infant phase; and Janet Ranganathan for providing advice throughout the project and presenting interim results at the 3rd IUCN World Conservation Congress in Bangkok.

We are deeply indebted to Ololtisatti Ole Kamuaro, Mohammed Said, Paul Okwi, Jaspat L. Agatsiva, Godfrey Ndeng'e, Anthony K.M. Kilele, Robin Reid, and Carlos Seré for their efforts to obtain the necessary support from the Permanent Secretaries. We would like to thank Chris Tuite, Peter Mungai, Peter Ndunda, Wanjira Mathai, and David Jhirad for assistance that opened the door for Professor Wangari Maathai's introductory statement.

Maps and the underlying spatial data sets are the first pillar upon which this atlas has been built. A special thank you goes to: Janet Nackoney for her enthusiasm, technical expertise, meticulous data compilation, multiple draft maps and reviews, and excellent presentations throughout the project; Florence Landsberg for bringing precision, beauty, and coordination skills to the final map-making process; Russ Kruska for sharing his extensive experience with spatial data and providing technical support despite multiple demands on his time; An Notenbaert for her skills in convincing our colleagues that livelihood maps could strengthen this atlas; and Geoffrey Kimathi and Alan Kute for their support that made the final version of the livelihood maps possible.

Words are the second pillar upon which this atlas has been built. The report has greatly benefited from the word-smithing skills of the following individuals. We would like to thank: Greg Mock for his ideas to improve this publication and his consistent final editing help from the first to the last word; Karen Holmes for translating general ideas and map patterns into readable prose and improving greatly upon the first drafts; Emily Cooper for sharpening the language of earlier drafts within tight deadlines; Linda Cotton for rewriting chapters and skillfully smoothing the styles of various authors when time was short; Hyacinth Billings for final copyediting and guiding the atlas through its final production stage; and Jo Tunstall for editorial advice and reviews when this publication was at a critical juncture.

We greatly appreciate Yumiko Kura's and Eriko Hoshino's early background research on water and fishery issues and Felix Murithi's help when we needed to locate more current data. Thank you to Phil Thornton for sharing and explaining unpublished rainfall data.

Lucille Kirori, Douglas Okwara, Margaret Githuni, and the ILRI transport staff provided impeccable logistical arrangements and support. Their contribution made Nairobi visits more efficient and pleasant.

We have benefited from the work of Elsie Velez-Whited, Chris Elias, Saima Zaman, Rina Rodriguez, and Dan Tunstall. They kept our funders interested in this project, shared our progress enthusiastically, and made sure that our reporting was timely and accurate.

It has been a pleasure to work with Kim Pollock and Holly Hagen from Dever Designs. We thank the staff at Jarboe Printing for a timely and efficient printing process.

We would like to express our gratitude to the following individuals who all made valuable contributions to this atlas by writing drafts, preparing maps, supplying data, reviewing methods for map indicators, or providing expert advice: Stephen Adam, Jaspat L. Agatsiva, Polly Akwanyi, Michael Arunga, Richard Bagine, Isabelle Baltenweck, Hyacinth Billings, Emily Cooper, Linda Cotton, Carolina de Rosas, Norbert Henninger, Karen Holmes, Moses Ikiara, Fred Kaigua, Eunice Kariuki, Shem Kifugo, Anthony K.M. Kilele, Geoffrey Kimathi, Evans Kituyi, Patti Kristjanson, Russ Kruska, Alan Kute, Florence Landsberg, Diana Lee-Smith, Bancy Mati, Greg Mock, Wycliffe Mutero, Stephen Mutimba, Nancy Mutunga, Janet Nackoney, An Notenbaert, Godfrey Ndeng'e, James Nyoro, Pamela Ochungo, Alex Odero, James Oduor, Paul Okwi, Dickson Ole Kaelo, Ololtisatti Ole Kamuaro, Dan Omolo, Daniel Prager, Robin Reid, Mohammed Said, Amos Situma, Sammy Towett, Dan Tunstall, Jo Tunstall, Sandra van Dijk, Ville Vuorio, Patrick Wargute, and Jamie Worms.

We would like to thank Carolina de Rosas and David Jhirad for organizing a smoothly run review process. We have greatly benefited from our reviewers who provided timely and detailed comments on various drafts of the text and the maps: Carmelle Terborgh at Environmental Systems Research Institute; Antonio Di Gregorio and Craig von Hagen at the Food and Agriculture Organization of the United Nations; Peter Ndunda at the Green Belt Movement; Faith Githui at the IGAD Climate Prediction and Applications Centre; Luis Navarro at the International Development Resource Centre; Wilbur Ottichilo at the Regional Centre for Mapping of Resources for Development; Shalini Vajjhala at Resources for the Future; Pat Naidoo at the Rockefeller Foundation; Carmen Revenga at The Nature Conservancy; Michael Colby and Carrie Stokes at the United States Agency for International Development; David Mungai at the University of Nairobi; Frank Place and Brent Swallow at the World Agroforestry Centre; Phil Angell, Lauretta Burke, Habiba Gitay, David Jhirad, Christian Layke, Susan Minnemeyer, Daniel Prager, and Janet Ranganathan at the World Resources Institute. Without implicating them in any way, we thank them for their comments that helped to improve this document. We retain full responsibility for any remaining errors of fact or interpretation.

The Authors

List of Abbreviations and Acronyms

Africover	Digital georeferenced database on land cover for Africa produced by FAO and country partners
ALRMP	Arid Lands Resource Management Project, Kenya
CBS	Central Bureau of Statistics, Kenya
DRSRS	Department of Resource Surveys and Remote Sensing, Kenya
EBA	Endemic Bird Area
ESDA	Energy for Sustainable Development Africa, Kenya
ESRI	Environmental Systems Research Institute
FAO	Food and Agriculture Organization of the United Nations
FEWS NET	Famine Early Warning System Network
GBM	Green Belt Movement
GDP	Gross Domestic Product
IBA	Important Bird Area
ICPAC	IGAD Climate Prediction and Applications Centre
ICRAF	World Agroforestry Centre
ICRISAT	International Crops Research Institute for the Semi-Arid Tropics
IDRC	International Development Research Centre
IGAD	Intergovernmental Authority on Development
ILRI	International Livestock Research Institute
IUCN	The World Conservation Union
IWMI	International Water Management Institute
KATO	Kenya Association of Tour Operators
kg	kilogram
KIPPRA	Kenya Institute for Public Policy Research and Analysis
km	kilometer
Ksh	Kenyan Shilling
KWS	Kenya Wildlife Service
MDGs	Millennium Development Goals
RCMRD	Regional Centre for Mapping of Resources for Development
RF	Rockefeller Foundation
RFF	Resources for the Future
sq. km	square kilometer
Tegemeo	Tegemeo Institute of Agricultural Policy and Development, Kenya
TLU	Tropical Livestock Unit (equal to an animal weight of 250 kilograms)
TNC	The Nature Conservancy
UoN	University of Nairobi
US$	US Dollar
USAID	United States Agency for International Development
WFP	United Nations World Food Programme
WRI	World Resources Institute

Sources

Chapter 1 Ecosystems and Ecosystem Services

Biggs, R., E. Bohensky, P.V. Desanker, C. Fabricius, T. Lynam, A. A. Misselhorn, C. Musvoto, M. Mutale, B. Reyers, R.J. Scholes, S. Shikongo, and A.S. van Jaarsveld. 2004. *Nature Supporting People, The Southern African Millennium Ecosystem Assessment, Integrated Report. A Contribution to the Millennium Ecosystem Assessment, Prepared by the Southern African Millennium Ecosystem Assessment Team.* Pretoria: Council for Scientific and Industrial Research. Online at www.maweb.org/safma

Central Bureau of Statistics (CBS), Ministry of Planning and National Development. 2001. *Population Distribution by Administrative Areas and Urban Centres, Kenya 1999 Population and Housing Census. Vol. 1.* Nairobi: CBS.

Central Bureau of Statistics (CBS), Ministry of Planning and National Development. 2002. *Kenya 1999 Population and Housing Census GIS Database.* Nairobi: CBS.

Central Bureau of Statistics (CBS), Ministry of Planning and National Development. 2003. *Geographic Dimensions of Well-Being in Kenya: Where are the Poor? From Districts to Locations. Vol. I.* Nairobi: CBS. Online at http://www.worldbank.org/research/povertymaps/kenya/volume_index.htm

Central Bureau of Statistics (CBS), Ministry of Planning and National Development. 2004. *Economic Survey 2004.* Nairobi: CBS.

Central Bureau of Statistics (CBS), Ministry of Planning and National Development. 2005. *Economic Survey 2005.* Nairobi: CBS.

Central Bureau of Statistics (CBS), Ministry of Planning and National Development. 2006. *Kenya Facts and Figures, 2006 Edition.* 2006. Nairobi: CBS. Online at http://www.cbs.go.ke/downloads/pdf/Kenyafacts2006.pdf

Department of Resource Surveys and Remote Sensing (DRSRS), Ministry of Environment and Natural Resources. 2003. *DRSRS GIS Database.* Nairobi: DRSRS.

Food and Agriculture Organization (FAO) of the United Nations. 2000. *Africover Multipurpose Land Cover Databases for Kenya.* Rome: FAO. Online at www.africover.org

Food and Agriculture Organization (FAO) of the United Nations. 2005. "Kenya Country Report." In: *Irrigation in Africa in Figures, AQUASTAT Survey 2005.* Rome: FAO. Online at http://www.fao.org/ag/agl/aglw/aquastat/countries/kenya/index.stm

Government of Kenya (GoK). 2003. *Economic Recovery Strategy for Wealth and Employment Creation 2003-2007.* Nairobi: GoK. Online at http://www.planning.go.ke/publication/economic_recovery_plan.pdf

Grunblatt, J., M. Said, P. Wargute, and S.C. Kifugo. 1995. *DRSRS Data Summary Report: Kenyan Rangelands 1977-1994.* Nairobi: Department of Resource Surveys and Remote Sensing (DRSRS), Ministry of Planning and National Development.

Grunblatt, J., M. Said, and P. Wargute. 1996. *DRSRS National Rangelands Report: Summary of Population Estimates for Wildlife and Livestock, Kenyan Rangelands 1977-1994.* Nairobi: Department of Resource Surveys and Remote Sensing (DRSRS), Ministry of Planning and National Development.

Hijmans, R.J., S. Cameron, and J. Parra. 2005. *WorldClim Climate Surfaces, Version 1.4.* Online at http://www.worldclim.org/

Millennium Ecosystem Assessment (MA). 2003. *Ecosystems and Human Well-Being: A Framework for Assessment.* Washington, DC: Island Press. Online at http://www.maweb.org/en/Framework.aspx

Millennium Ecosystem Assessment (MA). 2005. *Ecosystems and Human Well-Being: Synthesis.* Washington, DC: Island Press. Online at http://www.maweb.org/en/Synthesis.aspx

Ministry of Planning and National Development (MoPND), United Nations Development Programme, and Government of Finland. 2005. *Millennium Development Goals in Kenya, Needs & Costs.* Nairobi: Government of Kenya. Online at http://www.ke.undp.org/KenyaMDGNeedsAssessment.pdf

Ministry of Water Development (MoWD) and Japan International Cooperation Agency (JICA). 1992. *The Study on the National Water Master Plan. GIS Data of Drainage and Subdrainage Areas.* Nairobi, Kenya: MoWD, JICA, and International Livestock Research Institute.

National Imagery and Mapping Agency of the United States (NIMA). 1997. *VMAP_1V10 - Vector Map Level 0 (Digital Chart of the World).* Washington, DC: NIMA.

Republic of Kenya (RoK). 2006. *Medium Term Expenditure Framework 2006/2007-2008/2009. Report for the Agriculture and Rural Development (ARD) Sector. Final Draft. February 2006.* Nairobi: RoK Online at http://www.treasury.go.ke/sector_reports/ARD_Sector%20_Report_2007%20_Draft.pdf

Survey of Kenya (SoK). 2003. *National Atlas of Kenya, Fifth Edition.* Nairobi: Survey of Kenya.

Survey of Kenya (SoK) and International Livestock Research Institute (ILRI). 2000. *Cities, Towns, and Market Centers GIS Database.* Digitized by ILRI from Survey of Kenya's 1:250,000 topographic sheets (for northern Kenya) and 1:50,000 topographic sheets for the rest of the country. Nairobi: SoK and ILRI.

Survey of Kenya (SoK), Japan International Cooperation Agency (JICA) and International Livestock Research Institute (ILRI). 1996. *250-meter Digital Elevation Model.* Created by ILRI and JICA, derived from SoK 1:250,000 Topographic Map. Nairobi: ILRI.

United Nations Environment Programme (UNEP). 2001. *An Assessment of the Status of the World's Remaining Closed Forests.* Nairobi: UNEP. Online at http://grid2.cr.usgs.gov/publications/closedforest.pdf#search=%22an%20assessment%20of%20the%20worlds%20remaining%20closed%20forests%22

World Conservation Union (IUCN) and United Nations Environment Programme/World Conservation Monitoring Centre (UNEP/WCMC). 2006. *2006 World Database on Protected Areas.* Cambridge, United Kingdom: IUCN and UNEP/WCMC.

World Gazetteer. 2006. *Kenya: Largest Cities and Towns.* Online at http://www.world-gazetteer.com/wg.php?x=&men=gcis&lng=en&dat=32&srt=pnan&col=ahq&geo=-112

Chapter 2 Spatial Patterns of Poverty and Human Well-Being

Central Bureau of Statistics (CBS), Ministry of Planning and National Development. 1994. *Kenya Population Census 1989. Vol. 1.* Nairobi: CBS.

Central Bureau of Statistics (CBS), Ministry of Planning and National Development. 1995. *Kenya Population Census of 1989.* This GIS coverage was created jointly by International Livestock Research Institute (ILRI) GIS unit, Famine Early Warning Systems (FEWS), and Tropical Soil Biological and Fertility Programme (TSBF). Nairobi: CBS.

Central Bureau of Statistics (CBS), Ministry of Planning and National Development. 2001. *Population Distribution by Administrative Areas and Urban Centres, Kenya 1999 Population and Housing Census. Vol. 1.* Nairobi: CBS.

Central Bureau of Statistics (CBS), Ministry of Planning and National Development. 2002. *Kenya 1999 Population and Housing Census GIS Database.* Nairobi: CBS.

Central Bureau of Statistics (CBS), Ministry of Planning and National Development. 2003. *Geographic Dimensions of Well-Being in Kenya: Where are the Poor? From Districts to Locations. Vol. I.* Nairobi: CBS. Online at http://www.worldbank.org/research/povertymaps/kenya/volume_index.htm

Central Bureau of Statistics (CBS), Ministry of Planning and National Development. 2005. *Geographic Dimensions of Well-Being in Kenya: Who and Where are the Poor? A Constituency Level Profile. Vol. II.* Nairobi: CBS.

Central Bureau of Statistics (CBS), Ministry of Planning and National Development. 2006. *Kenya Facts and Figures, 2006 Edition.* Nairobi: CBS. Online at http://www.cbs.go.ke/downloads/pdf/Kenyafacts2006.pdf

Food and Agriculture Organization (FAO) of the United Nations. 2000. *Africover Multipurpose Land Cover Databases for Kenya.* Rome: FAO. Online at www.africover.org

Millennium Ecosystem Assessment (MA). 2005. *Ecosystems and Human Well-Being: Synthesis.* Washington, DC: Island Press. Online at http://www.maweb.org/en/Synthesis.aspx

Ravallion, M. 1992. *Poverty Comparisons, A Guide to Concepts and Methods, LSMS Working Paper Number 88.* Washington, DC: The World Bank.

Sen, A. 1999. *Development and Freedom.* Oxford: Oxford University Press.

Survey of Kenya (SoK) and International Livestock Research Institute (ILRI). 2000. *Cities, Towns, and Market Centers GIS Database.* Digitized by ILRI from Survey of Kenya's 1:250,000 topographic sheets (for northern Kenya) and 1:50,000 topographic sheets for the rest of the country. Nairobi: SoK and ILRI.

World Conservation Union (IUCN) and United Nations Environment Programme/World Conservation Monitoring Centre (UNEP/WCMC). 2006. *2006 World Database on Protected Areas.* Cambridge, United Kingdom: IUCN and UNEP/WCMC.

Chapter 3 Water

Akotsi E. and M. Gachanja. 2004. *Changes in Forest Cover in Kenya's Five Water Towers 2000-2003.* Nairobi: Department of Resource Surveys and Remote Sensing and Kenya Forest Working Group. Online at http://www.unep.org/dewa/assessments/EcoSystems/Land/mountain/pdf/forest_catchment_2003_report.pdf

Amissah-Arthur, A., S. Jagap, and C. Rosenzweig. 2002. "Spatio-Temporal Effects of El Niño Events on Rainfall and Maize Yields in Kenya." *International Journal of Climatology,* 22, 1849-1860.

Balla, P.T. 2006. National Study on Small Hydro Power Development: Status and Potential of Small Hydro Power Development in the Tea Industry in Kenya. Online at http://greeningtea.unep.org/downloads/

Brakenridge, G.R., E. Anderson, and S. Caquard. 2006. *Global Active Archive of Large Flood Events.* Data selected for Kenya, 2002-2006. Hanover, USA: Dartmouth Flood Observatory. Online at http://www.dartmouth.edu/~floods/2003282.html

Campbell, D.J., D.P. Lusch, T. Smucker, and E.E. Wangui. 2003. *Root Causes of Land Use Change in the Loitokitok Area, Kajiado District, Kenya,* LUCID Working Paper 19. East Lansing, Michigan: Michigan State University. Online at http://www.lucideastafrica.org/publications.htm

Central Bureau of Statistics (CBS), Ministry of Planning and National Development. 2002. *Kenya 1999 Population and Housing Census GIS Database.* Nairobi: CBS.

Central Bureau of Statistics (CBS), Ministry of Planning and National Development. 2003. *Geographic Dimensions of Well-Being in Kenya: Where are the Poor? From Districts to Locations. Vol. I.* Nairobi: CBS. Online at http://www.worldbank.org/research/povertymaps/kenya/volume_index.htm

Central Bureau of Statistics (CBS), Ministry of Planning and National Development. 2006. *Kenya Facts and Figures, 2006 Edition.* Nairobi: CBS. Online at http://www.cbs.go.ke/downloads/pdf/Kenyafacts2006.pdf

Central Bureau of Statistics (CBS), Ministry of Health (MOH), and ORC Macro. 2004. *Kenya Demographic and Health Survey 2003.* Calverton, Maryland: CBS, MOH, and ORC Macro. Online at http://www.statcompiler.com/pubs/pub_details.cfm?ID=462&srchTp=advanced

Department of Resource Surveys and Remote Sensing (DRSRS), Ministry of Environment and Natural Resources. 2003. *DRSRS GIS Database.* Nairobi: DRSRS.

Emerton, L. and F. Vorhies. 1998. *Wetlands Services - Getting Customers to Pay. Paper presented to the Second International Conference on Wetlands and Development.* Dakar, Senegal: World Conservation Union (IUCN), Economics Service Unit.

Food and Agriculture Organization (FAO) of the United Nations. 2000a. *Africover Multipurpose Land Cover Databases for Kenya.* Rome: FAO. Online at: www.africover.org

Food and Agriculture Organization (FAO) of the United Nations. 2000b. *Crop and Food Supply Situation in Kenya, FAO Global Information and Early Warning System on Food and Agriculture, Special Report, 10 July 2000.* Online at http://www.fao.org/docrep/004/x7697e/x7697e00.htm

Food and Agriculture Organization (FAO) of the United Nations. 2005. "Kenya Country Report." In: *Irrigation in Africa in Figures, AQUASTAT Survey 2005.* Rome: FAO. Online at http://www.fao.org/ag/agl/aglw/aquastat/countries/kenya/index.stm

Government of Kenya (GoK). 2003. *Economic Recovery Strategy for Wealth and Employment Creation 2003-2007.* Nairobi: GoK. Online at http://www.planning.go.ke/publication/economic_recovery_plan.pdf

Grunblatt, J., M. Said, P. Wargute, and S.C. Kifugo. 1995. *DRSRS Data Summary Report: Kenyan Rangelands 1977-1994.* Nairobi: Department of Resource Surveys and Remote Sensing (DRSRS), Ministry of Planning and National Development.

Grunblatt, J., M. Said, and P. Wargute. 1996. *DRSRS National Rangelands Report: Summary of Population Estimates for Wildlife and Livestock, Kenyan Rangelands 1977-1994.* Nairobi: Department of Resource Surveys and Remote Sensing (DRSRS), Ministry of Planning and National Development.

Hijmans, R.J., S. Cameron, and J. Parra. 2005. *WorldClim Climate Surfaces, Version 1.4.* Online at http://www.worldclim.org/

Intermediate Technology Development Group (ITDG) and Energy for Sustainable Development Africa (ESDA). 2005. "Table of Existing and Potential Micro-Hydropower Sites in Kenya." Nairobi: ITDG and ESDA.

Jones P.G. and P.K. Thornton. 1999. "Unpublished Maps and Data of Bimodal Climates in Africa, Latin America, and Asia."

Katui-Katua, M. 2004. *Drawers of Water II: Thirty Years of Change in Domestic Water Use and Environmental Health in East Africa-Kenya Country Study.* London: International Institute for Environment and Development (IIED). Online at http://www.iied.org/pubs/pdf/full/9050IIED.pdf

Kenya Electricity Generating Company (KenGen) Limited. 2006. *Prospectus.* Online at http://www.kengen.co.ke/IPO/full_prospectus.pdf

Kenya Food Security Steering Group (KFSSG). 2006. *Kenya Short Rains Assessment Report 2005, 9 February 2006, Consolidated Interagency Report.* Nairobi: KFSSG. Online at http://documents.wfp.org/stellent/groups/public/documents/ena/wfp087348.pdf

Millennium Ecosystem Assessment (MA). 2005. *Ecosystems and Human Well-Being: Wetlands and Water Synthesis.* Washington, DC: World Resources Institute. Online at http://www.maweb.org/en/Synthesis.aspx

Ministry of Agriculture and Livestock Development (MoALD). 1995. *District Profiles of Irrigation and Drainage in Kenya* (All Districts). Nairobi: MoALD, Irrigation and Drainage Branch.

Ministry of Water Development (MoWD) and Japan International Cooperation Agency (JICA). 1992a. *The Study on the National Water Master Plan. GIS Data of Drainage and Subdrainage Areas.* Nairobi: MoWD, JICA, and International Livestock Research Institute.

Ministry of Water Development (MoWD) and Japan International Cooperation Agency (JICA). 1992b. *The Study on the National Water Master Plan.* "Table 4.7.2: Balance between Water Demands and Potential Available Water." Nairobi: MoWD and JICA.

Ministry of Water Development (MoWD) and Japan International Cooperation Agency (JICA). 1992c. *The Study on the National Water Master Plan.* "Figure M10.1: Location Map of Potential and Existing Damsites." Nairobi: MoWD and JICA.

Ministry of Water Development (MoWD) and Japan International Cooperation Agency (JICA). 1992d. *The Study on the National Water Master Plan.* "Figure M10.2: Water Transfer Scheme." Nairobi: MoWD and JICA.

Ministry of Water Development (MoWD) and Japan International Cooperation Agency (JICA). 1992e. *The Study on the National Water Master Plan.* "Map U-1: Nairobi." Nairobi: MoWD and JICA.

Ministry of Water Development (MoWD) and Japan International Cooperation Agency (JICA). 1992f. *The Study on the National Water Master Plan.* "Map U-32: Mombasa." Nairobi: MoWD and JICA.

Ministry of Water Development (MoWD) and Japan International Cooperation Agency (JICA). 1992g. *The Study on the National Water Master Plan.* "Map U-25: Watamu." Nairobi: MoWD and JICA.

Ministry of Water Development (MoWD) and Japan International Cooperation Agency (JICA). 1992h. *The Study on the National Water Master Plan.* "Table 4.5: Hydropower Projects to be Implemented by 2010." Nairobi: MoWD and JICA.

Ministry of Water Development (MoWD) and Japan International Cooperation Agency (JICA). 1992i. *The Study on the National Water Master Plan.* "Figure 4.4: National Power Grid Map." Nairobi: MoWD and JICA.

Ministry of Water Development (MoWD) and Japan International Cooperation Agency (JICA). 1992j. *The Study on the National Water Master Plan.* "Table 4.3: Major Irrigation Projects to be Implemented by 2010." Nairobi: MoWD and JICA.

Ministry of Water Development (MoWD) and Japan International Cooperation Agency (JICA). 1992k. *The Study on the National Water Master Plan.* "Figure 4.3: Location of Selected Large-Scale Irrigation Schemes." Nairobi: MoWD and JICA.

Ministry of Water Development (MoWD) and Japan International Cooperation Agency (JICA). 1992l. *The Study on the National Water Master Plan.* "Section 2.4: Daily Water Consumptions for Major Species." Nairobi: MoWD and JICA.

Ministry of Water Development (MoWD) and Japan International Cooperation Agency (JICA). 1992m. *Main Report.* Nairobi: MoWD and JICA.

Munga, D., J.U. Kitheka, S.M. Mwaguni, H.S. Massa, B.M. Mwashote, S. Mwangi, H. Ong'anda, M.M. Muthuka, F. Mdoe, S.J. Chidagaya, and G. Opello. 2004. *Pollution and Vulnerability of Water Supply Aquifers in Mombasa, Kenya: Interim Progress Report.* Mombasa: Kenya Marine and Fisheries Research Institute, Coast Development Authority, Ministry of Water Resources Development and Management, Municipal Council of Mombasa, and Government Chemistry Department. Online at www.unep.org/groundwaterproject/Archives/Kenya-midReport.pdf

Nairobi City Council. 2006. "Water and Sewerage." Online at http://www.nairobicity.org/departments/default2.asp?search=sewerage

Nakagawa, H., S. Ikebuchi, S. Kohsaki, K.S. Makhanu, and K. Wakabayashi. 1994. *21st Century Water Challenges in Kenya. Proceedings from the 20th WEDC Conference on Affordable Water Supply and Sanitation, Colombo, Sri Lanka.* Online at http://www.lboro.ac.uk/departments/cv/wedc/papers/20/sessiong/nakagawa.pdf

National Imagery and Mapping Agency of the United States (NIMA). 1997. *VMAP_1V10 - Vector Map Level 0 (Digital Chart of the World).* Washington, DC: NIMA.

Njuguna-Githinji, W. 1991. "Kenya." In *ITT Industries Guidebook to Global Water Issues.* Online at http://www.itt.com/waterbook/

Nyaoro, J.R. 1999. "Case Study on Kenya River Watershed Management and Arising Conflicts." In *Report on the Conference on Cooperation or Conflict: Ways of Managing Scarce Natural Resources in Africa, 18-23 June 1999.* Maputo, Mozambique: Civil Society Conflict Prevention Network and Citizens' Security Council (KATU). Online at http://www.katunet.fi/suomi/pdf/NATURAL%20RESOURCE%20REPORT.pdf

Oindo B.O. and A.K. Skidmore. 2002. "Interannual Variability of NDVI and Species Richness in Kenya." *International Journal of Remote Sensing*, 23, 285-298.

Onjala, J. 2001. *Water Pricing Options in Kenya: Cases of Mwea and West Kano Irrigation Schemes.* CDR Working Paper 01.9, December 2001. Copenhagen: Centre for Development Research. Online at http://www.diis.dk/graphics/CDR_Publications/cdr_publications/working_papers/wp-01-9.pdf

Onjala, J. 2002. *Managing Water Scarcity in Kenya: Industrial Response to Tariffs and Regulatory Enforcement.* Ph.D. Dissertation, Roskilde University, Denmark. Online at http://diggy.ruc.dk/bitstream/1800/470/1/managing_water_scarcity.pdf

Owore, D. 2004. "The Water Distribution Factor in Karen." *Network for Water and Sanitation (NETWAS) Newsletter.* Online at http://www.netwas.org/newsletter/articles/2004/01/6

Peden D., G. Tadesse, and M. Mammo. 2003. "Improving the Water Productivity of Livestock: An Opportunity for Poverty Reduction." In P.G. McCornick, A.B. Kamara, and G. Tadesse (editors). *Integrated Water and Land Management Research and Capacity Building Priorities for Ethiopia. Proceedings of a MoWR/EARO/IWMI/ILRI International Workshop held at ILRI, Addis Ababa, Ethiopia, 2-4 December 2002.* Colombo, Sri Lanka and Nairobi: International Water Management Institute (IWMI) and International Livestock Research Institute (ILRI). Online at http://www.ilri.org/publications/cdrom/integratedwater/IWMI/Documents/Print_version/IWMI_Pro.pdf

Peden, D. 2004. *Water and Livestock, A Framework for Water and Livestock Investment and Initial Findings. Paper presented at the World Water Week Seminars on 19-20th August 2004.* Stockholm.

ReliefWeb. 2005. *International Federation of Red Cross and Red Crescent Societies (IFRC) Report Flooding in Kenya, 31 May 2005.* Online at http://www.reliefweb.int/rw/RWB.NSF/db900SID/EGUA-6CWKSN?OpenDocument

Survey of Kenya (SoK). 1971. *Directorate of Overseas Surveys: 1:50,000 Topographical Maps.* Series Y731. Nairobi: SoK.

Survey of Kenya (SoK). 2003. *National Atlas of Kenya, Fifth Edition.* Nairobi: SoK.

Survey of Kenya (SoK) and International Livestock Research Institute (ILRI). 1997. *Road Network GIS Database.* Digitized by ILRI from Survey of Kenya's 1:50,000 topographic sheets for 1978-1997. Nairobi: SoK and ILRI.

Survey of Kenya (SoK) and International Livestock Research Institute (ILRI). 2000. *Cities, Towns, and Market Centers GIS Database.* Digitized by ILRI from Survey of Kenya's 1:250,000 topographic sheets (for northern Kenya) and 1:50,000 topographic sheets for the rest of the country. Nairobi: SoK and ILRI.

Survey of Kenya (SoK), Japan International Cooperation Agency (JICA), and International Livestock Research Institute (ILRI). 1996. *250-meter Digital Elevation Model.* Created by ILRI and JICA, derived from SoK 1:250,000 Topographic Map. Nairobi: ILRI.

Thompson J., I. Porras, M. Katui-Katua, M. Mujwahuzi, J. Tumwine, N. Johnstone, and L. Wood. 2001. *Drawers of Water II: Thirty Years of Change in Domestic Water Use and Environmental Health in East Africa, Summary.* London: International Institute for Environment and Development (IIED). Online at http://www.iied.org/pubs/pdf/full/9049IIED.pdf

United Nations Development Programme (UNDP), Intermediate Technology Development Group (ITDG), and Energy for Sustainable Development Africa (ESDA). 2005. *Kenya Energy Atlas.* Nairobi: United Nations Development Programme (UNDP) and Global Village Energy Partnership. Online at http://www.ke.undp.org/KenyaEnrgyAtlasFinal-UN.pdf

United Nations Environment Programme (UNEP). 2001. *An Assessment of the Status of the World's Remaining Closed Forests.* Nairobi: UNEP. Online at http://grid2.cr.usgs.gov/publications/closedforest.pdf#search=%22an%20assessment%20of%20the%20worlds%20remaining%20closed%20forests%22

United Nations Environment Programme (UNEP). 2006. *Africa Environment Outlook 2 - 2006.* Nairobi: UNEP. Online at http://www.unep.org/DEWA/Africa/AEO2_Launch/index.asp

United States Geological Survey (USGS). 2004. *90-meter Digital Elevation Model based on Shuttle Radar Topography Mission (SRTM).* Online at http://www.landcover.org.

Wambua, S. 2003. "Yatta Canal: The 'Nile' of Northern Machakos." *Network for Water and Sanitation (NETWAS) Newsletter.* Online at http://www.netwas.org/newsletter/articles/2003/09/8

Were, E., B. Swallow, and J. Roy. *Water, Women and Local Social Organization in the Western Kenya Highlands. ICRAF Working Paper No. 12.* Nairobi: World Agroforestry Centre (ICRAF). Online at http://www.worldagroforestry.org/water/downloads/Watershed%20Publications/2006%20were,%20swallow%20and%20roy.pdf

World Bank. 2004. "Towards a Water-Secure Kenya: Water Resources Sector Memorandum" cited in D. Grey and C.W. Sadoff. *Water for Growth and Development.* Thematic Documents of the IV World Water Forum. Comision Nacional del Agua: Mexico City. 2006. Online at http://siteresources.worldbank.org/INTWRD/Resources/FINAL_0601_SUBMITTED_Water_for_Growth_and_Development.pdf

World Conservation Union (IUCN). 2003. *Tana River, Kenya: Integrating Downstream Values into Hydropower Planning. Case Studies in Wetland Valuation Number 6, May 2003.* IUCN. Online at http://www.iucn.org/themes/wani/econ/CaseStudy06Tana.pdf

World Conservation Union (IUCN) and United Nations Environment Programme/World Conservation Monitoring Centre (UNEP/WCMC). 2006. *2006 World Database on Protected Areas.* Cambridge, United Kingdom: IUCN and UNEP/WCMC.

Zecchini, A. 2000. "Clash of Interests in Masai Country: Kenya's Battle for Biodiversity." *Le Monde Diplomatique*, November 2000. Online at http://mondediplo.com/2000/11/21masai

Chapter 4 Food

Arid Lands Resource Management Project (ALRMP), United Nations World Food Programme/Vulnerability Analysis and Mapping (WFP/VAM), Famine Early Warning Systems Network (FEWS NET), and Ministry of Agriculture (MoA). 2006. *Kenya Livelihood Mapping GIS Database*, 2003-05. Nairobi: ALRMP, WFP/VAM, FEWS NET, and MoA.

Baltenweck I., S. Staal, L. Njoroge, F. Wanyoike, P. Ochungo, and E. Kariuki. 2005. *Targeting Pro-Poor Investment in the Kenyan Dairy Sub-Sector. Report prepared by the International Livestock Research Institute (ILRI) for IFAD and the Government of Kenya.* Nairobi, Kenya: ILRI.

Central Bureau of Statistics (CBS), Ministry of Planning and National Development. 1986. *1986 Statistical Abstract.* Nairobi: CBS.

Central Bureau of Statistics (CBS), Ministry of Planning and National Development. 1995. *1995 Statistical Abstract.* Nairobi: CBS.

Central Bureau of Statistics (CBS), Ministry of Planning and National Development. 2001. *2001 Statistical Abstract.* Nairobi: CBS.

Central Bureau of Statistics (CBS), Ministry of Planning and National Development. 2002. *Kenya 1999 Population and Housing Census GIS Database.* Nairobi: CBS.

Central Bureau of Statistics (CBS), Ministry of Planning and National Development. 2003. *Geographic Dimensions of Well-Being in Kenya: Where are the Poor? From Districts to Locations. Vol. I.* Nairobi: CBS. Online at http://www.worldbank.org/research/povertymaps/kenya/volume_index.htm

Central Bureau of Statistics (CBS), Ministry of Planning and National Development. 2004. *Economic Survey 2004.* Nairobi: CBS.

Central Bureau of Statistics (CBS), Ministry of Planning and National Development. 2005. *Economic Survey 2005.* Nairobi: CBS.

De Groote, H., G. Owuor, C. Doss, J. Ouma, L. Muhammad, and K. Danda. 2005. "The Maize Green Revolution in Kenya Revisited." *electronic Journal of Agricultural and Development Economics (eJADE)* 2: 32-49. Online at: ftp://ftp.fao.org/docrep/fao/008/ae692e/ae692e00.pdf

Delgado, C.L., A.N. Clare, and M.T. Marites. 2003. *Policy, Technical, and Environmental Determinants and Implications of the Scaling-Up of Livestock Production in Four Fast-Growing Developing Countries: A Synthesis.* Washington DC: IFPRI and FAO. Online at: http://www.fao.org/WAIRDOCS/LEAD/X6170E/X6170E00.HTM

Department of Resource Surveys and Remote Sensing (DRSRS), Ministry of Environment and Natural Resources. 2003. *DRSRS GIS Database.* Nairobi: DRSRS.

Food and Agriculture Organization (FAO) of the United Nations. 2000. *Africover Multipurpose Land Cover Databases for Kenya.* Rome: FAO. Online at: www.africover.org

Food and Agriculture Organization (FAO) of the United Nations. 2006. *FAO STAT.* Online at http://faostat.fao.org/site/291/default.aspx

Gebru, G. and J. McPeak. 2004. *Herd Accumulation: A Pastoral Strategy to Reduce Risk Exposure. Global Livestock Collaborative Research Support Program. PARIMA Research Brief 04-05.* Davis, California: University of California-Davis. Online at: http://glcrsp.ucdavis.edu/publications/parima/04-05-PARIMA.pdf

Government of Kenya (GoK). 2003. *Economic Recovery Strategy for Wealth and Employment Creation 2003-2007.* Nairobi: GoK. Online at http://www.planning.go.ke/publication/economic_recovery_plan.pdf

Grunblatt, J., M. Said, P. Wargute, and S.C. Kifugo. 1995. *DRSRS Data Summary Report: Kenyan Rangelands 1977-1994.* Nairobi: Department of Resource Surveys and Remote Sensing (DRSRS), Ministry of Planning and National Development.

Grunblatt, J., M. Said, and P. Wargute. 1996. *DRSRS National Rangelands Report: Summary of Population Estimates for Wildlife and Livestock, Kenyan Rangelands 1977-1994.* Nairobi: Department of Resource Surveys and Remote Sensing (DRSRS), Ministry of Planning and National Development.

Hijmans, R.J., S. Cameron, and J. Parra. 2005. *WorldClim Climate Surfaces, Version 1.4.* Online at http://www.worldclim.org/

Jayne, T. S., T. Yamano, J. Nyoro, and T. Awuor. 2000. *Do Farmers Really Benefit from High Food Prices? Balancing Rural Interests in Kenya's Maize Pricing and Marketing Policy. Tegemeo Institute of Agricultural Policy and Development. Policy Brief No.1.* Nairobi: Tegemeo. Online at: www.tegemeo.org

Kristjanson P., M. Radeny, D. Nkedianye, R. Kruska, R. Reid, H. Gichohi, F. Atieno, and R. Sanford. 2002. *Valuing Alternative Land Use Options in the Kitengela Wildlife Dispersal Area of Kenya. International Livestock Research Institute (ILRI) Impact Assessment Series 10.* Nairobi: ILRI. Online at: www.ilri.org

Lamprey, R. and R.S. Reid. 2004. "Expansion of Human Settlement in Kenya's Maasai Mara: What Future for Pastoralism and Wildlife?" *Journal of Biogeography* 31: 997-1032.

Matiru, V. 1999. *Forest Cover and Forest Reserves in Kenya: Policy and Practice.* Nairobi: World Conservation Union (IUCN), Eastern African Regional Office. Online at http://www.iucn.org/places/earo/pubs/forest/forestcover.pdf

Ministry of Livestock and Fisheries Development (MoLFD). 2001. *Fisheries Department Annual Report 2001.* Nairobi: MoLFD.

Ministry of Planning and National Development (MoPND), United Nations Development Programme, and Government of Finland. 2005. *Millennium Development Goals in Kenya, Needs & Costs.* Nairobi: Government of Kenya. Online at http://www.ke.undp.org/KenyaMDGNeedsAssessment.pdf

Muendo, K.M., D. Tschirley, and M.T. Weber. 2004. *Improving Kenya's Domestic Horticultural Production and Marketing System: Current Competitiveness, Forces of Change, and Challenges for the Future. Tegemeo Institute of Agricultural Policy and Development. Working Paper No. 8.* Nairobi: Tegemeo. Online at: www.tegemeo.org

Mutunga, N. 2006. Personal Communication, September 2006.

Nkedianye, D.O. 2003. *Testing the Attitudinal Impact of a Conservation Tool outside a Protected Area: The Case of the Kitengela Wildlife Conservation Lease Programme for Nairobi National Park. MA Thesis.* Nairobi: University of Nairobi, Department of Rural Sociology and Community Development.

Norton-Griffiths, M., M.Y. Said, S. Serneels, D.S. Kaelo, M. Coughenour, R.H. Lamprey, D.M. Thompson, and R.S. Reid. (in press). "Land Use Economics in the Mara Area of the Serengeti Ecosystem." In: A.R.E. Sinclair and J.F. Fryxell (editors). *Serengeti III: The Future of an Ecosystem.* Chicago: University of Chicago Press.

Nyoro, J.K., L. Kirimi, and T.S. Jayne. 2004. *Competitiveness of Kenyan and Ugandan Maize Production: Challenges for the Future. Tegemeo Institute of Agricultural Policy and Development. Working Paper No. 10.* Nairobi: Tegemeo. Online at: www.tegemeo.org

Pflanz, M. 2005. "Small Game, Big Problem as Poaching Grows in Kenya." *Christian Science Monitor, January 31, 2005.* Online at http://www.csmonitor.com/2005/0131/p07s01-woaf.html

Republic of Kenya (RoK). 2006. *Medium Term Expenditure Framework 2006. 2007-2008/2009. Report for the Agriculture and Rural Development (ARD) Sector. Final Draft. February 2006.* Nairobi: RoK Online at http://www.treasury.go.ke/sector_reports/ARD_Sector%20_Report_2007%20_Draft.pdf

Serneels, S. and E.F. Lambin. 2001. "Impact of Land-Use Changes on the Wildebeest Migration in the Northern Part of the Serengeti-Mara Ecosystem." *Journal of Biogeography* 28: 391-407.

Staal, S. 2004a. *The Demand for Dairy Products in Kenya. Smallholder Dairy Project Policy Brief 1.* Nairobi: ILRI. Online at: http://www.smallholderdairy.org/publications/Policy%20briefs/SDP%20BRIEF%201%20-FINAL%20R.pdf

Staal, S. 2004b. *Employment Generation in the Kenya Dairy Industry. Smallholder Dairy ProjectPolicy Brief 2.* Nairobi: ILRI. Online at: http://www.smallholderdairy.org/publications/Policy%20briefs/SDP%20BRIEF%202%20-UPDATE.pdf

Steve, S., J. Omiti, C. Delgado, F. Wanyoike, and L. Njoroge. 2006. *Differential Impacts of Policy and Scale Factors in Livestock Production in Kenya: The Case of Dairy and Poultry Farming.* Nairobi: ILRI and IFPRI.

Survey of Kenya (SoK). 2003. *National Atlas of Kenya, Fifth Edition.* Nairobi: Survey of Kenya.

Survey of Kenya (SoK) and International Livestock Research Institute (ILRI). 2000. *Cities, Towns, and Market Centers GIS Database.* Digitized by ILRI from Survey of Kenya's 1:250,000 topographic sheets (for northern Kenya) and 1:50,000 topographic sheets for the rest of the country. Nairobi: SoK and ILRI.

TRAFFIC East and Southern Africa. 2000. *The Utilization of Wild Meat in Eastern and Southern Africa.* Harare and Nairobi: TRAFFIC. Online at http://www.traffic.org/bushmeat/overview.pdf

Urban Harvest. 2004. *Urban Agriculture.* Online at http://www.cipotato.org/urbanharvest/home.htm

World Agroforestry Centre (ICRAF) and Department of Resource Surveys and Remote Sensing (DRSRS), Ministry of Environment and Natural Resources. 2001. *GIS Database of Land Use and Land Cover for Kenya's Medium to High Potential Districts.* DRSRS interpretation of 5,747 aerial photographs from transects flown in 1997. Approach summarized in report prepared for ICRAF and African Highlands Initiative (E.C. Njuguna. 2001. *Aerial Survey Sample Photo Interpretation, Report on Land Use and Land Cover For Medium to High Potential Districts of Kenya.* Nairobi: DRSRS). Nairobi: ICRAF and DRSRS.

World Conservation Union (IUCN) and United Nations Environment Programme/World Conservation Monitoring Centre (UNEP/WCMC). 2006. *2006 World Database on Protected Areas.* Cambridge, United Kingdom: IUCN and UNEP/WCMC.

Chapter 5 Biodiversity

Achieng, A.P. 1990. "The Impact of the Introduction of Nile Perch, *Lates niloticus* (L.), on the Fisheries of Lake Victoria." *Journal of Fish Biology* 37(Supplement A): 17-23.

Adanje, S.A. and W.K. Ottichilo. 1999. "Population Status and Feeding Habits of the Translocated Sub-Population of Hunter's Antelope or Hirola (*Beatragus hunteri*, Scalter, 1889) in Tsavo National Park, Kenya." *African Journal of Ecology*, 37: 38-48.

African Bird Club. 2006. *Important Bird Areas.* Online at http://www.africanbirdclub.org/countries/Kenya/ibas.html

Bennun, L. and P. Njoroge. 1999. *Important Bird Areas in Kenya.* Nairobi: Nature Kenya, The East Africa Natural History Society.

Bigalke, R.C. 1978. "Present-day Mammals of Africa." In V.J. Magio and H.B.S. Cooke (editors). *Evolution of African Mammals.* Cambridge, Massachusetts: Harvard University Press.

Biggs, R., E. Bohensky, P.V. Desanker, C. Fabricius, T. Lynam, A. A. Misselhorn, C. Musvoto, M. Mutale, B. Reyers, R.J. Scholes, S. Shikongo, and A.S. van Jaarsveld. 2004. *Nature Supporting People, The Southern African Millennium Ecosystem Assessment, Integrated Report. A Contribution to the Millennium Ecosystem Assessment, Prepared by the Southern African Millennium Ecosystem Assessment Team.* Pretoria: Council for Scientific and Industrial Research. Online at http://www.maweb.org/safma

BirdLife International. 2006. *EBAs in Kenya.* Online at http://www.birdlife.org/datazone/ebas/index.html?action=EbaHTMFindResults.asp&INam=&Cty=110&z=1

Central Bureau of Statistics (CBS), Ministry of Planning and National Development. 2001. *Population Distribution by Administrative Areas and Urban Centres, Kenya 1999 Population and Housing Census. Vol. 1.* Nairobi: CBS.

Central Bureau of Statistics (CBS), Ministry of Planning and National Development. 2002. *Kenya 1999 Population and Housing Census GIS Database.* Nairobi: CBS.

Central Bureau of Statistics (CBS), Ministry of Planning and National Development. 2003. *Geographic Dimensions of Well-Being in Kenya: Where are the Poor? From Districts to Locations. Vol. I.* Nairobi. CBS. Online at http://www.worldbank.org/research/poverty-maps/kenya/volume_index.htm

Central Bureau of Statistics (CBS), Ministry of Planning and National Development. 2005. *Geographic Dimensions of Well-Being in Kenya: Who and Where are the Poor? A Constituency Level Profile. Vol. II.* Nairobi: CBS.

Dawson, J. 2004. "Can the Lion Lie Down with the Lamb?" *Earth Island Journal* 19 (2). Online at http://www.earthisland.org/eijournal/new_articles.cfm?articleID=878&journalID=79

de Leeuw, J., M.N. Waweru, O.O. Okello, M. Maloba, P. Nguru, M.Y. Said, H.M. Aligula, I. M. Heitkonig, and R.S. Reid. 2001. "Distribution and Diversity of Wildlife in Northern Kenya in Relation to Livestock and Permanent Water Points." *Biological Conservation* 100: 297-306.

Department of Resource Surveys and Remote Sensing (DRSRS), Ministry of Environment and Natural Resources. 2003. *DRSRS GIS Database.* Nairobi: DRSRS.

Fishpool, L.D.C., and M.I. Evans. 2001. *Important Bird Areas in Africa and Associated Islands: Priority Sites for Conservation.* Cambridge, United Kingdom: BirdLife International.

Food and Agriculture Organization (FAO) of the United Nations. 2000. *Africover Multipurpose Land Cover Databases for Kenya.* Rome: FAO. Online at: www.africover.org

Georgiadis, N. and G. Ojwang'. 1997, 1999, and 2001. *Laikipia Sample Survey Report.* Unpublished reports.

Georgiadis, N., J.G.N. Olwero, and G. Ojwang'. 2003 *Laikipia Sample Survey Report.* Unpublished report.

German Technical Cooperation (GTZ). 1996. *The North Kenya GIS Database.* United Nations Environment Programme (UNEP)/GRID-Nairobi. Nairobi: UNEP/GRID-Nairobi.

Gichohi, H.W. 1996. *The Ecology of a Truncated Ecosystem: The Athi-Kapiti Plains. Ph.D. Thesis.* Leicester: University of Leicester.

Gichohi, H. 2003. *Direct Payments as a Mechanism for Conserving Important Wildlife Corridor Links Between Nairobi National Park and its Wider Ecosystem: The Wildlife Conservation Lease Program.* Presentation to the Fifth World Parks Congress, Durban, South Africa, September 2003. Online at http://www.conservationfinance.org/WPC/WPC_documents/Apps_09_Gichohi_v2.pdf

Grunblatt, J., M. Said, P. Wargute, and S.C. Kifugo. 1995. *DRSRS Data Summary Report: Kenyan Rangelands 1977-1994.* Nairobi: Department of Resource Surveys and Remote Sensing (DRSRS), Ministry of Planning and National Development.

Grunblatt, J., M. Said, and P. Wargute. 1996. *DRSRS National Rangelands Report: Summary of Population Estimates for Wildlife and Livestock, Kenyan Rangelands 1977-1994.* Nairobi: Department of Resource Surveys and Remote Sensing (DRSRS), Ministry of Planning and National Development.

Hansen, A.J., M.Y. Said, J.O. Ogutu, R.S. Reid, R. DeFries, J. Dempewolf, J.F. Robinson-Cox, S.A.R. Mduma, and J.G.C. Hopcraft. Submitted. "Land-Use Change Around Protected Areas: Effects on Large Mammal Populations in Maasailand of East Africa." *Landscape Ecology.*

Hansen, M., R. DeFries, J.R. Townshend, M. Carroll, C. Dimiceli, and R. Sohlberg. 2003. *500m MODIS Vegetation Continuous Fields Version 1.0.* College Park, Maryland: The Global Land Cover Facility.

Institute of Applied Ecology (IEA). 1998. *African Mammals Databank, A Databank for the Conservation and Management of the African Mammals. Vol. 1 and 2.* Brussels: European Commission Directorate.

Kaufman, L. 1992. "Catastrophic Change in Species-Rich Freshwater Ecosystems: The Lessons from Lake Victoria." *Bioscience* 42: 846-858.

Kennedy, T.A., S. Naeem, K.M. Howe, J.M.H. Knops, D. Tilman, and P. Reich. 2002. "Biodiversity as a Barrier to Ecological Invasion." *Nature* 417: 636-638.

Lamprey, R. and R.S. Reid. 2004. "Expansion of Human Settlement in Kenya's Maasai Mara: What Future for Pastoralism and Wildlife?" *Journal of Biogeography* 31: 997-1032.

Millennium Ecosystem Assessment (MA). 2003. *Ecosystems and Human Well-Being: A Framework for Assessment.* Washington, DC: Island Press. Online at http://www.maweb.org/en/Framework.aspx

Moehlman, P.D. (editor). 2002. *Equids: Zebra, Asses and Horses. Status Survey and Conservation Action Plan.* Gland, Switzerland and Cambridge, United Kingdom: IUCN/SSC Equid Specialist Group.

Ng'weno, F., P. Matiku, and N. Otieno (editors). 2004. *Kenya's Important Bird Areas, Status and Trends 2004.* Nairobi: Nature Kenya, the East Africa Natural History Society. Online at http://www.birdlife.org/action/science/indicators/pdfs/kenya_iba_trends.pdf

Norton-Griffiths, M. 1998. "The Economics of Wildlife Conservation Policy in Kenya." In: E. J. Milner-Gullard and R.P. Mace (editors). *Conservation of Biological Resources:* 279-293. Oxford: Blackwell Science Ltd.

Norton-Griffiths, M., M.Y. Said, S. Serneels, D.S. Kaelo, M. Coughenour, R.H. Lamprey, D.M. Thompson, and R.S. Reid. (in press). "Land Use Economics in the Mara Area of the Serengeti Ecosystem." In: A.R.E. Sinclair and J.F. Fryxell (editors). *Serengeti III: The Future of an Ecosystem.* Chicago: University of Chicago Press.

Ogari, J. 2001. *Impact of Exotic Fish Species and Invasive Water Weeds such as the Water Hyacinth on Lake Victoria Fisheries.* Paper presented at the Lake Victoria Environmental Management Project (LVEMP) conference, Kisumu, Kenya.

Reid, R.S., C.J. Wilson, R.L. Kruska, and W. Mulatu. 1997. "Impacts of Tsetse Control and Land-Use on Vegetative Structure and Tree Species Composition in Southwestern Ethiopia." *Journal of Applied Ecology* 34: 731-747.

Reid, R.S., M.E. Rainy, J. Ogutu, R. Kruska, L.K. Kimani, M. Nyabenge, M. McCartney, M. Kshatriya, J.S. Worden, L. Ng'ang'a, J. Owuor, J. Kinoti, E. Njuguna, C.J. Wilson, and R.H Lamprey. 2003. *People, Wildlife and Livestock in the Mara Ecosystem: the Mara Count 2002.* Nairobi: International Livestock Research Institute. Online at www.maasaimaracount.org

Reid, R.S., H. Gichohi, M.Y. Said, D. Nkedianye, J.O. Ogutu, M. Kshatriya, P. Kristjanson, S.C. Kifugo, J.L. Agatsiva, S.A. Adanje, and R. Bagine. 2006. "Fragmentation of a Periurban Savanna, Athi-Kaputiei Plains, Kenya." In: K.A. Galvin, R.S. Reid, R.H. Behnke, and N.T. Hobbs (editors). *Fragmentation in Semi-arid and Arid Landscapes: Consequences for Human and Natural Systems.* Dordrecht: Kluwer Academic.

Revenga, C., J. Brunner, N. Henninger, K. Kassem, and R. Payne. 2000. *Pilot Analysis of Global Ecosystems, Freshwater Systems.* Washington, DC: World Resources Institute. Online at http://pdf.wri.org/page_freshwater.pdf

Said, M.Y., R.N. Chunge, C.G Craig, C.R. Thouless, R.F Barnes, and H.T Dublin. 1995. *African Elephant Database 1995.* Gland, Switzerland: World Conservation Union (IUCN).

Serneels, S. and E.F. Lambin. 2001. "Impact of Land-Use Changes on the Wildebeest Migration in the Northern Part of the Serengeti-Mara Ecosystem." *Journal of Biogeography* 28: 391-407.

Stattersfield, A.J., M.J. Crosby, A.J. Long, and D.C. Wege. 1998. *Endemic Bird Areas of the World: Priorities for Biodiversity Conservation.* Cambridge, United Kingdom: BirdLife International.

Survey of Kenya (SoK). 2003. *National Atlas of Kenya, Fifth Edition.* Nairobi: Survey of Kenya.

Survey of Kenya (SoK) and International Livestock Research Institute (ILRI). 1997. *Road Network GIS Database.* Digitized by ILRI from Survey of Kenya's 1:50,000 topographic sheets for 1978-1997. Nairobi: SoK and ILRI.

Survey of Kenya (SoK) and International Livestock Research Institute (ILRI). 2000. *Cities, Towns, and Market Centers GIS Database.* Digitized by ILRI from Survey of Kenya's 1:250,000 topographic sheets (for northern Kenya) and 1:50,000 topographic sheets for the rest of the country. Nairobi: SoK and ILRI.

Tilman, D., D. Wedin, and J. Knops. 1996. "Productivity and Sustainability Influenced by Biodiversity in Grassland Ecosystems." *Nature* 379: 718-720.

United Nations Environment Programme (UNEP). 2001. *An Assessment of the Status of the World's Remaining Closed Forests.* Nairobi: UNEP. Online at http://grid2.cr.usgs.gov/publications/closedforest.pdf#search=%22an%20assessment%20of%20the%20worlds%20remaining%20closed%20forests%22

Wargute, P.W. and M.Y. Said. 1997. *The Population and Distribution Trend of Grevy's Zebra (Equus grevyi Oustalet, 1882). Department of Resource Surveys and Remote Sensing, Technical Report.* Nairobi: Department of Resource Surveys and Remote Sensing (DRSRS), Ministry of Planning and National Development.

White, F. 1983. *The Vegetation of Africa, a Descriptive Memoir to Accompany the UNESCO/AETFAT/UNSO Vegetation Map of Africa.* Paris: UNESCO.

Williams, S.D. 1998. *Grevy's Zebra: Ecology in a Heterogeneous Environment. Ph.D. Thesis.* London: University College.

Wilson, C.J., R.S. Reid, N.L. Stanton, and B.D. Perry. 1997. "Ecological Consequences of Controlling the Tsetse Fly in Southwestern Ethiopia: Effects of Land-Use on Bird Species Diversity." *Conservation Biology* 11: 435-447.

World Agroforestry Centre (ICRAF) and Department of Resource Surveys and Remote Sensing (DRSRS), Ministry of Environment and Natural Resources. 2001. *GIS Database of Land Use and Land Cover for Kenya's Medium to High Potential Districts.* DRSRS interpretation of 5,747 aerial photographs from transects flown in 1997. Approach summarized in report prepared for ICRAF and African Highlands Initiative (E.C. Njuguna. 2001. *Aerial Survey Sample Photo Interpretation, Report on Land Use Land Cover For Medium to High Potential Districts of Kenya.* Nairobi: DRSRS). Nairobi: ICRAF and DRSRS.

World Conservation Union (IUCN) and United Nations Environment Programme/World Conservation Monitoring Centre (UNEP/WCMC). 2006. *2006 World Database on Protected Areas.* Cambridge, United Kingdom: IUCN and UNEP/WCMC.

World Conservation Union (IUCN). 2006. *2006 IUCN Red List of Threatened Species.* Online at: www.iucnredlist.org.

Chapter 6 Tourism

Arabuko-Sokoke Forest Management Team. 2002. *Arabuko-Sokoke Strategic Forest Management Plan 2002-2007.* Available online at http://www.birdlife.org/action/ground/arabuko/arabuko_sokoke_plan.pdf

Belau, D. 2003. *The Impact of the 2001-2002 Crisis on the Hotel and Tourism Industry.* Geneva: International Labour Organization (ILO). Online at http://www.ilo.org/public/english/dialogue/sector/papers/tourism/impact.pdf

Bennun, L. and P. Njoroge. 1999. *Important Bird Areas in Kenya.* Nairobi: Nature Kenya, The East Africa Natural History Society.

Central Bureau of Statistics (CBS), Ministry of Planning and National Development. 2002. *Kenya 1999 Population and Housing Census GIS Database.* Nairobi: CBS.

Central Bureau of Statistics (CBS), Ministry of Planning and National Development. 2003. *Geographic Dimensions of Well-Being in Kenya: Where are the Poor? From Districts to Locations. Vol. I.* Nairobi. CBS. Online at http://www.worldbank.org/research/povertymaps/kenya/volume_index.htm

Central Bureau of Statistics (CBS), Ministry of Planning and National Development. 2004. *Economic Survey 2004.* Nairobi: CBS.

Central Bureau of Statistics (CBS), Ministry of Planning and National Development. 2006. *Economic Survey 2006.* Nairobi: CBS.

Department of Resource Surveys and Remote Sensing (DRSRS), Ministry of Environment and Natural Resources. 2003. *DRSRS GIS Database.* Nairobi: DRSRS.

Ecotourism Society of Kenya (ESOK). 2005. *National Inventory of Ecotourism Projects in Kenya 2003-2005.* Nairobi: ESOK.

Filion, F.L., J.P. Foley, and A.J. Jacquemot. 1994. "The Economics of Global Ecotourism." In M. Munasinghe and J. McNeely (editors). *Protected Area Economics and Policy: Linking Conservation and Sustainable Development.* Washington DC: World Bank.

Fishpool, L.D.C., and M.I. Evans. 2001. *Important Bird Areas in Africa and Associated Islands: Priority Sites for Conservation.* Cambridge, United Kingdom: BirdLife International.

Food and Agriculture Organization (FAO) of the United Nations. 2000. *Africover Multipurpose Land Cover Databases for Kenya.* Rome: FAO. Online at www.africover.org

Grunblatt, J., M. Said, P. Wargute, and S.C. Kifugo. 1995. *DRSRS Data Summary Report: Kenyan Rangelands 1977-1994.* Nairobi: Department of Resource Surveys and Remote Sensing (DRSRS), Ministry of Planning and National Development.

Grunblatt, J., M. Said, and P. Wargute. 1996. *DRSRS National Rangelands Report: Summary of Population Estimates for Wildlife and Livestock, Kenyan Rangelands 1977-1994.* Nairobi: Department of Resource Surveys and Remote Sensing (DRSRS), Ministry of Planning and National Development.

Government of Kenya (GoK). 1995. *The Study on the National Tourism Master Plan in the Republic of Kenya (Volume 1).* Nairobi: GoK.

Government of Kenya (GoK). 2003. *Economic Recovery Strategy for Wealth and Employment Creation 2003-2007.* Nairobi: GoK. Online at http://www.planning.go.ke/publication/economic_recovery_plan.pdf

iExplore. 2006. *"Kenya Destination Guide"* at http://pta.iexplore.com/dmap/Kenya/Overview and http://pta.iexplore.com/dmap/Kenya/Where+to+Go;$sessionid$HLWYVYQACEVI2P2MN5XCFEQ

Ikiara, M. 2001. *Policy Framework of Kenya's Tourism Sector since Independence and Emerging Policy Concerns.* Kenya Institute for Public Policy Research and Analysis (KIPPRA) Working Paper No.2. Nairobi: KIPPRA. Online at http://www.kippra.org/resources/publications.asp

Ikiara, M. and C. Okech. 2002. *Impact of Tourism on Environment in Kenya: Status and Policy.* Kenya Institute for Public Policy Research and Analysis (KIPPRA) Discussion Paper No. 19. Nairobi: KIPPRA. Online at http://www.kippra.org/resources/publications.asp

Kahata, J. and J. Imbanga. 2002. *Sustainable Tourism. A Report on the Civil Society Review of the Implementation of Agenda 21 in Kenya, Kenya NGO Earth Summit 2002 Forum.* Online at http://www.worldsummit2002.org/texts/KENYAtourism.pdf

Kenya Wildlife Service (KWS). 2005. "Major Parks Yearly Report" (KWS internal document) and Sammy Towett (KWS), personal communication.

Kenya.com. 2006. "Shimba Hill, Description." Online at http://www.kenya.com/shimbahills.html

Laikipia Wildlife Forum. 2006. "Description of Species, Tour Operators, and Ecotourism" Online at http://www.laikipia.org/safari_laikipia.htm

MacMillan Education Ltd. 1993. *Kenya Traveller's Map, Third Edition.* London and Oxford: MacMillan Education Ltd.

National Environment Management Authority (NEMA). 2003. *State of the Environment (SoE) Report for Kenya, 2003, Summary.* Nairobi: NEMA. Online at http://www.nema.go.ke/downloads/SOE03summary.pdf

National Imagery and Mapping Agency of the United States (NIMA). 1997. *VMAP_1V10 - Vector Map Level 0 (Digital Chart of the World).* Washington, DC: NIMA.

Ojwang, G., M.Y. Said, P.W. Wargute, and R. Reid. 2006. *Dry Season Census of Large Herbivores in the Mara Ecosystem and Adjoining Areas 2000-2002.* Department of Resource Surveys and Remote Sensing (DRSRS) Technical Report. Nairobi: DRSRS.

Ottichilo, W.K., J. de Leeuw, A.K. Skidmore, H.H.T. Prins, and M.Y. Said. 2001. "Population Trends of Large Non-Migratory Wild Herbivores and Livestock in the Masai Mara Ecosystem, Kenya, between 1977 and 1997." *African Journal of Ecology* 38: 202-216.

Republic of Kenya (RoK). 2003. *Gazette Notice. 3976. The Hotels and Restaurants (Classification of Hotels and Restaurants) Regulations, 1988. Vol. CV, No. 62.* Nairobi: RoK.

Scholes, R.J. and R. Biggs (editors). 2004. *Ecosystem Services in Southern Africa: A Regional Assessment. A Contribution to the Millennium Ecosystem Assessment, Prepared by the Regional-Scale Team of the Southern African Millennium Ecosystem Assessment 2004.* Pretoria: Council for Scientific and Industrial Research (CSIR). Online at http://www.maweb.org/safma

Serneels, S. and E.F. Lambin. 2001. "Impact of Land-Use Changes on the Wildebeest Migration in the Northern Part of the Serengeti-Mara Ecosystem." *Journal of Biogeography* 28: 391-407.

Survey of Kenya (SoK), Japan International Cooperation Agency (JICA), and International Livestock Research Institute (ILRI). 1996. *250-meter Digital Elevation Model.* Created by ILRI and JICA, derived from SoK 1:250,000 Topographic Map. Nairobi: ILRI.

Survey of Kenya (SoK) and International Livestock Research Institute (ILRI). 2000. *Cities, Towns, and Market Centers GIS Database.* Digitized by ILRI from Survey of Kenya's 1:250,000 topographic sheets (for northern Kenya) and 1:50,000 topographic sheets for the rest of the country. Nairobi: SoK and ILRI.

United Nations. 2002. *Kenya Country Profile, Johannesburg Summit 2002.* Online at www.un.org/esa/agenda21/natlinfo/wssd/kenya.pdf

United Nations Development Programme (UNDP), Kenya Marine and Fisheries Research Institute (KMFRI), and Geological Survey of Denmark and Greenland (GEUS). 2006. *KenSea, Environmental Sensitivity Atlas for Coastal Area of Kenya.* Copenhagen: UNDP, KMFRI, and GEUS. Online at http://www.geus.dk/program-areas/nature-environment/international/reports/kensea/kensea_rapport.pdf

United Nations Environment Programme (UNEP)/GRID-Nairobi. 1998. *Eastern Africa Coastal and Marine Resources Database.* Nairobi: UNEP/GRID-Nairobi. Project description online at http://www.unep.org/eafatlas/dbke.htm

World Conservation Union (IUCN) and United Nations Environment Programme/World Conservation Monitoring Centre (UNEP/WCMC). 2006. *2006 World Database on Protected Areas.* Cambridge, United Kingdom: IUCN and UNEP/WCMC.

Chapter 7 Wood

African Bird Club. 2006. *Important Bird Areas.* Online at http://www.africanbirdclub.org/countries/Kenya/ibas.html

Akotsi, E. and M. Gachanja. 2004. *Changes in Forest Cover in Kenya's Five Water Towers 2000-2003.* Nairobi: Department of Resource Surveys and Remote Sensing (DRSRS) and Kenya Forest Working Group (KFWG). Online at http://www.unep.org/dewa/assessments/EcoSystems/Land/mountain/pdf/forest_catchment_2003_report.pdf

Arid Lands Resource Management Project (ALRMP), United Nations World Food Programme/Vulnerability Analysis and Mapping (WFP/VAM), Famine Early Warning Systems Network (FEWS NET), and Ministry of Agriculture (MoA). 2006. *Kenya Livelihood Mapping GIS Database, 2003-05.* Nairobi: ALRMP, WFP/VAM, FEWS NET, and MoA.

Central Bureau of Statistics (CBS), Ministry of Planning and National Development. 2002. *Kenya 1999 Population and Housing Census GIS Database.* Nairobi: CBS.

Central Bureau of Statistics (CBS), Ministry of Planning and National Development. 2003. *Geographic Dimensions of Well-Being in Kenya: Where are the Poor? From Districts to Locations. Vol. I.* Nairobi: CBS. Online at http://www.worldbank.org/research/povertymaps/kenya/volume_index.htm

Central Bureau of Statistics (CBS), Ministry of Planning and National Development. 2004. *Economic Survey 2004.* Nairobi: CBS.

Central Bureau of Statistics (CBS), Ministry of Planning and National Development. 2006. *Economic Survey 2006.* Nairobi: CBS.

Choge, S.K., A.B. Cunningham, and F.W. Ellery. 2002. "Sustaining the Wood Carving in Kenya: Issues and Challenges." In *Proceedings: Natural Forests and Savanna Woodlands Symposium III: 6-9 May 2002, University of Natal, South Africa.* Online at http://dwafapp4.dwaf.gov.za/dwaf/download.asp?f=4160___section3sustaining+the+woodcarving+industry+336.pdf&docId=4160

Energy for Sustainable Development Africa (ESDA). 2005a. *National Charcoal Survey: Exploring the Potential for a Sustainable Charcoal Industry in Kenya.* Nairobi: ESDA.

Energy for Sustainable Development Africa (ESDA). 2005b. *National Charcoal Survey: Summary Report. Exploring the Potential for a Sustainable Charcoal Industry in Kenya.* Nairobi: ESDA.

Food and Agriculture Organization (FAO) of the United Nations. 2000. *Africover Multipurpose Land Cover Databases for Kenya.* Rome: FAO. Online at: www.africover.org

Food and Agriculture Organization (FAO) of the United Nations. 2005. *Global Forest Resources Assessment 2005 (FRA 2005). Progress Towards Sustainable Forest Management, FAO Forestry Paper 147.* Rome: FAO. Online at http://www.fao.org/docrep/008/a0400e/a0400e00.htm

Gathaara, G.N. 1999. *Aerial Survey of the Destruction of Mount Kenya, Imenti, and Ngare Ndare Forest Reserves, February-June 1999.* Nairobi: Kenya Wildlife Service. Online at http://www.kenyaforests.org/reports/MtKenya99.pdf

Hansen, M., R. DeFries, J.R. Townshend, M. Carroll, C. Dimiceli, and R. Sohlberg. 2003. *500 meter MODIS Vegetation Continuous Fields Version 1.0.* College Park, Maryland: The Global Land Cover Facility. Online at http://glcf.umiacs.umd.edu/data/vcf/

Holding Anyonge, C. and J.M. Roshetko. 2003. "Farm-Level Timber Production: Orienting Farmers Towards the Market". *Unasylva* 212: 48-56. Online at ftp://ftp.fao.org/docrep/fao/005/y4744e/y4744e11.pdf

Jayne, T. S., T. Yamano, J. Nyoro, and T. Awuor. 2000. *Do Farmers Really Benefit from High Food Prices? Balancing Rural Interests in Kenya's Maize Pricing and Marketing Policy. Tegemeo Institute of Agricultural Policy and Development. Policy Brief No.1.* Nairobi: Tegemeo. Online at: www.tegemeo.org

Kameri-Mbote, P. 2005. *Land Tenure, Land Use and Sustainability in Kenya: Towards Innovative Use of Property Rights in Wildlife Management. IELRC Working Paper 2005 - 4.* Geneva, Switzerland: International Environmental Law Research Centre (IELRC). Online at http://www.ielrc.org/content/w0504.pdf

Lambrechts C., B. Woodley, C. Church, and M. Gachanja. 2003. *Aerial Survey of the Destruction of the Aberdare Range Forests.* Nairobi: United Nations Environment Programme (UNEP). Online at http://www.kenyaforests.org/reports/report%20Final.pdf

Lamprey, R. and R.S. Reid. 2004. "Expansion of Human Settlement in Kenya's Maasai Mara: What Future for Pastoralism and Wildlife?" *Journal of Biogeography* 31: 997-1032.

Matiru, V. 1999. *Forest Cover and Forest Reserves in Kenya: Policy and Practice.* Nairobi: World Conservation Union (IUCN), Eastern African Regional Office. Online at http://www.iucn.org/places/earo/pubs/forest/forestcover.pdf

Matiru, V. and S. Mutimba. 2002. "Legalise It." *Ecoforum*, Long Rains, 34.

Mbugua, D. 2000. *Forest Outlook Studies in Africa (FOSA), Kenya.* Rome: Food and Agriculture Organization (FAO) of the United Nations. Online at ftp://ftp.fao.org/docrep/fao/003/AB569E/AB569E00.pdf

Ministry of Energy (MoE). 2002. *Study on Kenya's Energy Demand, Supply and Policy Strategy for Households, Small Scale Industries and Service Establishments, Final Report, September 2002.* Nairobi: MoE and Kamfor Company Limited.

National Imagery and Mapping Agency of the United States (NIMA). 1997. *VMAP_1V10 - Vector Map Level 0 (Digital Chart of the World).* Washington, DC: NIMA.

Norton-Griffiths, M., M.Y. Said, S. Serneels, D.S. Kaelo, M. Coughenour, R.H. Lamprey, D.M. Thompson, and R.S. Reid. (in press). "Land Use Economics in the Mara Area of the Serengeti Ecosystem." In: A.R.E. Sinclair and J.F. Fryxell (editors). *Serengeti III: The Future of an Ecosystem.* Chicago: University of Chicago Press.

Serneels, S. and E.F. Lambin. 2001. "Impact of Land-Use Changes on the Wildebeest Migration in the Northern Part of the Serengeti-Mara Ecosystem." *Journal of Biogeography* 28: 391-407.

Survey of Kenya (SoK) and International Livestock Research Institute (ILRI). 1997. *Road Network GIS Database.* Digitized by ILRI from Survey of Kenya's 1:50,000 topographic sheets for 1978-1997. Nairobi: SoK and ILRI.

Survey of Kenya (SoK) and International Livestock Research Institute (ILRI). 2000. *Cities, Towns, and Market Centers GIS Database.* Digitized by ILRI from Survey of Kenya's 1:250,000 topographic sheets (for northern Kenya) and 1:50,000 topographic sheets for the rest of the country. Nairobi: SoK and ILRI.

United Nations Environment Programme (UNEP). 2001. *An Assessment of the Status of the World's Remaining Closed Forests.* Nairobi: UNEP. Online at http://grid2.cr.usgs.gov/publications/closedforest.pdf#search=%22an%20assessment%20of%20the%20worlds%20remaining%20closed%20forests%22

Vanleeuwe, H.B., C. Woodley, M. Lambrechts, and M. Gachanja. 2003. *Changes in the State of Conservation of Mount Kenya Forests: 1999-2002. An Interim Report February 2003.* Nairobi: DICE - University of Canterbury, Kenya Wildlife Service, and United Nations Environment Programme. Online at http://www.kenyaforests.org/reports/Mt.Kenya.pdf

Wass, P. 1995. *Kenya's Indigenous Forests, Status, Management and Conservation.* Gland, Switzerland: World Conservation Union (IUCN).

Wass, P. 2000. *Kenya's Forest Resource Assessment, Data Collection and Analysis for Sustainable Forest Management in ACP Countries - Linking National and International Efforts.* Addis Ababa: European Commission Directorate-General VIII Development and Food and Agriculture Organization. Online at ftp://ftp.fao.org/docrep/fao/004/X6809E/X6809E00.pdf

World Agroforestry Centre (ICRAF) and Department of Resource Surveys and Remote Sensing (DRSRS), Ministry of Environment and Natural Resources. 2001. *GIS Database of Land Use and Land Cover for Kenya's Medium to High Potential Districts.* DRSRS interpretation of 5,747 aerial photographs from transects flown in 1997. Approach summarized in report prepared for ICRAF and African Highlands Initiative (E.C. Njuguna. 2001. *Aerial Survey Sample Photo Interpretation, Report on Land Use Land Cover For Medium to High Potential Districts of Kenya.* Nairobi: DRSRS). Nairobi: ICRAF and DRSRS.

World Bank and Government of Kenya (GoK). 1999. *The Status of Plantation Grown Timber and its Implications on Sustainable Management and Rational Utilization of the Resource, Implementation, and Completion Report.* Nairobi: World Bank and GoK.

World Conservation Union (IUCN) and United Nations Environment Programme/World Conservation Monitoring Centre (UNEP/WCMC). 2006. *2006 World Database on Protected Areas.* Cambridge, United Kingdom: IUCN and UNEP/WCMC.

Chapter 8 The Upper Tana: Patterns of Ecosystem Services and Poverty

Central Bureau of Statistics (CBS), Ministry of Planning and National Development. 2002. *Kenya 1999 Population and Housing Census GIS Database.* Nairobi: CBS.

Central Bureau of Statistics (CBS), Ministry of Planning and National Development. 2003. *Geographic Dimensions of Well-Being in Kenya: Where are the Poor? From Districts to Locations. Vol. I.* Nairobi: CBS. Online at http://www.worldbank.org/research/povertymaps/kenya/volume_index.htm

Food and Agriculture Organization (FAO) of the United Nations. 2000. *Africover Multipurpose Land Cover Databases for Kenya.* Rome: FAO. Online at: www.africover.org

Government of Kenya (GoK). 2003. *Economic Recovery Strategy for Wealth and Employment Creation 2003-2007.* Nairobi: GoK. Online at http://www.planning.go.ke/publication/economic_recovery_plan.pdf

Jayne, T.S., T. Yamano, J. Nyoro, and T. Awuor. 2000. *Do Farmers Really Benefit from High Food Prices? Balancing Rural Interests in Kenya's Maize Pricing and Marketing Policy. Tegemeo Institute of Agricultural Policy and Development. Policy Brief No.1.* Nairobi: Tegemeo. Online at: www.tegemeo.org

Kenya Forests Working Group (KFWG). 2001. "Kenya Government Gives Away 5 Percent of Its Forests." Press Release from the Kenya Forests Working Group. Nairobi: KFWG, 2 November 2001.

Ministry of Energy (MoE). 2002. *Study on Kenya's Energy Demand, Supply and Policy Strategy for Households, Small Scale Industries and Service Establishments, Final Report, September 2002.* Nairobi: MoE and Kamfor Company Limited.

Ministry of Water Development (MoWD) and Japan International Cooperation Agency (JICA). 1992. *The Study on the National Water Master Plan. GIS Data of Drainage and Subdrainage Areas.* Nairobi: MoWD, JICA, and International Livestock Research Institute.

National Imagery and Mapping Agency of the United States (NIMA). 1997. *VMAP_1V10 - Vector Map Level 0 (Digital Chart of the World).* Washington, DC: NIMA.

Olson, J.M. 2004. *Multi-Scale Analysis of Land Use and Management Change on the Eastern Slopes of Mount Kenya.*, LUCID Working Paper 20. East Lansing, Michigan: Michigan State University. Online at http://www.lucideastafrica.org/publications.htm

Survey of Kenya (SoK), Japan International Cooperation Agency (JICA), and International Livestock Research Institute (ILRI). 1996. *250-meter Digital Elevation Model.* Created by ILRI and JICA, derived from SoK 1:250,000 Topographic Map. Nairobi: ILRI.

Lessons Learned and Next Steps

Biggs, R., E. Bohensky, P.V. Desanker, C. Fabricius, T. Lynam, A. A. Misselhorn, C. Musvoto, M. Mutale, B. Reyers, R.J. Scholes, S. Shikongo, and A.S. van Jaarsveld. 2004. *Nature Supporting People, The Southern African Millennium Ecosystem Assessment, Integrated Report. A Contribution to the Millennium Ecosystem Assessment, Prepared by the Southern African Millennium Ecosystem Assessment Team.* Pretoria: Council for Scientific and Industrial Research. Online at http://www.maweb.org/safma

Central Bureau of Statistics (CBS), Ministry of Planning and National Development. 2005. *Geographic Dimensions of Well-Being in Kenya: Who and Where are the Poor? A Constituency Level Profile. Vol. II.* Nairobi: CBS.

Government of Kenya (GoK). 2003. *Economic Recovery Strategy for Wealth and Employment Creation 2003-2007.* Nairobi: GoK. Online at http://www.planning.go.ke/publication/economic_recovery_plan.pdf

Millennium Ecosystem Assessment (MA). 2005. *Ecosystems and Human Well-Being: Synthesis.* Washington, DC: Island Press. Online at http://www.maweb.org/en/Synthesis.aspx

Ministry of Planning and National Development (MoPND), United Nations Development Programme, and Government of Finland. 2005. *Millennium Development Goals in Kenya, Needs & Costs.* Nairobi: Government of Kenya. Online at http://www.ke.undp.org/KenyaMDGNeedsAssessment.pdf

Republic of Kenya (RoK). 2006. *Medium Term Expenditure Framework 2006. 2007-2008/2009. Report for the Agriculture and Rural Development (ARD) Sector. Final Draft. February 2006.* Nairobi: RoK Online at http://www.treasury.go.ke/sector_reports/ARD_Sector%20_Report_2007%20_Draft.pdf

Scholes, R.J. and R. Biggs (editors). 2004. *Ecosystem Services in Southern Africa: A Regional Assessment. A Contribution to the Millennium Ecosystem Assessment, Prepared by the Regional-Scale Team of the Southern African Millennium Ecosystem Assessment 2004.* Pretoria: Council for Scientific and Industrial Research (CSIR). Online at http://www.maweb.org/safma